Habitat Ecology and Analysis

Habitat Ecology and Analysis

Joseph A. Veech

Department of Biology,
Texas State University, USA

OXFORD
UNIVERSITY PRESS

Great Clarendon Street, Oxford, OX2 6DP,
United Kingdom

Oxford University Press is a department of the University of Oxford.
It furthers the University's objective of excellence in research, scholarship,
and education by publishing worldwide. Oxford is a registered trade mark of
Oxford University Press in the UK and in certain other countries

First Edition published in 2021

Impression: 2

Published in the United States of America by Oxford University Press
198 Madison Avenue, New York, NY 10016, United States of America

British Library Cataloguing in Publication Data

Data available

Library of Congress Control Number: 2020945816

ISBN 978–0–19–882928–7 (hbk.)
ISBN 978–0–19–882941–6 (pbk.)

DOI: 10.1093/oso/9780198829287.001.0001

Printed and bound by
CPI Group (UK) Ltd, Croydon, CR0 4YY

Preface

I have always been impressed that the best way to go about finding an animal (and perhaps a plant) is to go look for it in its habitat. My childhood was filled with me doing this, typically trying to capture the local reptiles and amphibians more so than other types of species. I often consulted field guides to get guidance on where to look. As I eventually transitioned into a formal education in ecology and being a working professional ecologist, I became even more interested in how animals associate with particular types of habitat. My dissertation and research endeavors early in my career did not have much to do with habitat, at least not explicitly. Broadly speaking, my research has always been directed at examining the ecological factors that determine the distribution and abundance of species in nature as well as those factors affecting patterns of biodiversity. I have used many different approaches to research and worked on species in various taxonomic groups without having any particular study species or system. In this context, I have conducted studies at various spatial scales from local ecological communities to landscapes to biogeographic regions. I have also always been interested in developing new statistical approaches and examining their performance. Over the years, my research has been both field-based and computer-based as in simulation modeling. A few years ago, it finally occurred to me that the common theme underlying my disparate research interests was the recognition that habitat and a species' habitat requirements are fundamentally important to a lot of what we study as ecologists.

Most of my formal training and education is within the large and encompassing academic field of ecology. However, early in my career, I also conducted studies that most accurately could be labelled as wildlife ecology, and this line of research continues today. I consider myself to be an ecologist as well as a wildlife ecologist, and there are distinctions between the two. Thus, I have written this book from a combined perspective, and I hope that it will be of interest and use to students and practitioners in both disciplines.

The book can be considered as two parts. In the first part (Chapters 1–4) I discuss the history of the habitat concept and differences between ecology and wildlife ecology, present a conceptual/probabilistic model of the overriding importance of habitat, and outline goals of studying and analyzing habitat. In the second part (Chapters 5–11) I explain some relatively common methods for analyzing habitat and discuss some related statistical issues. This part of the book is intended as a user's manual or reference guide for readers confronted with the particular task of figuring out the habitat associations of a species. In writing the statistical part of the book, I strove to use common language and explain the "mysterious" terminology that sometimes accompanies descriptions of some statistical techniques. My intention is that the reader need only have an introductory-level comprehension of statistics in order to get a practical understanding out of the second part of the book. Indeed, some of Part II may be "review" to readers who have a more sophisticated and experienced perspective on statistics.

Given that my native language is English, there is an unintended neglect of the literature written in other languages, particularly with regard to Chapter 1 where I trace the history of the study of habitat. I know I have missed including the early writings of ecologists from countries such as Germany, France, Japan, Italy, Spain, various Latin

American countries, and perhaps others. An early ecology textbook (Adams 1913, cited in Chapter 1) provides an excellent list of references by German naturalists and ecologists. Otherwise, in writing this book, I have attempted to draw on a wide variety of studies conducted by researchers from around the world studying all sorts of animals.

I very much appreciate colleagues and students who either reviewed parts of the book, provided me with important materials, or engaged in email conversation. They are Ivan Castro-Arellano, Jared Haney, Steve Jenkins, John Majoris, Curt Meine, Randy Simpson, Stan Temple, Jeff Troy, and graduate students in my habitat ecology course in fall 2019. As I was writing this book over a 2-year period, I was not as diligent to the needs of my graduate students as I would have been otherwise; however, they all remained patient with me and I thank them for that. It was a pleasure working with Ian Sherman and Charlie Bath (OUP) in putting together this book. Lastly, my wife and daughter allowed me a nice quiet room at home to work on the book, particularly in spring 2020 when those fortunate enough to work from home were doing so, for obvious reasons.

Joseph A. Veech
San Marcos, Texas, USA
June 2020

Contents

PART I

History, Concepts, Models, and Goals

CHAPTER 1

Introduction

Habitat is where an organism lives—that's the simplest definition. However, as a concept integral to many areas of ecological investigation and knowledge, habitat is a bit more complicated and comprehensive than simply where an organism lives. I accessed the immense ISI Web of Science literature database and did a keyword search on "habitat" and several other terms often used by ecologists. Of course in return I received many results—more papers than I could read in several lifetimes and a list of titles that would take several days or longer to go through, never mind reading the abstracts. My point: for ecologists, "habitat" is a very familiar term, concept, and real-world entity and has been for a very long time. Perhaps it could even be considered a level of ecological (if not biological) organization. As a term of common usage, it ranks right up there with several others (Fig. 1.1). Thus "habitat ecology," broadly defined as *the study of the habitat requirements of species and effects of habitat on individual survival, population persistence, and spatial distribution*, has had a prominent place in the development of ecology as an academic field and as a knowledge base for conserving and managing the planet's living natural resources.

1.1 History of the habitat concept

In the scientific literature, use of "habitat" predates "ecology" as the latter was not coined until 1866 by the famous German biologist and philosopher, Ernst Haeckel (Stauffer 1957). The habitat concept developed over a period of at least 200 years (Fig. 1.2). To my knowledge, no one has ever tracked down the first use of the word "habitat." This would be a difficult task anyway given that the word has shared etymology with other similar words (e.g.,

"inhabit," "habitant") that have a greater realm of usage than just in biology or ecology. Carl Linnaeus used the term in 1745 in *Flora Svecica* as did his student, Johan Gustaf Hallman, in his dissertation in the same year, titled *Dissertatio Botanica de Passiflora*. By the 1766 edition of *Systema Naturae*, Linnaeus was routinely using "habitat" to denote the general area where a species was from, such as "*Mari mediterraneo*" (Mediterranean Sea) for "*Testudo coriacea*," now known as the leatherback sea turtle (*Dermochelys coriacea*). This particular usage continued for well over 120 years and was relatively common in broad taxonomic treatises wherein each species was described in a consistent format of categories such as "appearance," "diet," "habits," and "habitat." The latter would typically be followed by only a few words indicating the geographical region (sometimes as a geopolitical label, e.g., "Mexico") or even a single locality where the species existed and was collected. This was hardly a description of habitat in the modern sense of the word. Incidentally, the word "habits" probably dates back even further and today is somewhat archaic with regard to usage in ecological literature; it generally meant the behavior of the species.

A slightly more modern usage of "habitat" appeared by 1791. The very first volume of *Transactions of the Linnean Society of London* (a journal that is no longer in print) included accounts of some lichen "species" collected from southern Europe. For each species account, there was a habitat category filled with a very short descriptor in Latin such as *rupibus calcareis* (limestone rock), *rupibus alpinis* (alpine rock), *ericetis alpinis* (alpine moor), *truncis arborum* (tree trunk), and *corticibus olearum* (bark of olive trees) (Smith 1791). In the same volume, the habitat of the buff ermine moth (*Phalaena bombyx*, now

Habitat Ecology and Analysis. Joseph A. Veech, Oxford University Press (2021). © Joseph A. Veech. DOI: 10.1093/oso/9780198829287.003.0001

Figure 1.1 Word cloud depicting usage of "habitat" and select other terms in the ecological literature from 1864–2018. Each word was searched in ISI Web of Science along with "ecolog*" so as to eliminate papers that were not ecological. Each term appeared in the title or abstract of the following number of papers (×1,000): species—1,657, population—1,576, behavior—750, soil—749, community—728, resource—726, habitat—695, environment—592, reproduction—536, organism—415, wildlife—392, climate—352, ecosystem—333, individual—322, biodiversity—223, adaptation—210, gene—210, foraging—145, landscape—117, competition—107, dispersal—106, predation—105, photosynthesis—77, trophic—67, parasitism—52, niche—52, succession—52, primary production—51, food web—38, decomposition—38, pollination—27, and mutualism—24.

Spilarctia luteum) was listed as *arboribus pomiferis* and *quercu* (fruit-bearing trees and oaks) (Marsham 1791). These examples represent a transition to a modern concept of habitat in that the habitat of each species was viewed as the type of vegetation or substrate where the species exists rather than as the geographical location where the specimen was collected.

Another notable advancement occurred in 1836. Early in his career, the well-known naturalist and biologist, Richard Owen, prepared and presented a table of the habitats of antelope species in India and Africa during the Proceedings of the Zoological Society of London (Owen 1836). For each of 60 species, the table gave a short descriptor of the habitat, such as hilly forests, open plains, stony plains and valleys, desert borders, rocky hills, thickets, reedy banks, and *Acacia* groves. Of particular importance,

Owen had a specific goal for doing his habitat analysis—he wanted to test a hypothesis. He proposed that the suborbital, maxillary, and inguinal glands found in some species, but absent or underdeveloped in others, had the purpose of facilitating the aggregation of individuals of a species. That is, individual antelopes would secrete onto vegetation or large rocks and these secretions "might serve to direct individuals of the same species to each other." Owen was looking to see whether species inhabiting open plains lacked the glands (due to the absence of shrubs and rocks to secrete on) whereas those of forest areas had the glands, and further whether gregarious species had the glands whereas solitary species did not. According to Owen, his table showed that some of the plains-inhabiting species lacked the glands and some had the glands irre-

Historical development of the habitat concept

(1) First explicit use of "habitat"; Linnaeus and Hallman in 1745

(2) Habitat used to indicate type of vegetation or substrate, Smith and Marsham in 1791

(3) Habitat discussed in an adaptive context; Owen in 1836

(4) Golden age of natural history writing (1850–1900) but little mention of habitat, various authors; see Table 1.1

(5) Clear separation between Linnaean and modern concepts begins to appear (e.g., Miller 1899, Bailey 1900)

(6) Ecology develops as a modern science (1900–1930), modern concept of habitat emerges through the writings of Grinnell, Shelford, Dice, Elton, Lack, and others

(7) First quantitative analysis of habitat associations; Gause in 1930

(8) Wildlife management emerges as academic discipline (1930–1950), emphasis on role of habitat in sustaining populations and active management of habitat for select species; Leopold, Errington, and others

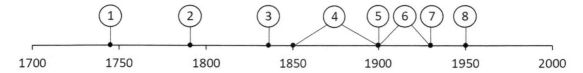

Figure 1.2 Historical timeline for the development of the habitat concept.

spective of whether the species was gregarious or solitary. A similar lack of pattern was evident in the forest-dwelling species. Hence, Owen rejected his own hypothesis. In addition to testing a hypothesis, his study is notable in several other respects. He had at least a dozen or so habitat types represented among the 60 species indicating a fairly detailed classification of habitat. Owen had a specific biological and ecological reason for identifying the habitat of each species. Presumably, he compiled the table from previously published literature—perhaps an early forerunner of studies that use a macroecological approach or the comparative method, both of which involve species as the units of analysis. Lastly, Owen implicitly recognized the possibility that a physical structure on an organism might relate to the habitat where it occurs and have a specific function related to behavior. Of course, in 1836 he did not use the word "adaptation"—the development of that concept would be left to someone a couple decades later.

Interestingly, Charles Darwin did not use "habitat" in *Voyage of the Beagle* (1839) or *On the Origin of Species* (1859) although there would have been ample opportunity to do so. Instead he used "station" and "habitation" in contexts where "habitat" could have fit, although both the former were also used in *Voyage* to denote a location of human settlement or occupation. In the 574 pages of *The Geographical Distribution of Animals* (1876), Alfred Russel Wallace used "habitat," "habitation," and "station" each a few times. However, he did attempt to define and distinguish "habitat" and "station" (p. 4). He defined "station" as a locality wherein two or more stations are separated by some distance but tend nonetheless to have the same habitat. In comparison with stations, habitats were recognized as having distinct vegetation or terrain. Wallace defined both terms in the context of changes in the species inhabiting each. In his words, "The whole area over which a particular animal is found may consist of any number of stations, but rarely of more than one habitat" (p. 4). With neither Darwin nor Wallace do we find a specific and detailed concept of habitat, nonetheless, Wallace's writings began to at least hint toward the modern concept of habitat.

In the early to mid-nineteenth century, published journals devoted to natural history began to emerge.

Prior to this, there had been scientific journals for at least 150 years, dating back to the *Philosophical Transactions of the Royal Society* established in 1665 by the Royal Society, which itself had been organized in England only 5 years earlier. Perhaps the earliest exclusively biological journals were *Transactions of the Linnean Society of London* (1791) and *Proceedings of the Zoological Society of London* (1830), and both of these often contained natural history contributions and species accounts. During the early years, scientific studies were read to members of the societies at regular meetings—this practice continued well into the nineteenth century as represented by the famous readings of Darwin and Wallace's theories about natural selection to members of the Linnean Society of London on July 1, 1858. Even relatively mundane papers were read at meetings, such as the previously described study by Owen presented at the meeting of the Zoological Society of London on March 22, 1836. Owen presented his hypothesis and assessment of antelope habitat as a follow-up to a paper read by Edward Bennett in which he described the facial glands of Indian antelopes. Eventually the "reading to the society members" route to publishing a scientific study became outdated. Henceforth, scientific journals could organize (perhaps under the auspices of a professional society or organization although not a necessity) without being tied to the physical process of a paper first being read at a meeting. The following journals emerged with a definite focus on natural history and in particular reporting on the discovery of new species: *Annals and Magazine of Natural History* (1838), *Ibis* (1859), *American Naturalist* (1867), *Bulletin of the American Museum of Natural History* (1881), and *The Auk* (1884).

The time period from about 1850–1900 may have been the golden age of natural history writing. Prior to this, much of the scientific biological literature was written in Latin or in a very obtuse and long-winded way. Into the nineteenth century there were still many species to be discovered and many regions to be explored and inventoried. Professional and amateur naturalists came along and began writing in a very easy-to-read conversational type of prose. Their writings often consisted of presenting behavioral and basic ecological information on a given species often obtained by direct observation of wild or even captive individuals. But there was either no mention of "habitat," or habitat was denoted as a region or locality where the species or individual originated, as though following the precedent set by Linnaeus more than 100 years earlier. Many of these natural history accounts appeared in the *American Naturalist* between 1867 and 1880 by authors such as Elliott Coues, Augustus Fowler, Samuel Lockwood, Alpheus Packard, and Charles Abbott (see Abbott 1860, 1870a, b, c, 1873; Packard 1867, 1871, 1876; Fowler 1868a, b; Lockwood 1875, 1876). Another form of natural history writing, which might now be thought of as early taxonomy, involved a systematic treatise of a group of species in which specific characteristics of each species, sometimes including "habitat," were presented in an orderly way. Again, habitat was conceived by these authors only in the Linnaean fashion of identifying the collecting location or geographic region where the species existed. Naturalists engaged in this type of writing typically had formal scientific training and advanced degrees and to a modern audience their papers sound more scientific than the narratives of the other type of natural history writing. These papers appeared in all the previously mentioned journals; particularly prolific writers included Edward Cope, David Jordan, Charles Abbott, Albert Günther, and Charles Cory (see Abbott 1860, 1870a, b, c, 1873; Cope 1869, 1896; Günther 1871, 1875; Jordan 1874, 1877; Jordan and Copeland 1877; Cory 1891, 1886). The authors of both types of paper often presented very comprehensive measurements and information on the morphological traits of the species with no description of habitat—again this reflects an early emphasis on taxonomy and classifying species (often by authors with close associations with museums) rather than on studying the species' ecological role in nature.

This inattention to habitat (as we conceive it from our modern perspective) was prevalent and pervasive in the period 1850–1900. Naturalists of various nationalities writing about all types of organisms in all regions of the world generally neglected to even mention the habitat of their subjects or defined the habitat in a very broad and often geographic context; that is, the Linnaean way (Table 1.1). There were occasional exceptions. Abbott (1870a) wrote about "mud-loving fishes" and even used that term for the title of his paper. He described how a group of nine fish species (of various genera) tended to be found in very shallow and sluggish streams near Trenton, New Jersey, United States. He even mentioned that the fishes "preferred such shallow water, with the mud, to that which was deeper, to which they had access, because it was over a stony bed" (p. 387). His implicit recognition of a species having a habitat preference foreshadows a major aspect of our modern perspective of habitat—that is, that species have preferences and can actively select habitat. Even very small organisms were seen by some as having a habitat. Stokes (1888) identified the habitats of various ciliate species respectively as ponds and then went on to give a very brief description of the water and sometimes whether the ciliate species was associated with certain aquatic plants. Williamson (1894) gave a brief but very direct statement on the habitat of abalone (*Haliotis* sp.): "the habitat of abalone is among rocks, where, at very low tide, they may be found huddled together in a corner of a rock pool, or hedged in between fissures of immense rocks" (p. 854). In a very comprehensive species account, Wilder (1899) wrote a detailed 300-word description of the habitat of two salamander species, *Desmognathus fusca* and *Spelerpes bilineatus*: "they are found in and about running brooks that are plentifully supplied with small stones, and they seem to prefer spots shaded by trees" and "perhaps the best brooks of all are the little mountain streams that run swiftly down quite steep inclines, forming miniature cascades alternating with small shallow basins" (p. 233). Interestingly, Wilder's habitat description is given in the context of instructing the reader on how to go about finding these secretive salamanders.

Around 1900, the definition of habitat and its use as an ecological term finally began to transition into our modern concept of habitat. A great example of this is Miller (1899). In his *Mammals of New York*, Gerrit Smith Miller gave a brief one-page account for each of 81 species known in the state of New York. Each account followed the same systematized format in which Miller provided information in seven different categories: synonymy, type locality, faunal position, habitat, distribution in New York, principal records, and remarks. Entries for the "habitat" category typically described the major type of vegetation or terrain where the species could be found,

such as forests, woods, woodlands, thickets, old fields, prairies, meadows, rocky hillsides, marshes, bogs, streams, and ponds, among others. Today these basic descriptors are so ingrained in both the public's recognition of habitat and scientific parlance that we could easily and rightly say that this is a list of *types of habitats*. Notably, Miller's species accounts included identification of the type locality (the place where the type specimen was collected), faunal position (a phrase Miller used to indicate a biome), and distribution in New York. So these three categories together represented the Linnaean definition of habitat (as a geographic place or region) and allowed Miller to implicitly and perhaps intentionally distinguish between that view of habitat and the emerging view of habitat as the type of vegetation, terrain, or environment where a species could be found. The latter is our modern concept of habitat in its simplest and most general form.

Vernon Bailey also recognized and described habitats in much the same way as Miller. In his 1900 monograph on voles (*Microtus*), Bailey provided a taxonomic key and extensive quantitative description of each species including measurements of particular morphological characteristics (Fig. 1.3; Bailey 1900). His monograph is thoroughly modern and scientifically detailed, more so than Miller's book on New York mammals published just a year earlier. Because of the scientific prose and content of his monograph on voles, Bailey further entrenched the new definition and concept of habitat within the emerging field of ecological science. The emergence of this new perspective on habitat was not limited solely to vertebrates; Hargitt (1901a, 1901b, 1901c) also separately distinguished habitat from type locality and geographic region in his three-part monograph on hydrozoans.

In 1907 Cora Daisy Reeves published a paper, "The breeding habits of the rainbow darter (*Etheostoma caeruleum*, Storer), a study in sexual selection." As the title suggests, Reeves' paper was primarily devoted to the reproductive behavior of darters. However, she also wrote about their habitat in detail. She described it as the rapids of swift-flowing streams and the pools in between rapids. She wrote about the darters being found among large stones and on a gravel sheet. Her description was also quantitative; she recorded that the water above the

Table 1.1 List of late-nineteenth-century naturalists who used the word "habitat" in a way now seen as archaic (to refer to a particular geographic region or locality) or not at all when writing accounts of particular species. List is selective and only includes authors that were actively writing at some time from 1850–1900. Naturalists are ordered chronologically by birth year.

Name	Birth—death	Species group	Region
John Edward Gray	1800–1875	Everything, corals to mammals	Worldwide
John Gould	1804–1881	Birds	UK, Australia, New Guinea, North, Central, and South America
Edward Blyth	1810–1873	Birds, mammals	India, Southeast Asia
Augustus Fowler	1812—unknown	Birds	Primarily eastern North America
Gustav Hartlaub	1814–1900	Birds	Europe, Africa
John Henry Gurney, Sr.	1819–1890	Birds	Europe, Africa
Samuel Lockwood	1819–1894	Reptiles, fish, invertebrates	North America
Henry Baker Tristram	1822–1906	Various groups but primarily birds	Pacific islands and Middle East
Thomas Vernon Wollaston	1822–1878	Insects, primarily beetles	Europe
Spencer Fullerton Baird	1823–1887	Reptiles, birds	North America
Edgar Leopold Layard	1824–1900	Birds	South Africa, Pacific islands
John Lawrence LeConte	1825–1883	Insects	North America
Octavius Pickard-Cambridge	1828–1917	Spiders	Worldwide
Philip Lutley Sclater	1829–1913	Birds	Central and South America, Pacific islands
Albert Günther	1830–1914	Reptiles	Worldwide
Daniel Giraud Elliot	1835–1915	Birds, mammals	North America
Edward Hargitt	1835–1895	Woodpeckers	Worldwide
Tomasso Salvadori	1835–1923	Birds	Italy, India
Joel Asaph Allen	1838–1921	Birds, mammals	North and South America, Caribbean
Alpheus Spring Packard	1839–1905	Insects	North America
Edward Drinker Cope	1840–1897	Primarily reptiles	Western Hemisphere
Elliott Coues	1842–1899	Primarily birds, some mammals	North America
Charles Conrad Abbott	1843–1919	Fish, invertebrates, amphibians	Primarily eastern North America
William Healey Dall	1845–1927	Primarily mollusks	North America
Richard Bowdler Sharpe	1847–1909	Birds	Great Britain, Africa, Asia
Robert Ridgway	1850–1929	Birds	North and South America
Hans von Berlepsch	1850–1915	Birds	North and South America
David Starr Jordan	1851–1931	Primarily fish	North America
Clinton Hart Merriam	1855–1942	Mammals, birds	North America
Charles Barney Cory	1857–1921	Birds	North and South America, Caribbean
George Albert Boulenger	1858–1937	Reptiles, amphibians, fish	Worldwide
Alfred Webster Anthony	1865–1939	Birds, mammals	Western United States, particularly Pacific Coast
Harry Church Oberholser	1870–1963	Birds	Primarily western United States, Mexico
Joseph Grinnell[1]	1877–1939	Birds, mammals	Primarily western North America

[1] More so than the other naturalists, Grinnell eventually became thoroughly modern in his view of habitat.

48   NORTH AMERICAN FAUNA.   [NO. 17.

MICROTUS LONGICAUDUS (Merriam).   Long-tailed Vole.

Arvicola (Mynomes) longicaudus Merriam, Am. Nat., XXII, 934–935, Oct., 1888.

Type locality.—Custer, S. Dak. (in the Black Hills at an altitude of about 5,500 feet).

Geographic distribution.—Boreal cap of the Black Hills and down some of the cold streams well into the Transition zone.

Habitat.—Banks of cold streams and in mountain meadows.

General characters.—Size of body about equal to that of *Microtus pennsylvanicus;* tail much longer; ears larger; colors grayer; skull flatter; braincase wider.

Color.—*Summer pelage:* Upperparts dull bister, darkened with numerous black-tipped hairs, becoming grayish on the sides and shading into dull, buffy gray on belly; feet plumbeous; tail dimly bicolor, blackish above, soiled whitish below.   *Winter pelage* (old and faded in a June specimen from Sundance, Wyo.): Upperparts grayish bister, mixed with blackish-tipped hairs, shading gradually into slightly paler sides and dull whitish belly; tail distinctly bicolor; feet soiled whitish.

Cranial characters.—Skull long and not much arched; rostrum long; nasals reaching to anterior plane of incisors; bullæ large and rounded; molar pattern similar to that of *pennsylvanicus,* except for absence of posterior loop in middle upper molar; m3 with 3 closed triangles, 3 outer and 4 inner salient angles; $\overline{m}1$ with anterior loop, 5 closed triangles, 4 outer and 5 inner salient angles.   From *mordax* it differs in slightly shorter, heavier rostrum and wider nasals; narrower interpterygoid fossa; wider expansion of jugal; shorter and wider condyloid ramus of mandible.

Measurements.—Type, ♀ ad.: Total length, 185; tail vertebræ, 65; hind foot, 21.   Topotype, ♀ ad.: 184; 61; 22.   *Skull* (of type): Basal length, 25; nasals, 7.8; zygomatic breadth, 15.2; mastoid breadth, 11.6; alveolar length of upper molar series, 6.3.

General remarks.—*Microtus longicaudus* stands as one of the few outlying and isolated forms, though the first-described species of its widely distributed group.   Its nearest neighbor is *M. mordax* of the Bighorn Mountains, Wyoming, between which range and the Black Hills neither species is known to occur.

Specimens examined.—Total number, 6, from the following localities:

   South Dakota: Custer, 2.
   Wyoming: Sundance (in the western edge of the Black Hills), 4.

Figure 1.3 Species description for the long-tailed vole (*Microtus longicaudus*) as presented by Vernon Bailey in *Revision of American Voles of the Genus Microtus*, North American Fauna Volume 17, 1900. This is the actual page from the monograph. Note that "type locality," "geographic description," and "habitat" are separate categories thereby indicating that Bailey explicitly recognized that habitat was more than just the geographic location where a species was found.

gravel sheet was 1.5–6 in. deep with a current velocity of 75 ft per min. The gravel pebbles averaged 0.5 in. in diameter. In addition to being one of the first quantitative descriptions of a species' habitat, Reeves' study also was one of the first to (1) recognize a distinction between the habitat of the species during the breeding and nonbreeding seasons, (2) provide an estimate of population density within the habitat, and (3) mention the presence of other similar species (fish) using the same habitat, namely creek chubs and stonerollers. To be thorough, the published paper also included an annotated photograph of the habitat. Reeves accomplished all this without habitat being the primary focus of her study. She presents very detailed information and interpretation of darter mating behavior—surely her study is also an early standard in the field of behavioral ecology. Although she never published a similar assessment of habitat (or reproductive behavior), Reeves went on to have a very interesting academic career nonetheless (Box 1.1).

In the early twentieth century, ecologists also began to use the term "habitat" in a theoretical context and not just as a category or label for the type of environment where a particular species lived. Leavitt (1907) reviewed the different theories regarding the necessity of geographic separation for the process of speciation. In so doing he recognized that the number of species of a given kind occurring within a region (he called it a "district") might closely correspond to the number of different types of habitat in the region. He also stated that "difference of local habitat" could provide the isolation needed to preclude interbreeding of incipient species and he referred to the need for individuals to be adapted to their habitat. Although Leavitt did not discuss habitat as a necessary and broad concept within ecology and evolution, he did clearly use the term in a modern way and in the context of the contentious debate on whether natural selection or mutation was more important in producing new evolutionary forms.

Box 1.1 Cora Daisy Reeves

Cora Reeves (1873–1953) was born in Rockford, Illinois, United States but eventually found her way to the Pacific Coast where she enrolled at the California State Normal College in Los Angeles (later to become UCLA). She graduated in 1894 and later returned east to complete a PhD at the University of Michigan in 1917. Her dissertation research experimentally examined how fish perceive the wavelength of light. Soon after completing her dissertation, she was hired to the faculty at Ginling College in Nanking (Nanjing), China. The college had been founded only a few years previous by missionaries from the United States with the purpose of providing advanced education to Chinese women. Today, the college is part of Nanjing Normal University. Reeves remained at Ginling College until 1941 and upon her retirement she returned to the United States. Although Reeves' academic career was primarily founded on teaching and administration, she routinely undertook collection of fish, reptile, and insect specimens as a graduate student in Michigan and then later as an academician in China. Some of her specimens still reside in the collections at the Museum of Zoology at the University of Michigan. She also wrote an 806-page "field guide," *Manual of the Vertebrate Animals of Northeastern and Central China, Exclusive of Birds*, published in 1933. Reeves' accomplishments are particularly remarkable given the time period of her life and career. Reeves (1907) is not only one of the first scientific accounts of a species habitat, but it is likely also one of the first ecology journal papers written in IMRAD format, an approach to written communication of research that would become the standard in nearly all fields of science. In this regard, Reeves was also thoroughly modern—her name deserves to be better recognized among the early pioneers of natural history and ecology.

Robert Leavitt was actually best known as a botanist and author of several popular textbooks, not as an ecologist. However, he was a contemporary of plant ecologists such as Henry Cowles, Eugen Warming, Frederic Clements, Arthur Tansley, Henry Gleason, and Volney Spalding. The research and writings of these scientists greatly influenced the early development of ecology as a scientific discipline. They wrote about the process of succession, plant communities, and ecosystems, essentially demonstrating that ecology could be more than just the autecological study of the natural history of single species. These founders of modern ecology used the term "habitat" extensively in their writings, particularly Clements and Tansley. In his classic, *Plant Succession*, Clements (1916) used "habitat" hundreds of times, but in a way that we (students of modern-day ecology) use the word "environment." Also, he never used "habitat" in reference to any particular plant or animal species. Plant ecologists of his day were primarily concerned with three main pursuits: figuring out ways of hierarchically grouping sets of co-occurring plant species, determining the abiotic and biotic factors affecting the distribution of plants, and studying natural temporal change in plant communities such as the process of succession. Their concept of habitat was compatible with each of these research arenas. To them, habitat was a physical place where all the environmental conditions were favorable for the development of certain plant communities or prohibitive in the development of other plant communities. Understandably, unlike animal ecologists they did not see the plant communities or vegetation as forming the habitat. Plant ecologists thought of habitat as a template of mostly abiotic physical factors upon which a plant community could develop and be maintained, and possibly transition to a different set of species through succession.

Clements' ideas about succession and habitat were formed by his direct experience spending time in the natural landscapes of his native Nebraska. To him, habitat was more than just a particular place where certain plants grow.

Ecology concerns the relation of plants to their surroundings, both physical and biological. The habitat of a plant is an aggregate of influences or factors acting upon the plant and causing it to exhibit certain phenomena and structures more or less peculiar to the habitat and plant in question. Each habitat is dominated by one or more controlling factors in the presence of which other factors are insignificant or ineffective. Taken as a whole, these factors constitute the physical environment of a plant, or its habitat. They are temperature, light, water-content, soil, atmosphere, precipitation, and physiography. (Paraphrased from Pound and Clements 1900, p.161)

[At the time, Roscoe Pound, only a few years older than Clements, was director of the state botanical survey in Nebraska, hence the order of authorship. Pound went on to have more of a career as a legal scholar than as an ecologist.]

Clements expressed a relatively modern concept of habitat and one that was a long way from the early Linnaean depiction of habitat as a specific geographic region or locality. Indeed, Tansley (1920) wrote about the impracticality of physically delimiting any single habitat and even suggested that reference to particular habitat types (especially as synonymous with plant formations or associations) was misguided. Beginning with Cowles and Warming, and continuing into modern times, plant ecologists have always used a very comprehensive and heuristic concept of habitat.

In the early twentieth century, there was a divide between plant ecologists and animal ecologists. Indeed, *ecology* was *plant ecology* or what was sometimes called phytogeography or botanical geography. Perhaps because of this divide, animal ecologists were a bit slower to develop a concept of habitat that was as theoretical and mechanistic in subscribing to habitat a central role in determining distribution and abundance of species. In essence, that was the key feature of "habitat" in plant ecology: habitat subsumes all the various environmental factors affecting the persistence of a species and communities. This mechanistic aspect of habitat was initially missing in how animal ecologists used the term "habitat."

In the separate academic field of animal ecology, Joseph Grinnell is probably the person who deserves the most credit in bringing about a modern concept of habitat (Box 1.2). At first, Grinnell (1897) used the Linnaean definition in identifying the habitat of a new species of towhee as San Clemente Island off the coast of California; by the way, it is presently considered to be a subspecies *Pipilo maculatus clementae*. Interestingly, in this species account, Grinnell also identified a type locality with a little more geographic precision (Smuggler's

Box 1.2 Joseph Grinnell

Joseph Grinnell (1877–1939) was born in the region that was known as "Indian Territory" at the time and which later became the state of Oklahoma. His father was a medical doctor assigned by the US federal government to serve groups of Native Americans at Fort Sill. When he was a child, his family moved to California where he would remain throughout his life. He received a PhD from Stanford University in 1913. Shortly prior to that, he became the first director of the newly created Museum of Vertebrate Zoology at the University of California, Berkeley, a position that he held for his entire career. Grinnell was a keen observer of vertebrate fauna in the field, writing extensively about how various species are adapted to the conditions of their habitats. He published 554 research papers, popular articles, and books between 1893 and his death in 1939 (Alagona 2012). He also was an early proponent of the need to preserve species and their natural habitats. This included, among many conservation accomplishments, a critical visionary role in establishing the University of California's system of natural reserves dedicated to scientific research, education, and conservation (Alagona 2012). More so than any other person, Grinnell should be credited with bringing about our modern multi-faceted concept of wildlife habitat. He recognized habitat as a general ecological concept as well as an extrinsic characteristic of a species. Further, his conservation advocacy also provided a physical reality to habitat. He saw that habitats were indeed places in nature that needed protection for their own sake as well as providing the places of residence for various wildlife species.

Cove) where a male and female had been collected just a few months prior. He did not otherwise describe the habitat. However, by 1904 he was transitioning. In a species account of the chestnut-backed chickadee (originally *Parus rufescens* but very recently *Poecile rufescens*), he describes the habitat as a long and narrow region along the Pacific Coast from southeastern Alaska to northern California, and he presents a map of this geographic range. But in the same paper, he also writes about its habitat as being dense, well-shaded, humid coniferous forest with substantial cloudiness and precipitation. Thus he was using "habitat" in both the old-fashioned sense of the word and its newly emerging usage as a descriptor for the vegetation and other environmental conditions where a species is found. Incidentally, he also suggested that the dorsal brown plumage coloration of the bird was an adaptation for the humid environment (though he does not give a mechanism). He further described geographic variation within the species, this included descriptions of several subspecies.

Lastly, Grinnell also suggested a very well thought-out hypothesis on the origin of *P. rufescens* from another nearby species, *P. hudsonicus*, which involved isolation due to habitat differences. In most respects, Grinnell (1904) was a very modern paper on the ecology and evolution of a species.

By 1917 Grinnell was exclusively using "habitat" in its modern sense. In his study of the California thrasher (*Toxostoma redivivum*), Grinnell thoroughly described the chaparral habitat of the bird, including a detailed description of the breeding habitat and ways in which the foraging behavior of thrashers relates to the physical structure of the vegetation. More remarkably, Grinnell clearly recognized that a set of environmental factors (i.e., the habitat) was important in restricting the geographic distribution of the species. According to Grinnell (1917), "An explanation of this restricted distribution is probably to be found in the close adjustment of the bird in various physiological and psychological respects to a narrow range of environmental conditions. The nature of these critical conditions is to be learned through an examination of the bird's habitat" (p. 428). In recognizing habitat as a factor affecting the occurrence and spatial distribution of a species, it is easy to conceive that Grinnell may have read and been influenced by the studies of the plant ecologists that were emerging at the time. Grinnell (1917) is also the very first paper to use the term "niche" and present a description of a species' niche; ironically although the word appears in the title of the paper, Grinnell only used it a few times at the very end of the paper. Nonetheless, the relevance here is that "habitat" and "niche" were used in the same paper and thus distinguished from one another. In the ecological literature, "niche" is not nearly as widely used as "habitat" (Fig. 1.1), but it is a very familiar term and concept to all ecologists. As a concept, niche is more complex, intangible, and theoretical than is habitat (Section 1.2).

The early divide between plant and animal ecologists was bridged somewhat by Victor Shelford (Box 1.3). There was a direct link: Shelford had been a student of Henry Cowles at the University of Chicago. As such, Shelford was influenced by Cowles' pioneering study of plant succession in the Indiana Dunes, an extensive area of dune formation on the south shore of Lake Michigan. Perhaps due to a desire to carve out his own scientific career independent of Cowles, Shelford did not study plants. His organisms of choice were tiger beetles (family *Cicindelidae*). Shelford was one of the first ecologists to frequently discuss "habitat selection" and "habitat preference." In his dissertation, Shelford (1907, 1908) described how female tiger beetles must find (i.e., select) the correct soil type and moisture conditions for egg-laying. This focus on studying the factors affecting habitat selection and preferences

---

Box 1.3 Victor Ernest Shelford

Victor Shelford (1877–1968) was born in the small town of Chemung, New York. He received his PhD from the University of Chicago in 1907. He was then hired to the faculty at the University of Illinois in 1914 where he remained for his entire career. Shelford had an immense range of ecological research interests. He conducted field work and published research on mammals, birds, fishes, and insects in forests, grasslands, dune complexes, tundra, swamps, streams, rivers, and lakes. From 1914–1930 he spent his summers on the coast of the state of Washington studying various fish species as well as the pelagic and near-shore communities of marine invertebrates in Puget Sound (e.g., Shelford and Powers 1915; Shelford et al. 1935). He was also concerned with advancing the experimental study of animals in their natural habitats and laboratory settings. As such, he wrote papers on the proper methodology for and inferences from such studies (e.g., Shelford 1926, 1934, 1954; Shelford and Eddy 1929). Shelford even wrote textbooks on ecology. From our present-day perspective, his research contributions can be categorized within a wide range of subdisciplines: physiological ecology, community ecology, landscape ecology, population ecology, biogeography, and conservation biology. Shelford retained a lifelong interest in ecological succession, perhaps a holdover from his early days with Cowles. However, above all else, he was an experimentalist. He very much believed that an organism's adaptation to its abiotic environment affected its daily performance (survival and reproduction) as well as the habitats that it occupied and the geographic distribution of the species. He investigated his view of ecology and nature with carefully designed experiments that were thoroughly modern and scientific. Lastly, Shelford shared with Grinnell not only a birth year, but also a desire to protect native flora and fauna, which included advocating for the permanent protection of natural habitat (e.g., Shelford 1933, 1941).

would continue throughout his career, although he quickly moved on from insects to fishes and other vertebrates.

Considering that Shelford studied tiger beetles, it is easy to imagine how he began to think about animals actively selecting habitat. Tiger beetles are very fast runners, they fly well, and they have excellent vision. They are generally active for long periods of the day, moving about, always searching for something. Much of the time the searching is the hunting of prey; tiger beetles are aggressive predators, hence the name. Shelford observed this natural behavior of tiger beetles in the dune fields as well as the precise and deliberate egg-laying behavior of females—this involves careful construction of a burrow chamber, once the appropriate soil habitat is found (Shelford 1908). He must have been impressed by the seemingly deliberate and careful behavior of the beetles.

In 1911 and 1912, Shelford published a five-part series of papers in the *Biological Bulletin*. This was likely the first attempt to apply the process of succession to animal communities, both aquatic and terrestrial. In part I, Shelford discussed how a stream might have different sets of fish species at different locations as the stream aged becoming deeper and wider through erosion and incision; that is, successional change in a stream. He also presented data on the collecting localities of numerous fish species in several creeks that drained into Lake Michigan near Chicago. He clearly saw habitat preferences and the implied capacity for fish to select habitat as being important to their distribution, writing "Fishes have definite habitat preferences which cause them to be definitely arranged in streams which have a graded series of conditions from mouth to source" (Shelford 1911, p. 33). Further, in this paper and others, he stressed that animals must be physiologically and structurally adapted to the environmental conditions of their specific habitat. This led him to measure particular environmental factors in different habitats, such as the penetration of light through the canopies of forests at different successional stages (Shelford 1912a).

The *rate of evaporation*—early on in his scientific career Shelford identified this as a crucially important factor in all habitats, and he measured, studied, and wrote about it his entire life. He was the first animal ecologist to identify a particular abiotic

(edaphic) habitat factor and attribute importance to it from both a theoretical standpoint and with regard to empirical observation (Shelford 1911, 1912a). Again, he was influenced by plant ecologists. According to Shelford (1912a), over decades of study, plant ecologists had come to realize that rate of evaporation was the best descriptor of plant environments. In the same paper, he also noted that spiders and insects captured on the prairies will quickly die when placed in screened cages in the laboratory but will survive if kept in glass-enclosed containers. He attributed the mortality to evaporation of water from the organisms' bodies, that is, desiccation. To Shelford, rate of evaporation was a common measuring stick to compare among different habitat types. He regarded the "evaporating power of the air" as the most inclusive and best index of the environmental conditions that an animal was exposed to (Shelford 1912a). Perhaps because he was able to a priori identify factors presumed to be important in habitat selection, Shelford went so far as to perform experiments on how animals select habitat (Shelford 1915). Some of these experiments even involved captive organisms. Due to his research on organism–environment interaction, Victor Shelford is also considered a founder of physiological ecology.

He can also be considered a founder of community ecology. In the 1912 series of papers, he thoroughly establishes the idea that different animal species at a given location constitute a community just as much as do the plant species. Shelford also set the stage for thinking about vegetation (i.e., plant communities) as habitat for animal communities, an implicit notion of habitat that still persists today. Much of Shelford (1912b) is devoted to an attempt to find ways of classifying habitats and putting "habitat" as a real unit of nature into the hierarchy of biological organization. Plant ecologists had been up to this task for several decades and they were following the lead of Linnaean taxonomists to essentially classify nature. Of importance, Shelford (1912b) was also attempting to define the emerging academic field of ecology and set it apart from natural history. As an example, he wrote "The work of naturalists is important though it is defective mainly in that one often has difficulty in determining what habitat is meant" (p. 336).

Victor Shelford was extremely productive. In 1913, he published *Animal Communities in Temperate*

America, a 362-page book that would go on to influence the intellectual development of ecology for decades. The book further developed all of Shelford's ideas on the importance of habitat, among other things. In these past six paragraphs on Victor Shelford, I have unintentionally short-changed him a bit. He contributed enormously to the growth of ecology as an academic and research discipline; and Shelford helped advance many of its different subdisciplines—this cannot be overstated. He even saw this as a duty and often referred to the need to "make progress" in ecology. Most of his papers and books still have great relevance today. For the particular purpose of this book, his greatest contribution was in bringing about a very multi-faceted concept of habitat that is integral to the many ecological processes and patterns that we still study today.

Lee Raymond Dice was another important ecologist who contributed to the development of our modern concept of habitat (Box 1.4). For his disser-

tation at the University of California, Berkeley, Dice studied the vertebrate communities in his home state of Washington, particularly the southeastern corner. He spent much time in the field, observing birds during the day and trapping mammals by night—in a way reminiscent of the early naturalists. However, his intent was very ecological and modern. He wanted to understand the extent to which habitat determined the spatial and geographic distribution of species, a pursuit that ecologists still work on today. Dice (1916) described the "ecological method" as a way of describing "animal distributions" based on studying the adaptive relationships of organisms with their environment. Dice provided lists of various mammal, bird, reptile, and amphibian species found in different major habitat types (vegetation associations). In addition to thoroughly describing the vegetation in each habitat type, he presented detailed climatological information for the habitats. In so doing, Dice (1916)

Box 1.4 Lee Raymond Dice

Lee Dice (1887–1977) was born in Savannah, Georgia, but moved with his family to Washington State at an early age. As an undergraduate student, he briefly attended Washington State Agricultural College (now Washington State University), the University of Chicago, and then Stanford University. While at the University of Chicago, he enrolled in Victor Shelford's course in animal ecology. He was immediately hooked on the subject matter; the course was an early and profound influence in his career development (Evans 1978). Dice obtained his PhD in 1915 from the University of California, Berkeley partially under the direction of Joseph Grinnell. Upon completing his doctoral degree, Dice was employed at various universities and government agencies until 1919 when he accepted a permanent position in the Department of Zoology at the University of Michigan where he remained employed until retiring in 1957. Although Dice started his career as a mammalogist and field ecologist, his career eventually came to include an eclectic mix of research topics. He conducted research on the genetics of deer mice (*Permoyscus*) and this eventually led to an interest in the genetics of human epilepsy (Evans 1978). He had early and lifelong interests in paleontology and biogeography (Evans 1978). He even developed a statistical index to measure the amount of "ecological association" between species (Dice 1945). Despite seeming to have many admin-

istrative duties at the University of Michigan, Dice managed to remain engaged in field research, often in relatively far-off places such as Arizona and New Mexico (Evans 1978). From the very beginning of his career, Dice recognized and used the modern concept of habitat. His contributions to the historical development of ecology and wildlife ecology are not as well-known as those of Grinnell and Shelford, but they should be.

furthered the idea of vegetation as habitat for animals and recognized a very comprehensive concept of habitat that took into account abiotic and biotic conditions. Several years later, he conducted a similar study in Montana (Dice 1923). He was very interested in the extent to which structurally- and climatologically similar habitats in different regions might harbor similar animal species. He considered the ecological method of studying species–habitat relationships as transcendent among habitats and regions, applicable to any place on the planet.

In a synthesis paper, Dice (1922) more closely linked animal communities to the habitat concept and he described the need for a new system of ecological classification based on habitats and ecological communities. He appealed to field biologists (museum collectors) to record detailed information on the habitats where their specimens were collected. Dice saw such information as being extremely valuable to ecologists in learning more about the multiple factors that affect (restrict) the distribution of species in nature. He also emphatically commented on the urgency of this task. Even way back in 1922, he recognized that natural habitats were disappearing fast due to humankind's activities and a rapidly modernizing world:

There is pressing need that the work of describing the biotic areas and habitats of the world should be speedily done . . . in our more settled districts it is now difficult or impossible to find even small areas of the original habitats. It is important to determine quickly the habitat preferences of the native plants and animals. It behooves us to record all we can of natural habitats and habitat preferences before it is too late. (p. 338)

This was prescient writing by Dice and crystal-clear recognition of a concern that still exists today.

Meanwhile, the occupational divide between plant and animal ecologists continued into the 1920s. Yapp (1922) discussed different definitions or concepts of habitat but did not cite any of the papers of the animal ecologists and he even claimed that animal ecology was lagging behind plant ecology. He presented a concept of habitat that was spatially abstract and multi-level. Yapp saw a difference between the habitat of an individual plant, a community of plants, and even groups of related plant communities (as in succession). He also said that it was

impossible to physically delimit the outer spatial boundaries of any habitat. To Yapp, habitat consisted of the factors or conditions of the environment that the physiology of an individual plant (and by extension plant communities) must be adapted to, or else the plant would not exist where it does. According to Yapp, "Apart from the factors, the habitat, from the point of view of ecology, is little more than an abstraction" (p. 11)—this was a concept that did not catch on entirely. Nonetheless, Yapp advanced our modern concept of habitat by suggesting that habitat could be viewed and studied in the abstract without reference to a spatial location or the physical structure of vegetation or anything else. Although Yapp separated the concept of a plant association from habitat as a concept, he did acknowledge that animal species might be associated with certain plant communities and such associations were worthy of further study. Eventually, the gap between plant and animal ecologists was closed somewhat with the joint authorship of a book by Clements and Shelford in 1939. To some extent, the early notions of habitat as held by animal ecologists were too much tied to the process of succession. That is, animal ecologists were interested in discovering the animal equivalent to plant succession. Of course, the succession of animal communities is a real process in nature. But the habitat relationships of animal species can certainly be studied without reference to succession.

As early as 1905, Clements had called for a quantitative analysis of habitats, but what he really meant by habitats was plant formations or associations (see also Clements 1913). He desired statistical methods for elaborating the process of succession, regardless of learning more about vegetation as habitat for animals. Further, Shelford and other animal ecologists had introduced an experimental and quantitative approach to the study of habitat selection, although it was very rudimentary by today's standards. Through the 1920s, the modern concept of habitat was becoming better established even though a quantitatively rigorous assessment of a species–habitat relationship had yet to appear.

Perhaps the first quantitative and statistical analysis of the habitat associations of species was Georgy Gause's study of grasshoppers near the city of Sochi in the USSR. Of course, Gause is otherwise well-known to ecologists for his experimental research and writing on the Principle of Competitive Exclusion. Prior to becoming a population biologist and antibiotics researcher, Gause was studying the habitat associations of grasshoppers (Box 1.5). He was only 19 years of age when he wrote "Studies on the Ecology of the Orthoptera" and managed to get it published in *Ecology*, one of the very first ecological journals. The year was 1930 and this was one of Gause's first scientific papers and the first written in English. Although Gause does not mention any practical motivation for studying grasshoppers, we can assume there was one. Grasshoppers were major pests of the vast grain crops under cultivation in the burgeoning USSR, so Gause or his superiors were probably tasked with learning more about the ecology of these organisms. Gause's analysis of the habitat associations of 15 grasshopper species was remarkably sophisticated for the day. Even by today's standards, his analysis is thorough and appropriate; it would be publishable in any number of ecological or entomological journals.

Gause recorded the abundances of the grasshopper species in 400 m² plots in each of 21 sampling localities (or "habitats"). Based on previous studies of other scientists, Gause knew that the different grasshopper species differed in the level of humidity that they required or tolerated. Further, he reasoned that the vegetation in each plot produced a certain microclimate (and yes he used that term) with a particular level of humidity. Therefore for each of the habitats he measured plant biomass and height in 1 m² plots. Then he placed each habitat into one of six categories that represented a progression from xerophytic conditions with little vegetative cover to mesophytic conditions with dense cover to a hydrophilic condition represented by a damp meadow. The latter habitat category had on average about 20 times the vegetative biomass of the most xerophytic category.

Gause described his grasshopper habitat analysis as "a study of the distribution of organisms in relation to the factors of the environment" (p. 307), which he regarded as a fundamental pursuit in the still relatively new science of ecology. In addition to ecological interests, Gause evidently had excellent training in math and statistics. In general terms he described a plot of data in which an environmental factor was on the x-axis and species abundance on the y-axis. He then stated that from the plot, "we will obtain a curve representing the relation which exists between abundance and the given factor" (p. 307). Gause wanted to find the mathematical

---

Box 1.5 Georgy Gause

Georgy Gause (1910–1986) was born in Moscow, Russia. He was admitted to Moscow State University where he received his undergraduate degree in 1931 and PhD in 1940. Prior to achieving his doctoral degree, Gause was already conducting research and writing about the mathematical underpinnings of population biology. For example, he published his ecological classic, *The Struggle for Existence*, in 1934 (and in English!). This book and similar research papers in the 1930s helped lay the early foundation for population ecology and indeed fostered the introduction of mathematics into the historical development of ecology. In ecology, Gause's contributions are equally acknowledged along with those of other early ecologists who studied population dynamics; for example, Vito Volterra, Alfred Lotka, and Raymond Pearl. However, most ecologists are not aware that Gause eventually went on to conduct very important research on antibiotics (Kodash and Fischer 2018) and somewhat left ecology behind. Nonetheless, his simple yet elegant experiments on the coexistence (or lack thereof) of *Paramecium* remain as the classic empirical demonstration of competitive exclusion. Gause was a gifted ecologist and mathematician. If his research interests had remained within ecology, no doubt he would have continued to make meaningful contributions.

expression for the connection between the two variables, the environmental factor and species abundance. He suggested that the abundance data for a given species spread over the habitat categories could be described as a binomial curve (Fig. 1.4). He further stated that so-called ecological curves follow the Law of Gauss (a reference to Carl Friedrich

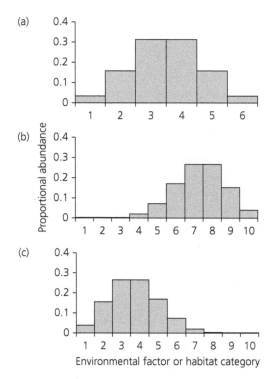

Figure 1.4 Binomial distributions derive from a binomial function that gives the probability of x "successes" for a given number of "trials" and a given probability of success during each trial, $P(x) = \binom{n}{x}p^x(1-p)^{n-x}$, where n = number of trials and p = probability of success. The establishment and existence of an individual of a given species in a sampling plot of a given habitat category can be conceived as a binomial process. In that scenario, a "trial" is a given level of an environmental factor and a "success" is the individual being able to exist in a plot given that level of the environmental factor. In (a) there are six habitat categories (or levels) and $p = 0.5$. Category 1 is a default category and represents zero successes out of six trials. Category 2 includes individuals that had one success. Category 3 represents individuals that had two successes, and so on. In this example, the proportional abundance of the species is identical to $P(x = X)$ for each $X = 0$ to 5, representing the six habitat categories. Any number of habitat categories or levels of an environmental factor could be described by a binomial distribution and p could be a value other than 0.5. In (b) there are 10 levels and $p = 0.7$, in (c) there are 10 levels and $p = 0.3$.

Gauss, the great nineteenth-century mathematician). Gause was likely referring to the fact that binomial distributions can sometimes be approximated by normal (i.e., Gaussian) distributions, even though the former are discrete and the latter are continuous. Binomial distributions derive from binomial processes wherein there are only two possible outcomes (e.g., yes/no, pass/fail, present/absent). Modeling the species–environment relationship as the result of a binomial process is very reasonable in that an individual of the species either is present in or absent from a sampling plot that has a given value(s) of the environmental factor(s)—and presumably presence/absence might be due to one or more of the environmental factors, although Gause did not explain it like this. By extension, the proportion of individuals of the given species in each habitat would follow a binomial distribution (Fig. 1.4). Thus, Gause was envisioning the six habitat categories arranged along the x-axis (as a single comprehensive environmental variable) and observed proportional abundance of each species plotted on the y-axis. In this way, an abundance curve for each species could be derived and analyzed as well as species compared with one another (Fig. 1.5). Of note, Harris et al. (1929) had also suggested a similar statistical comparative approach but without applying it to a particular species.

Somewhat ironically, despite the elegant statistical framework, Gause did not actually make much use of the curves. In principle, he could have tested to see if each curve departed from a random distribution (which would be manifested as a symmetrical hump-shaped curve) or perfectly even distribution (which would be represented as a horizontal line) (Fig. 1.5). A relatively simple goodness-of-fit test (e.g., chi-square test) would have sufficed for this task. To his credit, Gause may have recognized the limits to his data—he had only six habitat categories and hence his environmental variable was very discrete (not continuous) and thus perhaps not very suitable to discriminating a normal distribution. Gause actually alluded to this issue briefly in stating that "the approach to the binomial curve of distribution depends on the number of variants" (or habitat categories) (p. 317).

Nonetheless, Gause made good use of his data. For each species, he calculated the weighted mean

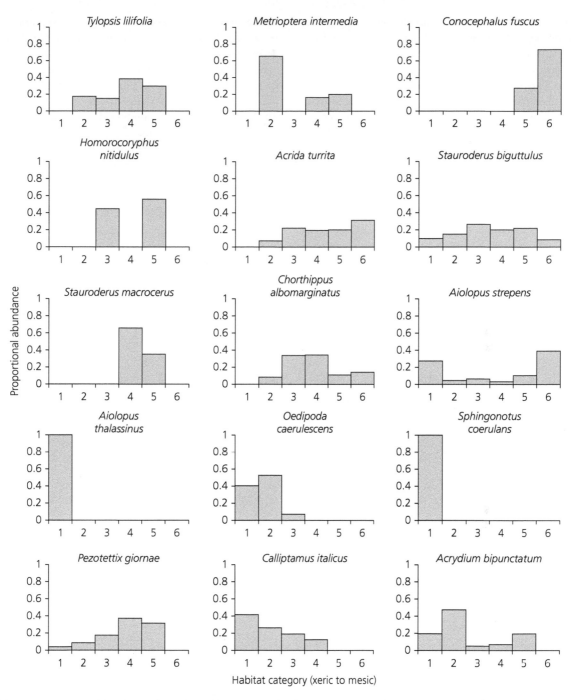

Figure 1.5 Proportional abundances of Gause's grasshopper species distributed among the six habitat categories ranging from xeric (1) to mesic (6) conditions. Gause (1930) did not present the data in this way although he describes "abundance curves." Some of the species have an approximately normal distribution (e.g., *Stauroderus buguttulus, Chorthippus albomarginatus*), being found throughout most of the moisture gradient with a peak in abundance in habitat category 3 or 4. Other species clearly seem to associate with either xeric (e.g., *Aiolopus thalassinus*) or mesic (e.g., *Conocephalus fuscus*) conditions. No species has a uniform distribution that would be evident as the same proportional abundance (1/6 = 0.167) in each habitat category.

of its abundance among the six habitat categories ordered from dry to wet, $M = \Sigma(J \times w_j)$ for $J = 1$ to 6 and w_j = the proportional abundance of the species in habitat category J. Thus, species with low M values were taken to be xerophytic and those with high values were considered mesophytic. Gause also calculated the standard deviations of the M values, in the typical way for a weighted mean, $\mathrm{SD} = \Sigma(m_j - M)^2$ with $m_j = J \times w_j$. This allowed him to compare species and discover that most species were different from one another in where they fell along the continuum of dry to wet environmental conditions (Fig. 1.5). Further he considered the standard deviation to also be a measure of the ecological plasticity of each species or the extent to which the species is found in multiple habitat types and can survive in a wide range of environmental conditions. To compare the ecological plasticity of each species, Gause determined the standard error (i.e., the standard deviation of the $|m_j - M|$ values) of each species plasticity value. Again he found the species were different in their plasticity values or tolerances. Thus, his statistical analysis of the data enabled him to conclude that "the mean value, which characterizes the average conditions, and the dispersion, which is the indicator of ecological plasticity, are on the whole very characteristic for each species" (p. 320). By taking an analytical approach, Gause discovered distinct habitat associations of the grasshopper species that might not have been as evident by merely observing the species in the field and writing a narrative description in the manner of the old-time naturalists. Again, Gause accomplished this in 1930, many years prior to the advent of the advanced statistical procedures that we have today.

Through the 1930s, our modern concept of habitat became more comprehensive and its central role in understanding many ecological processes became more widely recognized. As an example, the well-known ornithologist and evolutionary ecologist David Lack conducted pioneering studies of habitat selection in birds (Lack 1933, 1934) (Box 1.6). He

---

Box 1.6　David Lambert Lack

David Lack (1910–1973) was born in London, England. From an early age, he had an interest in birds and even published studies of bird behavior while an undergraduate student at Cambridge University. As a young man, Lack actually crossed paths with Joseph Grinnell, meeting him in August 1935 when Lack visited the Museum of Vertebrate Zoology at the University of California, Berkeley after a cross-country road trip (Anderson 2013). Lack received his ScD (PhD) from Cambridge University in 1948. His entire career was spent as director of the Edward Grey Institute of Field Ornithology at Oxford University. Lack's main interest in habitat was concerned with the role that habitat differences played in the speciation process as well as permitting coexistence of similar and potentially competing species in the same general region. By the 1940s he was clearly influenced by Gause and the Principle of Competitive Exclusion (Anderson 2013). Lack also studied the life histories of various bird species and the ways in which density-dependent processes regulated population size. He often saw natural selection at work in molding the behavior and traits of birds. He is probably best known for the idea that clutch size is a species-specific trait that maximizes the parent birds' fitness when the number of eggs laid corresponds to the number of nestlings that the parents will have the greatest success in fledging. This hypothesis is now widely known in behavioral ecology and ornithology as Lack's Principle. It is difficult to overestimate David Lack's contributions to ecology, ornithology, and behavioral ecology, particularly with regard to the evolutionary context he imbued in each of these disciplines.

sought evolutionarily based explanations for how and why certain species preferred specific habitats and in particular he attributed much of the selection process to innate behavior or what he often called the "psychological factor" (Lack 1933, 1937). Ultimately, Lack was interested in the factors that limited bird distribution and abundance, a pursuit of ecological understanding that he shared with many other fellow animal ecologists. However, to Lack, the limiting factor could be something as complex as a direct behavioral response to the physical structure of the habitat rather than an automatic physiological response of the animal to a physico-chemical factor in the environment. The latter had been the focus of many previous animal ecologists, particularly those studying invertebrates and fishes. Thus, Lack pushed forward the need to understand and thoroughly characterize the habitat of a species—it was a necessary step toward investigating ecological and behavioral processes. In short, during the 1930s, ecologists began to recognize the benefits of studying the relationship of the organism (species) itself to its habitat in addition to merely identifying and describing its habitat (Klauber 1931; Pearson 1933; Boycott 1934; Edge 1935; Mosauer 1935).

The 1930s also saw the birth and development of a new academic discipline, wildlife ecology. Initially this discipline was better known as wildlife management. Aldo Leopold is widely acknowledged as its founder, with wildlife management becoming recognized as an academic discipline when Leopold published *Game Management* in 1933 and acquired a professorship of game management at the University of Wisconsin in the same year. The title aptly described the book—it was about managing game species (notably upland bird species and deer) for the purpose of maintaining populations that could be harvested or hunted. Somewhat ironically, the book did not have any discussion of habitat (Leopold used the word "habitat" only six times in the entire book) although habitat management has become a huge part of wildlife ecology and conservation in the past five decades or so (see Chapter 2). Leopold's main goal was to discuss how the productivity (i.e., population growth rate) of a game species could be increased by identifying and then manipulating the factors that were restricting the population. These factors included food supply,

water, nesting sites (for birds), predators, disease, and vegetative cover to hide from predators and to shelter from weather. Leopold referred to "coverts" as patches of vegetative cover and discussed how these could be created or modified—essentially he was advocating for habitat management. Leopold certainly recognized the broader ecological concept of habitat; he simply did not use the word much.

Herbert Stoddard's *Bobwhite Quail: Its Habits, Preservation, and Increase*, published in 1931, was perhaps the first discussion of habitat in the context of conservation and management of a harvested species. Stoddard's book is remarkable for its thoroughness in covering all aspects of bobwhite biology and ecology, including the species economic and cultural importance as a game species that had been hunted in America since European settlement. In *Game Management*, Leopold discussed Stoddard's research on bobwhite quail at length, and clearly admired the holistic and comprehensive approach of Stoddard. In addition to bobwhite quail, wildlife ecologists conducted habitat studies and implemented management for other wildlife species during this time period, such as muskrat (Errington 1937, 1940; Hamerstrom and Blake 1939; Bellrose and Brown 1941; Lay and O'Neil 1942), gray and fox squirrels (Baumgartner 1939; Goodrum 1940; Baker 1944; Brown and Yeager 1945), white-tailed deer (Hosley and Ziebarth 1935; Morton and Sedam 1938; Buechner 1944; Cook 1945; Allen 1948), and various waterfowl species (Girard 1941; Low 1941; Lynch 1941; Wiebe 1946; Mendall 1949). Although these and other wildlife studies rarely included an analysis of habitat that was as quantitative as Gause's (1930) study of grasshoppers, they did accomplish one crucial step forward. By focusing on the management of habitat as a way to maintain and increase populations, these studies attempted to discover the *habitat requirements* of species. Therefore in a sometimes subtle way, the early wildlife ecologists implicitly reinforced the idea that suitable habitat is required for a population's persistence and species' existence—even today this remains a main reason for conducting an analysis of habitat.

In general, by the 1940s, the modern perspective on habitat was firmly established in ecology and becoming more important in wildlife ecology. Several comprehensive ecological textbooks had appeared in the previous decades. Textbooks often have a subtle but widespread effect on the development and growth of a concept within an academic field. When something makes its way into a textbook then it is definitely knowledge that the elders want to pass on to the students of the discipline. As examples, we have Adams (1913) *Guide to the Study of Animal Ecology*, Pearse (1926) *Animal Ecology*, Elton (1927) *Animal Ecology*, Clements and Shelford (1939) *Bio-ecology*, and Allee et al. (1949) *Principles of Animal Ecology*. These books are still relevant today and worth browsing through—you will be surprised at how long we've been aware of some ecological concepts and processes that we still study nearly a century later.

1.2 Definitions and terminology

In any field of study or academic topic, precise terminology is important. Up to now, I have intentionally avoided giving a verbatim definition of "habitat." Rather, by describing the habitat concept and putting usage of the term "habitat" in a historical context, my hope was that a definition of habitat would implicitly emerge in the reader's mind (granted, most readers come to this book with some notion of habitat anyway). Nonetheless, I will now be more precise. However, I first comment on some definitions by other authors. Hall et al. (1997) defined "habitat" (paraphrasing) as "the resources and conditions in an area that lead to survival and reproduction, relating the presence of a species, population, or individual to an area's physical and biological characteristics, implying more than vegetation and vegetation structure, the sum of the specific resources that are needed by organisms" (p. 175). Morrison et al. (2006) then refined this definition and made it more detailed with regard to identifying the resources and conditions. They stated "Habitat is an area with a combination of resources (like food, cover, water) and environmental conditions (temperature, precipitation, presence or absence of predators and competitors) that promotes occupancy by individuals of a given species (or population) and allows those individuals to survive and reproduce" (p. 10); this definition was also advocated more recently by Mathewson and Morrison (2015) and Krausman and Morrison

(2016). It is a thorough and comprehensive definition but it has some fairly subtle implications that could limit its usefulness.

A careful reading of the above definitions implies that once a habitat loses a particular required resource (e.g., food) or gains a negative influence (e.g., a predator species) and thus conditions do not lead to survival and reproduction, then the area ceases to be habitat. These definitions do not allow (explicitly) for habitat to be unoccupied and by extension they do not recognize that species often do not saturate their habitat in a given region. A habitat that has a diminished food supply, excessive predator density, or that is difficult to reach or colonize is still habitat for a prey species, it simply is not occupied for obvious reasons. The definition provided by Morrison et al. (2006) fits very well with what we might call *realized habitat*, or the area where the species is *actually found* due to the area's combination of resources and conditions allowing survival and reproduction (i.e., occupancy).

Block and Brennan (1993) provided a very succinct definition of habitat as "the subset of physical environmental factors that a species requires for its survival and reproduction" (p. 36). Note that their definition implies that the habitat contains the resources for survival and reproduction without stating whether the species actually does survive and reproduce or even occupy the habitat—this is an improvement on the Hall et al. (1997) and Morrison et al. (2006) definitions in that it does not require occupancy. In most modern definitions of habitat, it is implied that the conditions and factors of a species' habitat typically facilitate and perhaps are required for survival and reproduction of individuals. Hence, the phrase "survival and reproduction" need not be a formal part of the definition of habitat. Indeed, intentionally omitting the phrase then is a way around the paradox identified by Van Horne (1983)—that is, the density of a species in a particular area may not necessarily indicate that the area consists of habitat that allows for survival and reproduction. Further, Mitchell (2005) criticized abundance-based definitions of habitat in that (according to him) they are only valid when abundance of the species is maximized; that is, the habitat is saturated. Otherwise, it is impossible to know whether a given area truly is habitat in the sense of restricting the distribution of the species, other non-habitat factors may be restricting a species' abundance in areas that we think (incorrectly) are habitat (Mitchell 2005). By using a habitat definition that does not directly refer to abundance, survival, and reproduction, we can avoid the conundrum described by Mitchell (2005).

In this book, I use an operational definition of habitat, defining it as *the set of physical environmental variables (excluding food resources, climate, and predators/competitors) that a species associates with in a nonrandom way and the spatial locations where all or some minimal subset of those variables occur*. This definition is closely tied to conducting an actual habitat analysis (Chapters 4–9). Clearly, organisms need food (energy) for survival and reproduction, and often the habitat of a species (particularly the foraging habitat) will have or be capable of having the food resource. In that sense, food may be a component of the habitat, just as weather, climate, and the presence (or absence) of other species are also components. Importantly these components do not define the habitat and thus the habitat may either have or not have these components (or certain conditions of the components) and still be considered habitat. The presence/absence of these other components may greatly influence whether a particular area of habitat has the species and at what density. For example, the food supply can presumably be diminished enough in the habitat that the species becomes extirpated or exists at very low density, in which case habitat is not as important (not the limiting factor) as food supply. My definition points to habitat being primarily characterized based upon the physical structure of the place (area) where the species resides (see Chapters 3 and 5). Again, the definition is intended to be operational in the context of an analysis of habitat (Chapter 5) and allow the effect of habitat to be separately compared with the effect of other factors (food resources, predators/competitors, climate, and disease) that also may determine species distribution and abundance (Chapter 3).

This definition of habitat based on relating a species occurrence to environmental conditions is similar to what Mitchell (2005) referred to as the distributional concept of habitat. Further, as pointed out by Mitchell (2005), this concept of habitat implies that some type of correlation analysis needs to be conducted to truly define the habitat of a species. Mitchell (2005) argued that there are conceptual and statistical weaknesses in trying to correlate species abundance to environmental factors. Most notably, Mitchell criticized the distributional concept because it lacks any kind of mechanistic underpinning; the concept alone does not answer how and why a species is associated with particular environmental conditions (i.e., habitat factors). As such, he argued that habitat was likely to have very limited usefulness in explaining the distribution and abundance of species. Much of Mitchell's critique of habitat concepts is centered on issues pertaining to the *analysis* of habitat; therefore I revisit his arguments in Chapter 5.

Kearney (2006) suggested that the distributional (or descriptive) concept of habitat is appropriate and that Mitchell's plea for a mechanistic concept was best considered as a concept or definition of the niche. This point might also apply to the very inclusive habitat definition offered by Hall et al. (1997) and Morrison et al. (2006). Interestingly, Hall et al. (1997) did not use the term "niche" and thus by default did not distinguish it from habitat. Morrison et al. (2006) presented a thorough discussion of the distinction between habitat and niche, with niche primarily referring to how an individual animal uses its habitat and the resources it contains whereas habitat is the description of the resources. They stated, "niche factors include various resources, such as the type and size of food required, and constraints on the acquisition of those resources, such as the activity of predators and competitors" (p. 421), thus there is some overlap in their concepts of niche and habitat. Nonetheless, they emphasized the importance of distinguishing between niche and habitat when examining the factors limiting the distribution and abundance of a species, as have other authors (Mitchell 2005; Kearney 2006). In this book, I regard the niche as almost any ecological textbook would. I also follow the above-mentioned authors in conceiving of niche as how a species uses resources as well as how it is adapted to a particular habitat and interacts with other species in the habitat. To me, the habitat is the where and what, the niche is the how and why. Habitat is tangible; niche is abstract (although real). I could take the reader outdoors and show the habitat of a given species. I could not show or point to the niche of the species.

In writing a book on habitat analysis, it is worthwhile to mention a few more terms related to habitat. These are selection, preference, use, requirements, suitability, and quality (insert the word "habitat" in front of each term). Rather than offer a precise definition of each, I will refer the reader to other sources that thoroughly discuss these terms in a literal sense, a conceptual sense, and as actual processes or observable characteristics of a species or the environment (Hall et al. 1997; Jones 2001; Morrison et al. 2006; Johnson 2007; Krausman and Morrison 2016; Kirk et al. 2018). The reader is forewarned that some of these authors (e.g., Hall et al. 1997; Morrison et al. 2006; Krausman and Morrison 2016) are strict in how these terms should be used. Hall et al. (1997) presented a very thorough discussion of these terms and their *possible* misuse in the literature. However, perceived misuse of these terms may be overstated if the context of their usage makes clear the author's intent and thereby does not lead to misunderstanding by the reader (Hodges 2008, 2014). Further, rigid definitions for some terms and concepts (and highly standardized ecological terminology in general) could actually hinder progress in understanding the underlying ecology if the pattern or process being studied does not fit neatly under one of the definitions (Hodges 2008). In using "habitat" and habitat-related terms the responsibility is on the writer to clearly define the intended meaning, but that meaning (or definition) need not exactly match the ways in which other authors have used the term. "Habitat" is a polysemous term, but that does not entail confusion or misunderstanding (Hodges 2008).

1.3 Conclusion

"Habitat" as a concept and descriptive label has a long history in ecology (Fig. 1.2). It dates back to at least the 1750s and the time of Linnaeus. Scientists, naturalists, and philosophers at that time were trying to make sense of the natural world by finding pattern and order. Associated with this pursuit, there was a desire to inventory nature. Thus, we had the Linnaean view of species as discrete items from nature that could be collected, described, and cataloged. This was a perspective that viewed the natural world as a library, with species being one

type of book and individuals the pages of the book. However, a habitat is not a discrete item of nature that can be collected as a physical sample in its entirety and brought into a museum. Hence, the first usage of "habitat" (as in the writings of Linnaeus) was as a descriptor to identify the locality or region where a species is found. Eventually, "habitat" went from being the actual locality where an organism lives to the *kind of locality*.

Habitat and a species requirement of or tolerance of particular environmental factors came to be seen as a determinant of species abundance and distribution. Ecologists became interested in how species select habitats and the evolutionary and ecological consequences of such selection with regard to individuals being adapted to certain habitats. Gradually, the concept of habitat became more comprehensive and integral to the development of ecological theory. At the same time, the practical implications of understanding a species habitat took on great importance—the need arose to conserve natural habitats that were quickly vanishing and to manage the habitat of wildlife species that were harvested.

As with any intellectually rich and broad arena of scientific investigation, the study and analysis of habitat comes with many terms and definitions. It is useful to recognize the distinctions of the terminology and use terms in a consistent way (Hall et al. 1997); however, unnecessary rigidity in definitions should not be allowed to interfere with the main goal of analyzing habitat and learning more about species–habitat relationships. Our modern concept of habitat was set by the 1940s and yet since then its usefulness in understanding species distribution and abundance and other ecological patterns and processes has been somewhat ignored. Chapter 3 is my attempt to re-energize the habitat concept for a central role in understanding the ecology of nature.

References

Abbott, C.C. (1860). Descriptions of two new species of *Pimelodus* from Kansas. *Proceedings of the Academy of Natural Sciences of Philadelphia*, 12, 568–69.

Abbott, C.C. (1870a). Mud-loving fishes. *American Naturalist*, 4, 385–91.

Abbott, C.C. (1870b). Notes on certain inland birds of New Jersey. *American Naturalist*, 4, 536–50.

Abbott, C.C. (1870c). Notes on freshwater fishes of New Jersey. *American Naturalist*, 4, 99–117.

Abbott, C.C. (1873). Notes on the habits of certain crawfish. *American Naturalist*, 7, 80–84.

Alagona, P.S. (2012). A sanctuary for science: the Hastings Natural History Reservation and the origins of the University of California's Natural Reserve System. *Journal of the History of Biology*, 45, 651–80.

Allen, G.W. (1948). The management of Georgia deer. *Journal of Wildlife Management*, 12, 428–32.

Anderson, T.R. (2013). *The Life of David Lack: Father of Evolutionary Ecology*. Oxford University Press, Oxford.

Bailey, V. (1900). Revision of American voles of the genus *Microtus*. North American Fauna, Number 17, United States Department of Agriculture, Washington, DC.

Baker, R.H. (1944). An ecological study of tree squirrels in eastern Texas. *Journal of Mammalogy*, 25, 8–24.

Baumgartner, L.L. (1939). Fox squirrel dens. *Journal of Mammalogy*, 20, 456–65.

Bellrose, F.C. and Brown, L.G. (1941). The effect of fluctuating water levels on the muskrat population of the Illinois River Valley. *Journal of Wildlife Management*, 5, 206–12.

Block, W.M. and Brennan, L.A. (1993). The habitat concept in ornithology, theory and applications. Pages 35–91 in *Current Ornithology*, Vol. 11, Power, D.P. (editor), Plenum Press, New York.

Boycott, A.E. (1934). The habitats of land Mollusca in Britain. *Journal of Ecology*, 22, 1–38.

Brown, L.G. and Yeager, L.E. (1945). Fox squirrels and gray squirrels in Illinois. *Illinois Natural History Survey Bulletin*, 23, 448–36.

Buechner, H.K. (1944). The range vegetation of Kerr County, Texas, in relation to livestock and white-tailed deer. *American Midland Naturalist*, 31, 697–43.

Clements, F.E. (1905). *Research Methods in Ecology*. University Publishing Company, Lincoln, Neb.

Clements, F.E. (1913). The alpine laboratory. *Science*, 37, 327–28.

Clements, F.E. (1916). *Plant Succession, an Analysis of the Development of Vegetation*. Carnegie Institution, Washington, DC.

Cook, F.W. (1945). White-tailed deer in the Great Plains region. *Journal of Wildlife Management*, 9, 237–42.

Cope, E.D. (1869). A review of the species of the Plethodontidae and Desmognathidae. *Proceedings of the Academy of Natural Sciences of Philadelphia*, 21, 93–118.

Cope, C.B. (1896). The geographical distribution of Batrachia and Reptilia in North America. *American Naturalist*, 30, 886–902.

Cory, C.B. (1891). Notes on West Indian birds. *The Auk*, 8, 41–46.

Cory, C.B. (1886). Descriptions of thirteen new species of birds from the island of Grand Cayman, West Indies. *The Auk*, 3, 497–501.

Dice, L.R. (1916). Distribution of the land vertebrates of southeastern Washington. *University of California Publications in Zoology*, 16, 293–48.

Dice, L.R. (1922). Biotic areas and ecologic habitats as units for the statement of animal and plant distribution. *Science*, 55, 335–38.

Dice, L.R. (1923). Mammal associations and habitats of the Flathead Lake Region, Montana. *Ecology*, 4, 247–60.

Dice, L.R. (1945). Measures of the amount of ecologic association between species. *Ecology*, 26, 297–302.

Edge, E.R. (1935). A study of the relation of the Douglas ground squirrel to the vegetation and other ecological factors in western Oregon. *American Midland Naturalist*, 16, 949–59.

Errington, P.L. (1937). Habitat requirements of stream-dwelling muskrats. *Transactions of the North American Wildlife Conference*, 2, 411–16.

Errington, P.L. (1940). Natural restocking of muskrat-vacant habitats. *Journal of Wildlife Management*, 4, 173–85.

Evans, F.C. (1978). Lee Raymond Dice (1887–1977). *Journal of Mammalogy*, 59, 635–44.

Fowler, A. (1868a). The belted kingfisher. *American Naturalist*, 1, 403–405.

Fowler, A. (1868b). The chickadee. *American Naturalist*, 1, 584–87.

Fowler, A. (1869). The golden-winged woodpecker. *American Naturalist*, 3, 422–27.

Gause, G.F. (1930). Studies on the ecology of the Orthoptera. *Ecology*, 11, 307–25.

Gause, G.F. (1934). *The Struggle for Existence*. William and Wilkins Publishing Company, Baltimore, Md.

Girard, G.L. (1941). The mallard: its management in western Montana. *Journal of Wildlife Management*, 5, 233–59.

Goodrum, P.D. (1940). A population study of the gray squirrel in eastern Texas. Texas Agricultural Experiment Station Bulletin 591. College Station, Tex.

Grinnell, J. (1897). Description of a new towhee from California. *The Auk*, 14, 294–96.

Grinnell, J. (1904). The origin and distribution of the chestnut-backed chickadee. *The Auk*, 21, 364–82.

Grinnell, J. (1917). The niche-relationships of the California thrasher. *The Auk*, 34, 427–33.

Günther, A. (1871). Description of *Ceratodus*, a genus of ganoid fishes, recently discovered in rivers of Queensland, Australia. *Philosophical Transactions of the Royal Society of London*, 161, 511–84.

Günther, A. (1875). Description of the living and extinct races of gigantic land-tortoises—Parts I. and II. Introduction, and the tortoises of the Galapagos Islands.

Philosophical Transactions of the Royal Society of London, 165, 251–84.

Hall, L.S., Krausman, P.R., and Morrison, M.L. (1997). The habitat concept and a plea for standard terminology. *Wildlife Society Bulletin*, 25, 173–82.

Hamerstrom, F.N. and Blake, J. (1939). Central Wisconsin muskrat study. *American Midland Naturalist*, 21, 514–20.

Hargitt, C.W. (1901a). Synopses of North American invertebrates, The Hydromedusae—Part I. *American Naturalist*, 35, 301–15.

Hargitt, C.W. (1901b). Synopses of North American invertebrates, The Hydromedusae—Part II. *American Naturalist*, 35, 379–95.

Hargitt, C.W. (1901c). Synopses of North American invertebrates, The Hydromedusae—Part III. *American Naturalist*, 35, 575–95.

Harris, J.A., Kuenzel, J., and Cooper, W.S. (1929). Comparison of the physical factors of habitats. *Ecology*, 10, 47–66.

Hodges, K.E. (2008). Defining the problem: terminology and progress in ecology. *Frontiers in Ecology and the Environment*, 6, 35–42.

Hodges, K.E. (2014). Clarity in ecology: terminological prescription is the wrong route. *Bioscience*, 64, 373.

Hosley, N.W. and Ziebarth, R.K. (1935). Some winter relations of the white-tailed deer to the forests in north central Massachusetts. *Ecology*, 16, 535–53.

Johnson, M.D. (2007). Measuring habitat quality: a review. *Condor*, 109, 489–504.

Jones, J. (2001). Habitat selection studies in avian ecology: a critical review. *The Auk*, 118, 557–62.

Jordan, D.S. (1874). A key to the higher algae of the Atlantic Coast, between Newfoundland and Florida. *American Naturalist*, 8, 479–93.

Jordan, D.S. (1877). On the fishes of northern Indiana. *Proceedings of the Academy of Natural Sciences of Philadelphia*, 29, 42–82.

Jordan, D.S. and Copeland, H.E. (1877). The sand darter. *American Naturalist*, 11, 86–88.

Kearney, M. (2006). Habitat, environment and niche: what are we modeling? *Oikos*, 115, 186–91.

Kirk, D.A., Park, A.C., Smith, A.C., et al. (2018). Our use, misuse, and abandonment of a concept: whither habitat? *Ecology and Evolution*, 8, 4197–208.

Klauber, L.M. (1931). A statistical survey of the snakes of the southern border of California. *Bulletins of the Zoological Society of San Diego*, No. 8.

Kodash, N. and Fischer, M. (2018). Georgy Gause's shift from ecology and evolutionary biology to antibiotics research: reasons, objectives, circumstances. *Theory in Biosciences*, 137, 79–83.

Krausman, P.R. and Morrison, M.L. (2016). Another plea for standard terminology. *Journal of Wildlife Management*, 80, 1143–44.

Lack, D. (1933). Habitat selection in birds, with special reference to the effects of afforestation on the Breckland avifauna. *Journal of Animal Ecology*, 2, 239–62.

Lack, D. (1934). Habitat distribution in certain Icelandic birds. *Journal of Animal Ecology*, 3, 81–90.

Lack, D. (1937). The psychological factor in bird distribution. *British Birds*, 31, 130–36.

Lay, D.W. and O'Neil, T. (1942). Muskrats on the Texas coast. *Journal of Wildlife Management*, 6, 301–11.

Leavitt, R.G. (1907). The distribution of closely related species. *American Naturalist*, 41, 207–40.

Lockwood, S. (1875). The pine snake of New Jersey. *American Naturalist*, 9, 1–14.

Lockwood, S. (1876). The Florida chameleon. *American Naturalist*, 10, 4–16

Low, J.B. (1941). Nesting of the ruddy duck in Iowa. *The Auk*, 58, 506–17.

Lynch, J.J. (1941). The place of burning in management of the Gulf Coast wildlife refuges. *Journal of Wildlife Management*, 5, 454–57.

Marsham, T. (1791). Observations on the *Phalaena Bombyx lubricipeda* of Linnaeus and some other moths allied to it. *Transactions of the Linnean Society*, 1, 67–75.

Mathewson, H.A. and Morrison, M.L. (2015). The misunderstanding of habitat. Pages 3–8 in *Wildlife Habitat Conservation*, Morrison, M.L. and Mathewson, H.A. (editors), Johns Hopkins University Press, Baltimore, MD.

Mendall, H.L. (1949). Food habits in relation to black duck management in Maine. *Journal of Wildlife Management*, 13, 64–101.

Miller, G.S. (1899). Mammals of New York. *Bulletin of the New York State Museum*, 6, 273–85.

Mitchell, S.C. (2005). How useful is the concept of habitat?—a critique. *Oikos*, 110, 634–38.

Morrison, M.L., Marcot, B.G., and Mannan, R.W. (2006). *Wildlife–Habitat Relationships: Concepts and Applications*. Third edition. Island Press, Washington, DC.

Morton, J.N. and Sedam, J.B. (1938). Cutting operations to improve wildlife environment on forest areas. *Journal of Wildlife Management*, 2, 206–14.

Mosauer, W. (1935). The reptiles of a sand dune area and its surroundings in the Colorado Desert, California: a study in habitat preference. *Ecology*, 16, 13–27.

Owen, R. (1836). Indian antelope (*Antilope cervicapra*, Pall.) with a tabular view of the relations between the habits and habitats of the several species of antelopes and their suborbital, maxillary, post-auditory, and inguinal glands. *Proceedings of the Zoological Society of London*, 4, 34–40.

Packard, A.S. (1867). The dragon-fly. *American Naturalist*, 1, 304–13.

Packard, A.S. (1871). Bristle-tails and spring-tails. *American Naturalist*, 5, 91–107.

Packard, A.S. (1876). The cave beetles of Kentucky. *American Naturalist*, 10, 282–87.

Pearson, J.F.W. (1933). Studies on the ecological relations of bees in the Chicago region. *Ecological Monographs*, 3, 373–41.

Pound, R. and Clements, F.E. (1900). *The Phytogeography of Nebraska*. University Publishing Company, Lincoln, Neb.

Reeves, C.D. (1907). The breeding habits of the rainbow darter (*Etheostoma coeruleum*, Storer), a study in sexual selection. *Biological Bulletin*, 14, 35–59.

Shelford, V.E. (1907). Preliminary note on the distribution of the tiger beetles (*Cicindela*) and its relation to plant succession. *Biological Bulletin*, 14, 9–14.

Shelford, V.E. (1908). Life-histories and larval habits of the tiger beetles (Cicindelidae). *Journal of the Linnean Society of London—Zoology*, 30, 157–84.

Shelford, V.E. (1911). Ecological succession, I. Stream fishes and the method of physiographic analysis. *Biological Bulletin*, 21, 9–35.

Shelford, V.E. (1912a). Ecological succession, IV. Vegetation and the control of land animal communities. *Biological Bulletin*, 23, 59–99.

Shelford, V.E. (1912b). Ecological succession, V. Aspects of physiological classification. *Biological Bulletin*, 23, 331–70.

Shelford, V.E. (1915). Principles and problems of ecology as illustrated by animals. *Journal of Ecology*, 3, 1–23.

Shelford, V.E. (1926). Methods for the experimental study of the relations of insects to weather. *Journal of Economic Entomology*, 19, 251–61.

Shelford, V.E. (1933). Conservation versus preservation. *Science*, 77, 535.

Shelford, V.E. (1934). Faith in the results of controlled laboratory experiments as applied in nature. *Ecological Monographs*, 4, 491–98.

Shelford, V.E. (1941). List of reserves that may serve as nature sanctuaries of national and international importance, in Canada, the United States, and Mexico. *Ecology*, 22, 100–107.

Shelford, V.E. (1954). An experimental approach to the study of bird populations. *Wilson Bulletin*, 66, 253–58.

Shelford, V.E. and Eddy, S. (1929). Methods for the study of stream communities. *Ecology*, 10, 382–91.

Shelford, V.E. and Powers, E.B. (1915). An experimental study of the movements of herring and other marine fishes. *Biological Bulletin*, 28, 315–34.

Shelford, V.E., Weese, A.O., Rice, L.A., Rasmussen, D.I., and MacLean, A. (1935). Some marine biotic communities of the Pacific Coast of North America, Part I. General survey of the communities. *Ecological Monographs*, 5, 249–92.

Smith, J.E. (1791). Descriptions of ten species of lichen, collected in the south of Europe. *Transactions of the Linnean Society*, 1, 81–85.

Stauffer, R.C. (1957). Haeckel, Darwin, and Ecology. *Quarterly Review of Biology*, 32, 138–44.

Stokes, A.C. (1888). Fresh-water infusoria. *Journal of the Trenton Natural History Society*, 1, 71–344.

Tansley, A.G. (1920). The classification of vegetation and the concept of development. *Journal of Ecology*, 8, 118–49.

Van Horne, B. (1983). Density as a misleading indicator of habitat quality. *Journal of Wildlife Management*, 47, 893–01.

Wiebe, A.H. (1946). Improving conditions for migratory waterfowl on TVA impoundments. *Journal of Wildlife Management*, 10, 4–8.

Wilder, H.H. (1899). *Desmognathus fusca* (Rafinesque) and *Spelerpes bilineatus* (Green). *American Naturalist*, 33, 231–46.

Williamson, M.B. (1894). Abalone or *Haliotis* shells of the Californian coast. *American Naturalist*, 28, 849–58.

Yapp, R.H. (1922). The concept of habitat. *Journal of Ecology*, 10, 1–17.

CHAPTER 2

Ecology and Wildlife Ecology as Distinct Academic Disciplines

As was evident from Chapter 1, wildlife ecology came into being as an academic discipline and applied science somewhat separately from ecology. This chapter examines and discusses their separate trajectories. Of course, they have at least one common interest which is also the raison d'être for this book; that is, a recognition that habitat matters to individuals, populations, and species. This chapter is not intended as a thorough review of the historical development of wildlife ecology, nor as a discussion of all the similarities and differences between wildlife ecology and ecology. Rather I focus on the extent to which practitioners in each discipline have directly studied habitat or invoked habitat in their study of various processes and patterns.

Ecology and wildlife ecology are distinct academic disciplines. As students, practitioners in each discipline may receive somewhat different training and education inasmuch as at many universities (particularly in the United States) there are separate graduate programs in the two disciplines, even though there are increasingly more interdisciplinary programs. There are scientific journals that fit best under the label of "ecology" and others that are best described as covering "wildlife ecology." I also suspect that there is not tremendous overlap in membership of professional societies, with few practitioners being dues-paying members of the Ecological Society of America (and/or the British Ecological Society) and the Wildlife Society. In the United States, the academic distinction between ecology and wildlife ecology is most pronounced at the land-grant universities (i.e., those universities designated as beneficiaries of the Morrill Acts of

1862 and 1890 with the mission of educating students in agriculture and engineering). There is typically one land-grant university per state and many of the land-grant universities have administrative structures in which there is a college of science separate from a college of agriculture, natural resources, or applied science. Each college has academic departments wherein departments focused on wildlife ecology are usually within the natural resources college, whereas departments providing training and education in ecology are within the college of science. The University of Wisconsin—Madison is the land-grant university for that state. When the university hired Aldo Leopold in 1933, he was assigned to the Department of Agricultural Economics. Shortly thereafter, administrators created a new academic department (initially the Department of Game Management and then renamed Wildlife Management a few years later and finally Wildlife Ecology in 1967) and placed it within the College of Agriculture with Leopold as the founding member of the department (Meine 2010). Placement of the department within the College of Agriculture was likely due to game management at that time being equated with game-cropping or the production of wildlife biomass for eventual harvest. Indeed, early on, wildlife management was about restoring game (wildlife) populations on private lands (farms). Thus, the separate historical trajectories and research emphases of ecology and wildlife ecology may partly be traced back to a decision made by administrators at a single university, as well as the view that wildlife ecologists would be providing a service (consultation) to

Habitat Ecology and Analysis. Joseph A. Veech, Oxford University Press (2021). © Joseph A. Veech. DOI: 10.1093/oso/9780198829287.003.0002

farmers. Certainly, this precedent was there when other land-grant universities sought to establish similar applied science departments focused on wildlife ecology and natural resources management, and assign those departments to the college of agriculture, not science.

Beyond the practical, academic, and occupational distinctions between ecology and wildlife ecology, the two disciplines differ with regard to research paradigms (Table 2.1). Wildlife ecology is narrower in subject matter and can be defined as the study of the factors (broadly considered) that affect persistence of populations of vertebrate species, whereas ecology is traditionally defined in either of two ways: study of the distribution and abundance of organisms or study of the interaction between organism and environment. Either way, ecology has had and continues to have a much more inclusive foundation of subject matter and is more concerned with the first principles that govern natural systems than is wildlife ecology. However, the differences between ecology and wildlife ecology (as listed in Table 2.1) are starting to diminish. Indeed, within contemporaneous wildlife programs and departments, there are certainly some faculty whose research and teaching activities would fit within the category of ecology more so than wildlife ecology, or perhaps fit equally well in either. In addition, the relationship between ecology and wildlife ecology is complicated a bit more by a third separate and distinct academic discipline, conservation biology. It emerged in the late 1980s as a highly interdisciplinary arena that includes aspects of applied ecology, economics, philosophy, geography, and even sometimes the social sciences. Thus, there are at least three academic and scientific disciplines that can involve the study and analysis of habitat.

The three disciplines differ in the extent to which practitioners actually do study and analyze habitat. I conducted a keyword search on "habitat" and related terms as they appeared in select journals from the three disciplines. Papers in journals covering wildlife ecology used "habitat" and related terms much more often than those covering ecology and conservation biology (Table 2.2). On average, about half the papers published in the six main wildlife ecology journals had "habitat" appear in either their title or abstract. Less than a third of the papers in eight ecological journals included "habitat" (Table 2.2). The wildlife ecology journals also had the greatest percentage of papers using "habitat" conjoined with another term (e.g., "habitat selection") indicating that those journals were more likely to publish papers addressing some aspect of habitat ecology rather than just using "habitat" as a common term for where an individual or species resides (Table 2.2). This keyword analysis revealed an interesting historical aspect. The modern concept of habitat emerged in ecology prior to the birth of wildlife ecology (Chapter 1) and yet, over the

Table 2.1 Comparison of ecology and wildlife ecology as academic disciplines and research arenas.

Point of comparison	Ecology	Wildlife ecology
Fundamental area of study	Distribution and abundance of organisms	Survival, reproduction, and habitat requirements of species
Derived area of study	Interrelationships between organisms and their environment	Environmental factors and demographic characteristics that affect population size and growth
Emphasis	Studies of communities and ecosystems (primarily synecological)	Single-species studies (primarily autecological)
Taxonomic focus	Wide range of species including microbes, protozoa, invertebrates, vertebrates, and plants	Vertebrate species, initially and primarily mammal and bird species that were harvested from nature
Role of theory[1]	Typically prominent role although often indirect; development and testing of theory is encouraged	Not as prominent; minimal amount of emergent theory
Role of habitat	Less recognition of the importance of habitat	Past and on-going recognition of the importance of habitat

[1] Note that this applies to the discipline in general, not the study of habitat in particular.

Table 2.2 Use of the word "habitat" and related terms in select journals from the first year of journal publication to 2017. Data were compiled from keyword searches in ISI Web of Science. Search was based on terms occurring in the title or abstract.

Publication	First year[1]	Number of papers	Percentage of papers using the given keyword				
			Habitat	Habitat selection	Habitat use	Habitat preference	Habitat management
Journals covering wildlife ecology							
Journal of Wildlife Management[2]	1937	10,167	34.4	6.6	7.3	5.0	6.6
Journal of Wildlife Management (1980–2017)	1980	6,358	49.5	10.4	11.3	7.7	10.5
Wildlife Research	1956	2,022	54.4	6.4	7.2	6.0	4.1
European Journal of Wildlife Research	2004	1,248	42.9	8.0	8.7	5.9	4.6
Wildlife Biology	1995	1,050	53.0	12.0	10.0	6.6	4.8
Journal of Fish and Wildlife Management	2010	327	55.0	3.1	6.7	6.4	10.7
Wildlife Monographs	1958	311	51.1	9.3	11.3	6.1	5.1
All combined			**50.2**	**9.3**	**10.0**	**7.0**	**8.0**
Journals covering ecology							
Ecology	1920	16,583	29.0	2.1	1.2	1.6	0.2
Oecologia	1968	12,762	33.9	2.4	1.7	2.2	0.3
Oikos	1949	9,060	32.8	2.7	1.4	2.0	0.3
Journal of Animal Ecology	1932	5,768	36.6	3.5	2.0	3.0	0.3
Functional Ecology	1987	3,929	19.2	1.0	0.8	1.3	0.2
Ecology Letters	1998	2,702	29.8	1.4	0.7	0.8	0.9
Ecological Monographs	1931	1,789	34.2	2.5	1.3	1.5	0.1
Evolutionary Ecology	1977	1,766	26.3	4.1	1.6	3.7	0
All combined			**31.0**	**2.4**	**1.4**	**2.0**	**0.3**
Journals covering conservation biology							
Biological Conservation	1968	8,598	58.3	4.0	3.9	4.4	8.4
Journal of Applied Ecology	1964	5,620	41.9	2.5	2.5	2.6	6.6
Conservation Biology	1987	5,504	40.6	1.4	1.6	1.1	5.3
Ecological Applications	1991	4,209	44.1	2.4	2.3	1.7	6.6
Restoration Ecology	1993	1,989	42.8	1.2	1.2	0.5	17.7
Animal Conservation	1998	1,294	51.5	8.3	4.7	4.5	7
All combined			**47.7**	**2.9**	**2.7**	**2.6**	**7.7**

[1] First year that the journal was published and hence first year included in the literature search.

[2] Not included in the percentages for "all combined."

years, wildlife ecologists have done more with the modern concept than have ecologists.

Since the early 1990s, attention to habitat has been much more common in research within wildlife ecology than in ecology, as evidenced by the percentage of published papers that use the word "habitat" (Fig. 2.1, top panel). In the decades prior to about 1974, approximately 10–20 percent of papers published annually in *Ecology* and the *Journal of Animal Ecology* included "habitat" in the title or abstract, whereas only 5–10 percent of papers published in the *Journal of Wildlife Management* did so. This changed throughout the 1980s and 1990s; percentages of "habitat"-citing papers in all three journals increased dramatically to near 50 percent. Most noticeably, the percentage for *Ecology* and the *Journal of Animal Ecology* has remained between 40 and 50 percent for the past two decades, while the per-

centage for the *Journal of Wildlife Management* has steadily increased such that about 60 percent of papers published annually in the journal mention "habitat" (Fig. 2.1, top panel). Simply put, wildlife ecologists seem to study and analyze habitat more than do ecologists. Granted, many of the "habitat"-citing papers in the *Journal of Wildlife Management* likely use the word in a descriptive sense simply to denote the place where an individual, population, or species exists. In such cases, the published research may not have literally studied or analyzed habitat. However, if we examine temporal usage of a phrase more in line with an actual analysis of habitat or invoking habitat in a more meaningful way (e.g., "habitat selection"), then we still see the same difference between the two academic disciplines (Fig. 2.1, bottom panel). The interesting question here may not be why wildlife ecologists have given

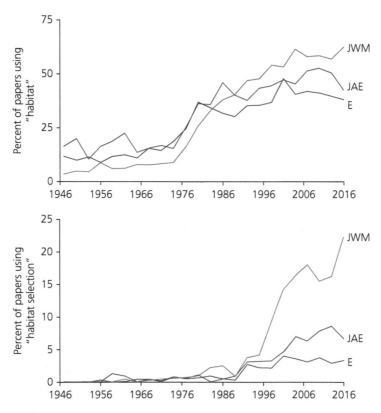

Figure 2.1 Increase in the percentage of published papers using the word "habitat" (top panel) or the phrase "habitat selection" (bottom panel) in either the title or abstract and over the time period 1946–2017 for three major journals, the *Journal of Wildlife Management* (JWM), the *Journal of Animal Ecology* (JAE), and *Ecology* (E). The plotted values are based on 3-year tallies of the number of papers as a percentage of all papers published in the journal during the 3-year period. Data were compiled from keyword searches in ISI Web of Science.

so much attention to habitat but rather why it has been relatively ignored in ecological research.

Part of the answer is that there was never a central and crucial role for habitat in any of the theory developed early on in ecology. From the 1920s to 1970s, much of the theory in ecology dealt with population growth, species interactions, coexistence of species, the flow of energy and nutrients in ecosystems, and maintenance of local species diversity (Real and Brown 1991). There are a few exceptions. David Lack identified habitat differences as being potentially important in the past speciation of closely related bird species. He clearly saw habitat selection as a species trait and referred to birds selecting their ancestral habitat; that is, the type of habitat that the ancestors of the bird had used and for which the species was adapted (Lack 1933). His hypothesis proposed that birds recognize the ancestral habitat of their species because it is visually conspicuous. Moreover birds are "faithful" to the habitat and of course breed there. Thus, at some point in the past, differences in selected habitat could have provided the spatial isolation necessary for speciation (Lack 1933). In a later paper, Lack pulled back from this idea and suggested that habitat selection was only important in maintaining the current spatial segregation of ecologically similar species and not necessarily a factor in the origination of the species (Lack 1940). Nonetheless, Lack's papers are early examples of habitat being explicitly incorporated into an eco-evolutionary concept; that is, habitat selection as a behavioral trait leading to the origin and maintenance of unique species. Papers by Robert MacArthur and Richard Levins are examples as well. There was an implicit role for habitat selection in the early and influential coexistence theory that they wrote about. Ecologically similar and potentially competing species could evolve to have different habitat preferences. This was an explanation for species coexistence as an outcome of niche divergence that led to the selection of different habitats and hence avoidance of interspecific competition (MacArthur 1958; MacArthur and Levins 1964, 1967). In addition, mathematical and conceptual models of habitat selection began to emerge in the 1960s (see Chapter 3). Such models were clearly theoretical and of course involved reference to habitat. However, the models primarily sought to explain the behavioral aspects of habitat selection in the context of an organism's fitness, rather than to invoke and ascribe a critical role for habitat in fundamental ecological theory concerning the distribution and abundance of species.

As wildlife ecology emerged as an academic discipline, research area, and profession, it was generally lacking in the development of any type of proprietary theory. There are no theories exclusive to wildlife ecology. This makes sense and is not surprising in that wildlife ecology began as and is still partly an applied science. Certainly some of the very first wildlife ecologists were aware of and read about the ecological theories and principles that were emerging simultaneously with the growth of their own academic field. For example, most of Chapters 2 and 3 in Aldo Leopold's *Game Management* book are ecological. When discussing population growth (or what he called "productivity") in mule deer (*Odocoileus hemionus*) and bobwhite quail (*Colinus virginianus*), Leopold referred to "unimpeded increase" and he described a wide variety of factors that could affect productivity and hence limit mule deer and quail populations. By "unimpeded increase," Leopold surely meant exponential population growth. Further, the figures in his book (e.g., figures 2–4) indicate a knowledge of logistic population growth (Box 2.1). He referred to the "potential" or "theoretical maximum rate of increase," which population ecologists now call the per-capita or intrinsic rate of increase and often symbolize it as r_{max}. He saw this rate of increase as being both a concept and a theoretical or hypothetical benchmark for comparison as well as a characteristic of a species. That is, mule deer and quail populations have different potential rates of increase. As mentioned in Chapter 1, *Game Management* was published in 1933, and it set in motion the development of wildlife ecology as an academic discipline and a knowledge base for wildlife management. Leopold discussed population fluctuations and correctly viewed various types of fluctuation as the norm rather than the exception (also Leopold 1931). He knew that populations do not continually exist at a static level exactly equal to the carrying capacity (and he used the term "carrying capacity" extensively). However, when discussing population fluctuations in *Game Management*,

Leopold again did not explain or even reference the relevant papers of the day that had presented the mathematical and quantitative foundation of predator–prey and competitor oscillations and cycles. Nonetheless, to his credit, Leopold essentially adhered to what became the perspective of modern-day population ecology: exponential and logistic population growth equations are models best used only as theoretical or hypothetical benchmarks for comparison and assessment of real population data.

My purpose in demonstrating a deficiency of ecological theory in *Game Management* is not to be overly critical of Leopold but rather just to make the point that this was an early missed opportunity for wildlife ecology as an academic discipline to more thoroughly connect with ecology. Granted, at the time that Leopold published *Game Management*, the theoretical foundation of ecology was still in its infancy, and among game managers Leopold actually stood out as being unusually knowledgeable about ecology (Meine 2010). But the question arises, in the historical development of wildlife ecology, when and what was the first published attempt to thoroughly incorporate ecological concepts, prin-

ciples, and theories into the pure and applied knowledge base of wildlife ecology? In particular, for the purposes of this book, what was the first such attempt within the realm of analyzing and managing habitat? With this question in mind, I attempted to find the first such published paper that discussed the need to understand a species' complete habitat requirements as they pertain to individual fitness, population growth and persistence, and a species' geographic distribution, and further how all that could be used to effectively manage habitat. I also intended (expected) that such a paper would build from ecological theory, particularly with regard to population growth and the niche, and that the paper would be a synthesis across species and habitats.

After thoroughly searching, I concluded that this paper does not exist. I searched the Transactions of the North American Wildlife and Natural Resources Conference between 1936 (first year) and 1958. This conference was and still is the annual meeting of the Wildlife Management Institute, a professional organization that works to improve the practice of wildlife management primarily in the United States and Canada. Between these years, no one attending

---

Box 2.1 Aldo Leopold's knowledge of population growth models

In reading *Game Management*, it is difficult to assess how aware or knowledgeable Aldo Leopold was about the mathematical population growth models that had been published since the early 1920s. Leopold did not use the words "exponential" and "logistic" anywhere in *Game Management* even though Raymond Pearl had previously used these terms in describing population growth (Pearl and Reed 1920; Pearl 1925, 1927) and Pearl was a well-known biologist at the time. The term "logistic" as applied to population growth actually traces back all the way to the 1830s when Belgian mathematician Pierre François Verhulst first applied it to describing the growth of human populations. Pearl also studied human populations and did not apply the terms to non-human populations until his 1927 paper that discussed population growth in yeast and *Drosophila* fruit flies. Perhaps because these papers discussed human population growth they were not well known to wildlife ecologists (or any type of ecologist) of the day. Georgy Gause was probably the first person to introduce exponential and logistic population growth models into the ecological literature (Gause 1931, 1932, 1934) and at about the time

Leopold would have been wrapping up the writing of *Game Management*. Clearly, Leopold had read Charles Elton's *Animal Ecology* by 1931 (Fig. 2.2) and he mentions Elton's book in *Game Management* writing "If the reader has never read a good text on ecology (such as Elton, 1927), he may pause here to do so, because game management, like every other form of land-cropping, is applied ecology." However, Elton (1927) also did not use the terms "exponential" or "logistic" and did not cite any of Pearl's papers except for a relatively obscure paper on the effects of population density on longevity of *Drosophila*. Therefore, Leopold would have acquired his knowledge of population growth models from a source other than Elton's book—although that source could have been Elton himself. Leopold and Elton first met at a conference in July 1931 and continued their friendship and professional correspondence throughout their careers (Meine 2010). Or perhaps Leopold had read some of Gause's papers. Regardless, even by 1933, Leopold clearly saw how a quantitative understanding of population dynamics could be useful in the management of wildlife species.

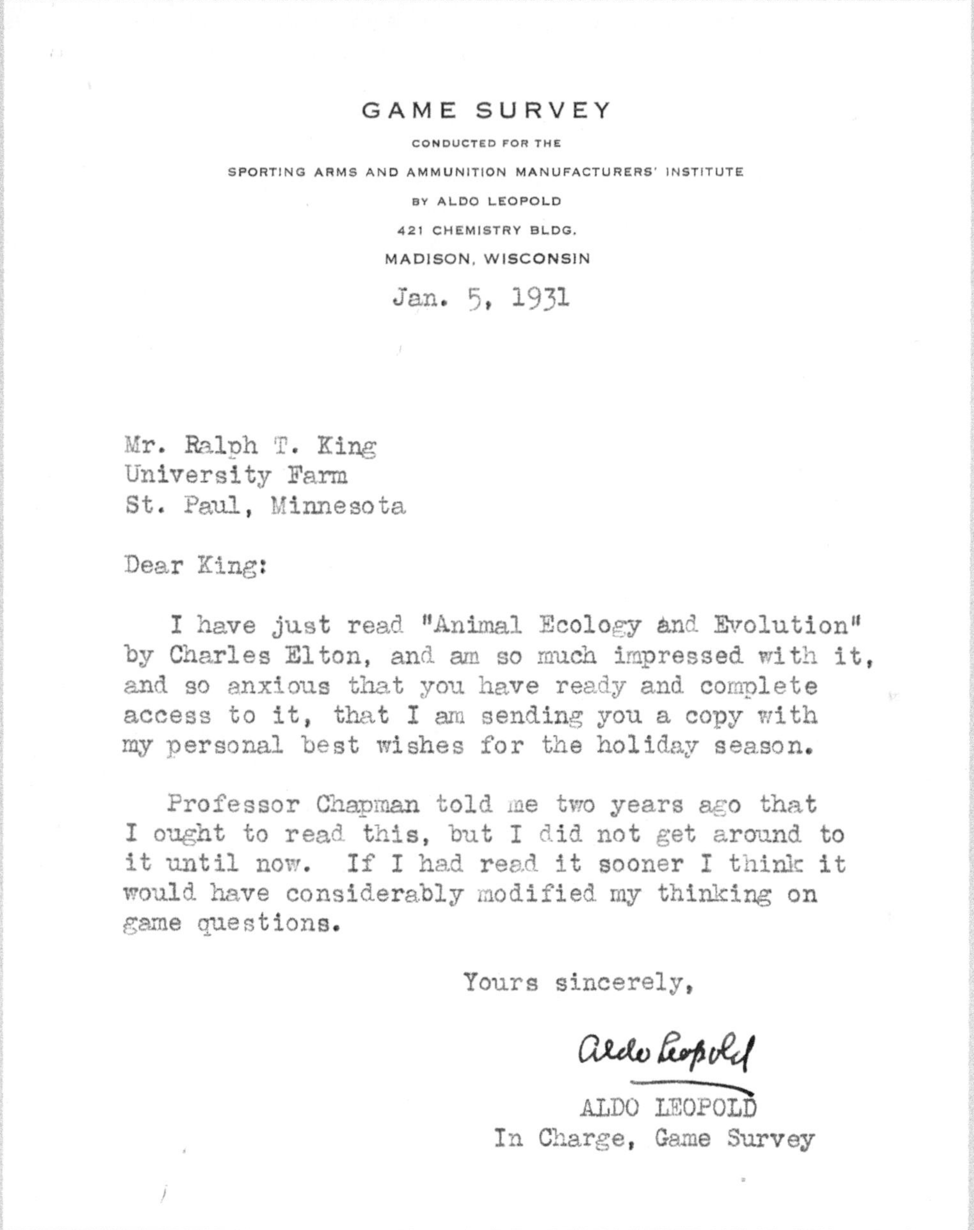

GAME SURVEY

CONDUCTED FOR THE

SPORTING ARMS AND AMMUNITION MANUFACTURERS' INSTITUTE

BY ALDO LEOPOLD

421 CHEMISTRY BLDG.

MADISON, WISCONSIN

Jan. 5, 1931

Mr. Ralph T. King
University Farm
St. Paul, Minnesota

Dear King:

    I have just read "Animal Ecology and Evolution"
by Charles Elton, and am so much impressed with it,
and so anxious that you have ready and complete
access to it, that I am sending you a copy with
my personal best wishes for the holiday season.

    Professor Chapman told me two years ago that
I ought to read this, but I did not get around to
it until now. If I had read it sooner I think it
would have considerably modified my thinking on
game questions.

                        Yours sincerely,

                        Aldo Leopold

                        ALDO LEOPOLD
                     In Charge, Game Survey

Figure 2.2 Letter from Aldo Leopold to Ralph King (a colleague at the University of Minnesota) informing him of Charles Elton's *Animal Ecology* published in 1927. "Professor Chapman" is probably referring to Herman Chapman, a professor at the School of Forestry at Yale University where Leopold had obtained an MSc degree in 1909. My thanks to Stan Temple for informing me about this letter. Letter is reprinted with permission from the Aldo Leopold Foundation.

the meeting presented (and hence published) any type of holistic perspective on habitat management deeply rooted in population ecology. There were presentations (papers) that discussed the habitat requirements of individual species including practical information on proper management, but these were primarily developed from knowledge of the natural history of the species, not from an analysis of the effect of a given factor on population growth rate. I also searched for the hypothetical paper using the Web of Science database, focusing on the years from the 1930s to 1970s, and did not find it. Lastly, such a paper is not among the collection in the *Essential Readings in Wildlife Management and Conservation* compiled, edited, and published by Paul Krausman and Bruce Leopold (no relation to AL) in 2013 (Krausman and Leopold 2013). Of the 42 papers that are included, many are classics from ecology and the collection is overall very comprehensive. The fact that the hypothetical paper is missing from this collection is fairly good evidence that it was never written.

If one were to thoroughly and intensively review the wildlife ecology literature in the early decades then it might be possible to detect some gradual transition toward research and management becoming more ecological. However, wildlife ecology in these decades generally followed from Leopold's *Game Management* in viewing wildlife as a stock or natural resource to be harvested. Leopold sometimes referred to "game crops." (Of note, Leopold himself viewed *Game Management* as being outdated by the end of his career [Meine 2010].) In retrospect, we can refer to this time period (roughly 1930s to 1960s) as the animal husbandry era of wildlife management. Habitat, often viewed synonymously with physical cover, was seen as a necessary resource along with food and water in producing or maintaining a harvestable population (or crop) of a wildlife species. Of course, later in his career, Leopold went on to develop a more holistic and ecological perspective on land use and the need to understand the inter-relations of plant and animal species in ecological communities. His land ethic as exemplified in *A Sand County Almanac* (published in 1949 right after Leopold's untimely death) brought attention to the direct role and responsibility that humans had in preserving natural landscapes, restoring degraded landscapes, and properly managing anthropogenic landscapes, all while providing habitat to wildlife.

Fifty years after the publication of *Game Management*, ecological theory finally appeared definitively in the wildlife ecology literature—and it was a textbook. Of course this is not to say that wildlife ecologists were completely unaware of ecological theory for 50 years; there simply had not been any published paper on the role of theory (particularly regarding habitat selection and use) in helping us better understand the dynamics of wildlife populations. In 1984, William Robinson and Eric Bolen published what I regard as the first modern wildlife ecology textbook with an ecological foundation, aptly titled *Wildlife Ecology and Management* (Robinson and Bolen 1984). In particular, chapters 4 (Ecosystems and Natural Communities), 5 (Population Ecology), and 13 (Predators and Predation) invoked quantitative ecological theory and principles. This textbook was evidently influential and popular having been republished several times since the first edition, most recently 2003. Fryxell et al. (2014) is another good example of a wildlife ecology textbook with an ecological foundation.

To conclude this chapter, I will pose this question to the reader: do wildlife ecology and ecology presently remain as separate academic disciplines? If "yes," do we need some type of disciplinary unification with regard to the study of habitat? I think the answer to the first question is "yes." However, the present separation of ecology and wildlife ecology is not so much a matter of differences in subject matter, research questions and methodology, and the over-arching goals of each discipline, but rather simply differences in tradition and training of practitioners. That is, the differences are superficial and not likely to hinder progress in either discipline. As such, we are not in need of any type of formal unification of the two disciplines, particularly if practitioners in both disciplines communicate with one another. Perhaps the most relevant questions are whether wildlife ecology will continue to reinforce itself with ecological theory, principles, and concepts—and whether ecology (and ecologists) will continue to build upon a pure knowledge base that is useful and applicable to the real-world conservation goals of wildlife ecology.

References

Elton, C. (1927). *Animal Ecology*. Macmillan Publishing Company, New York.

Fryxell, J.M., Sinclair, A.R.E., and Caughley, G. (2014). *Wildlife Ecology, Conservation, and Management*. Third edition. Wiley-Blackwell, Oxford.

Gause, G.F. (1931). The influence of ecological factors on the size of a population. *American Naturalist*, 65, 70–76.

Gause, G.F. (1932). Ecology of populations. *Quarterly Review of Biology*, 7, 27–46.

Gause, G.F. (1934). *The Struggle for Existence*. Williams and Wilkins Publishing Company, Baltimore, Md.

Krausman, P.R. and Leopold, B.D. (2013). *Essential Readings in Wildlife Management and Conservation*. Johns Hopkins University Press, Baltimore, Md.

Lack, D. (1933). Habitat selection in birds, with special reference to the effects of afforestation on the breckland avifauna. *Journal of Animal Ecology*, 2, 239–62.

Lack, D. (1940). Habitat selection and speciation in birds. *British Birds*, 34, 80–84.

Leopold, A. (1931). *Report on a game survey of the north central states*. Sporting Arms and Ammunition Manufacturers' Institute, Madison, Wis.

MacArthur, R.H. (1958). Population ecology of some warblers of northeastern coniferous forests. *Ecology*, 39, 599–19.

MacArthur, R. and Levins, R. (1964). Competition, habitat selection, and character displacement in a patchy environment. *Proceedings of the National Academy of Sciences USA*, 51, 1207–10.

MacArthur, R. and Levins, R. (1967). The limiting similarity, convergence, and divergence of coexisting species. *American Naturalist*, 101, 377–85.

Meine, C. (2010). *Aldo Leopold, His Life and Work*. University of Wisconsin Press, Madison, Wis.

Pearl, R. (1925). *The Biology of Population Growth*. Alfred A. Knopf Publishing, New York.

Pearl, R. (1927). The growth of populations. *Quarterly Review of Biology*, 2, 532–48.

Pearl, R. and Reed, L.J. (1920). On the rate of growth of the population of the United Stastes since 1790 and its mathematical representation. *Proceedings of the National Academy of Sciences USA*, 6, 275–88.

Real, L.A. and Brown, J.H. (1991). *Foundations of Ecology, Classic Papers with Commentaries*. University of Chicago Press, Chicago, Ill.

Robinson, W.L. and Bolen, E.G. (1984). *Wildlife Ecology and Management*. Macmillan Publishing Company, New York.

CHAPTER 3

The Primacy of Habitat

In Chapter 2, I briefly reviewed the history of the study of habitat in ecology and in wildlife biology. A core and fundamental pursuit in ecology is to explain the distribution and abundance of species (Krebs 2009). Yet, in my opinion, ecologists have historically overlooked the importance of habitat to species distribution and abundance and this neglect continues today (although see Table 3.1). To some extent, ecologists take it for granted that a species will (or can) occur where there is suitable habitat. Perhaps habitat-based explanations of distribution and abundance are not seen to be interesting enough for in-depth study. Instead, ecologists have expended considerable effort in developing and testing explanations based on species interactions and physiological adaptation to climatic conditions. However, attention is beginning to shift. The relatively new idea of habitat filtering implicitly recognizes that some habitat characteristics in combination with particular species traits (adaptations) can determine which species occur in a local community (Cornwell et al. 2006; Ulrich et al. 2010; Fichaux et al 2019). I believe further progress can be made in understanding distribution and abundance by studying the role of habitat in determining where individual organisms settle during the dispersal phase of their life cycle. Indeed, one way of viewing or defining habitat is through the act of a dispersing organism selecting and settling at a given location based upon its environmental properties (Mannan and Steidl 2013). Studies of habitat can be facilitated by a formal framework and analytical/conceptual model of the environmental cues that influence a dispersing individual's settlement at a particular location.

This framework is based on a scenario in which an individual organism leaves its natal location perhaps at a young age and searches for a new location in which to settle. The search need not involve directed movement (it could be random), nor necessarily require the individual to settle in a location currently unoccupied by other members of the species. The main idea is that the individual is moving over the landscape or seascape (at some scale relevant to its body size and dispersal speed) receiving environmental cues that will influence where it settles. By "settle" I mean that the individual enters into a potentially longer period of time in which it no longer moves as much as it did prior to settling (Fig. 3.1). If it and other individuals of its species can subsequently survive and reproduce at the particular settlement location, then a new population of the species can establish at the location (assuming the location was previously unoccupied by the species), or the settling individuals might augment a population already present. Either way, the cues affecting settlement maintain the spatial distribution and local abundance of the species over some spatial extent (Raimondi 1991; Booth and Wellington 1998; Larsson and Jonsson 2006; Vallès et al. 2008). Note that this framework can also explain why some individuals might remain at their natal location. The environmental cues that were strong enough to attract the parents or their ancestors remain strong enough (relative to other factors) to preclude the young individuals from dispersing.

The dispersal–settlement–establishment (DSE) process can be described by a series of probabilities. The intrinsic settlement probability, $P(S_{int})$, is the

Habitat Ecology and Analysis. Joseph A. Veech, Oxford University Press (2021). © Joseph A. Veech. DOI: 10.1093/oso/9780198829287.003.0003

Table 3.1 Quotes from selected papers that allude to the influence of habitat on species distribution and abundance. Although each paper suggested the importance of habitat selection to species distribution, none of the authors developed a conceptual/analytical framework to formally explain how distribution depends on habitat selection.

Quote[1]	Species and habitat	Source
Larval behavior [in selecting habitat] can be an important determinant of larval and adult distribution and abundance	Various sedentary marine invertebrates, particularly barnacles	Grossberg and Levitan 1992
These results indicate that **preferential habitat selection at settlement, . . ., may play a deciding role in determining distributions** of ecologically similar species of coral reef fish	Damselfish (*Stegastes leucostictus* and *S. variabilis*) on coral reefs	Wellington 1992
The focus on recruitment has been directed primarily to its role in regulating the abundance of fish populations. Little attention has been paid to its function in the **establishment of local patterns of distribution; for example, through habitat selection**	Damselfish (*Stegastes dorsopunicans* and *S. planifrons*) on coral reefs	Gutiérrez 1998
Habitat selection at settlement has been shown to be an important contributor to the distribution and abundance of coral reef fish within zones, among zones, and among reefs	Various coral reef fish of *Gobiodon*	Munday 2000
Our results indicate that **landscape-level patterns of distribution and species diversity can be driven to a large extent by habitat selection behavior,** a critical, but largely overlooked, mechanism of community and metacommunity assembly	Various aquatic beetle species inhabiting ponds	Binckley and Resetarits 2005
Larval choice ensured settlement differed substantially over this environmental gradient and ultimately dictated the pattern of early adult distribution	Barnacles (*Chthamalus montagui* and *C. stellatus*) on rocky shores	Jenkins 2005
Complete tests of habitat selection need to evaluate whether dispersers actively sample multiple habitats and choose the best available, and thus **whether individual-level processes determine large-scale patterns of distribution and abundance**	Treecreepers (birds, *Cormobates leucophaeus* and *Climacteris picumnus*) in eucalypt woodland	Doerr et al. 2006
In the community that we studied, **habitat appeared to be more influential than biological interactions in determining the distribution and abundance** of most avian species	Various bird species in forests	Dorazio et al. 2015

[1] Bolded part of each quote represents particular emphasis that I have added.

probability that an individual organism will detect a given settlement cue and successfully settle at a location with the particular cue given the perceptual ability of the individual and the time required to find the cue:

$$P(S_{int}) = (1 - T_p) \times (1 - E). \qquad (3.1)$$

Perception time (T_p) indicates the search time (during dispersal) that the individual needs to find and detect the cue. Given that no organism can disperse and search for a settlement location indefinitely, T_p is a proportion of the total amount of time available to the organism before it dies having been unable to find a settlement location. T_p is specific to a given environmental condition or feature; it is also determined by the intrinsic ability of the individual in

sensing a cue associated with the condition. For example, a dispersing juvenile organism may have 100 days to settle before dying. If it requires 20 days to find and detect the cue associated with a given factor or condition, then $T_p = 0.2$. Dispersing organisms sometimes make mistakes. Hence, E is the error inherent in using the given environmental factor as a cue for settling. This error represents the probability that the individual dies prior to reproducing because it erred in detecting or "reading" the cue. Thus E is determined by both the environmental cue and the fidelity of the organism's perceptual abilities. A given organism (species) could potentially have evolved to recognize several different cues, each with its own $P(S_{int})$. Other environmental factors could be conducive to the species

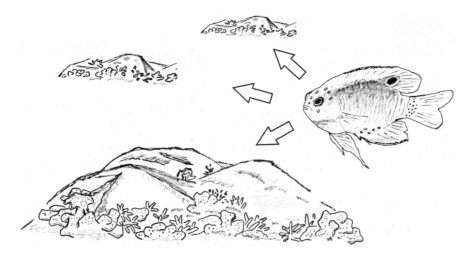

Figure 3.1 Many reef-dwelling fish species (e.g., *Stegastes* damselfishes) hatch from eggs laid in the parent's territory on the coral reef and then become free-swimming "larvae" during a pelagic phase lasting a few weeks to a few months. As they grow, the young fish eventually need to find a reef where they can settle and establish a territory. During this dispersal and search process, the fish receive and respond to environmental cues with regard to finding an appropriate settlement location. The figure depicts a juvenile *S. adustus* searching for a coral reef (not drawn to scale). More specifically, *S. adustus* individuals appear to cue in on the rocky substrate found at the top of the reef (shallow water), whereas a close congener, *S. planifrons*, uses the deeper coral substrate as a settlement cue (Gutiérrez 1998). Illustration by the author.

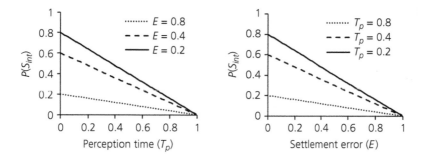

Figure 3.2 Effects of perception time and settlement error on the intrinsic settlement probability, $P(S_{int})$. In the equation, $P(S_{int}) = (1 - T_p) \times (1 - E)$, perception time and error have equal weight. An increase in either leads to the same decrease in $P(S_{int})$ for the same given value of the other.

survival but not used as settlement cues, either because they are difficult (time-consuming) to locate and detect, or they have a lot of inherent error. Note that in Equation 3.1, if either T_p or $E = 1$, then $P(S_{int}) = 0$ and hence the dispersing individual cannot (does not) successfully settle anywhere. In such a case the given environmental factor is thus not a settlement cue. As T_p or E increases, intrinsic settlement probability decreases. For a given value of T_p (or E), $P(S_{int})$ decreases as the same monotonic function of E (or T_p) (Fig. 3.2).

The intrinsic settlement probability describes the innate ability of the organism to search, find, and respond to an environmental cue. As such, the probability is not affected by specific conditions of a potential settlement location. However, those conditions matter to where an individual *actually will* settle. The site-specific settlement probability, $P(S_{ext})$, is based on the intrinsic settlement probabilities and strength of each cue at the given site i,

$$P(S_{ext})_i = \Sigma[P(S_{int})_j' \times Q_{ij}] \qquad (3.2)$$

for all j cues with strength Q_{ij} represented as a variable between 0 and 1. In this equation, the intrinsic settlement probabilities are represented as relative proportions (and indicated with the prime symbol) rather than as the raw values. For example, if a species can potentially respond to three settlement cues with intrinsic probabilities, $P(S_{int})_1 = 0.8$, $P(S_{int})_2 = 0.4$, and $P(S_{int})_3 = 0.1$, then the proportions are $P(S_{int})_1' = 0.62$, $P(S_{int})_2' = 0.31$, and $P(S_{int})_3' = 0.07$. Each of these proportions can be interpreted as the extent that a dispersing individual will use that cue to settle relative to the other cues. Further, the DSE model assumes that each cue can potentially "contribute" to an individual's settlement decision (or response) in an additive way, hence the summation operator in Equation 3.2.

Once an organism has dispersed and settled in a particular location, it must find a mate and survive long enough to reproduce and hence establish a population at the location. At every location there will be some site-specific mortality lowering the probability of an individual's survival post-settlement. This mortality is separate from the error (E) associated with the settlement event. The probability that a species establishes itself and hence exists at the location is given by

$$P(est)_i = P(S_{ext})_i \times (1 - m) \qquad (3.3)$$

where m is the probability of mortality. Further, recruitment limitation is common in nature. This represents a deficit of dispersing individuals available to establish new populations or augment existing ones. Recruitment limitation can be included in the model by subtracting a specified proportion such that

$$P(est)_i = \left[P(S_{ext})_i \times (1 - m) \right] - r \qquad (3.4)$$

where r is a value between 0 and 1. Larger values of r correspond to greater recruitment limitation. In some cases, large enough values of r will make $P(est)_i < 0$ and hence effectively 0 representing a situation where recruitment limitation precludes the species from occurring at the location.

The DSE model of cue-based settlement of a species is intended to provoke greater research on the importance of habitat in determining where a species exists and at what abundance. More practically, the model makes predictions about patterns of species occurrence in nature. The probability that a given species successfully establishes at a specific location, $P(est)_i$, can also be taken as the probability that the species will be detected and recorded as "present" at the location during a survey. To be exact, $P(species\ present)_i$ is actually less than $P(est)_i$ given that no survey method has perfect detection. For any given survey and species, there is a detection probability that takes into account that the survey might miss detecting the species even when it is present and available to be detected (MacKenzie et al. 2002; Gu and Swihart 2004) (Section 7.3). However, to the extent that detection probability is relatively constant over all survey sites then $P(species\ present)$ is directly proportional to $P(est)$. (If detection probability is not constant, then any site-specific empirical use of Equations 3.3 and 3.4 would need to multiply the right-hand side by $P(detect)_i$, assuming it can be estimated.) Of most importance, the DSE model can be used to generate expected relationships between an environmental factor and species presence (or expected occurrence). If the factor is a habitat characteristic that varies among the survey sites, then the model serves as an analytical and statistical tool to investigate species–habitat relationships.

The DSE model is general and flexible. It can be used to investigate the effect of any environmental cue (Q) on the settlement of individuals and establishment of a population. Given that this chapter focuses on the primacy of habitat, I present some results from the model wherein a habitat structural characteristic is assumed to be the most important environmental cue in having the shortest perception time and least error. Assume that a dispersing individual can potentially settle at a given location based on detecting and responding to any of three different environmental cues: habitat characteristic, perceived food supply, and absence of an antagonistic species (predator or competitor). Instead of strict presence/absence, the latter cue could be the inverse of the density of the antagonistic species such that higher Q_{ant} values represent a lower density of the antagonistic species. In accord with the mathematical foundation of the model, Q ranges from 0 to 1.

Figure 3.3 shows the relationship between the strength of each environmental cue and the *extrinsic*

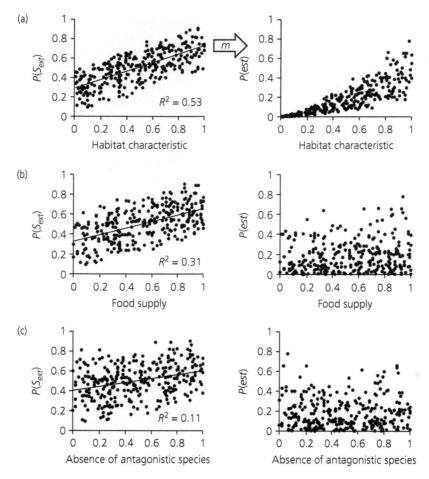

Figure 3.3 Relationship between extrinsic site-specific settlement probability, $P(S_{ext})$, and the environmental cue as a habitat characteristic, food supply, or absence of an antagonistic species (scatterplots on left). These relationships then lead to the relationships between probability of successful species establishment, $P(est)$, and the environmental factor (scatterplots on right) when a mortality function is included in the model. The mortality function, $m = 1 - [(Q_{hab} \times \varepsilon(0.2 \text{ to } 1)]$, represents a decrease in mortality probability as the habitat characteristic increases (i.e., quality of the habitat improves). ε represents stochasticity in the mortality function as a random number between the specified bounds. Values for the intrinsic settle probabilities, $P(S_{int})$, are 0.86, 0.64, and 0.3, respectively, for habitat, food supply, and absence of antagonistic species. These correspond to summative proportions of 0.48, 0.36, and 0.17. For all panels, $N = 300$.

site-specific settlement probability when *intrinsic* settlement probabilities equal 0.86, 0.64, and 0.3, respectively, for habitat, food supply, and antagonistic species. These probabilities derive from the following specified perception times and errors: habitat ($T_p = 0.1$, $E = 0.05$), food supply ($T_p = 0.2$, $E = 0.2$), and antagonistic species ($T_p = 0.5$, $E = 0.4$). Thus, these specified values represent a scenario where the habitat environmental cue can be located relatively quickly and with minimal detection error, whereas the environmental cues provided by food

supply and lack of (or low density of) an antagonistic species take longer to find and are more prone to error. As seen in Fig. 3.3 (left panels), there is a relatively strong relationship between the habitat environmental cue and extrinsic settlement probability, whereas food supply and absence of an antagonistic species each have a weaker effect on extrinsic settlement probability. After the individual has settled, mortality may occur with a specified probability. However, assuming that natural selection has led to the evolution of a reliable environmental cue (see below) the mortality should be an inverse function of the strength of the environmental cue for which the species has the greatest intrinsic settlement probability. Therefore, in the example of Fig. 3.3, I simulated mortality as a function of the habitat cue to thus examine the relationship between each environmental cue as the probability of the species establishing at a site (Fig. 3.3, right panels), recalling Equation 3.3. There is a clear and definite pattern of increasing probability of species establishment as the habitat characteristic gets stronger (Q_{hab} approaching 1) (Fig. 3.3A, right panel). The patterns for food supply and absence of an antagonistic species are essentially random (Fig. 3.3B and C, right panels). For each of the three types of cue, I intentionally specified values of T_p and E such that intrinsic settlement probability would be greatest for the cue based on a habitat feature. Obviously, this need not always be the case. However, the purpose of this chapter is to advance the hypothesis and conceptual framework that habitat structural features provide the strongest cue for settlement of a dispersing individual.

As previously stated, the relationship between the probability of species establishment and the environmental cue (e.g., habitat characteristic) is essentially a prediction or depiction of how the species should be distributed in nature among a set of survey or sampling sites. Of course, in practice, putting this prediction to the test requires that the empirical data for the environmental cues (Q_{ij} values) can be converted to a scale of 0 to 1. But this is no great obstacle as ecologists routinely rescale or reformat data for use in models and statistical analyses. The predicted relationship between an environmental cue and species establishment can be made more realistic by including recruitment limitation (r), recalling Equation 3.4. This has the effect of adding scatter (noise) to the pattern (Fig. 3.4), but this scatter is realistic because recruitment limitation likely occurs for many species (Tilman 1997; Chesson 1998; Wright 2002). The lack of habitat saturation or inability of a species to occupy all suitable habitat patches may arise when the number of dispersing individuals (offspring) is limited.

The particular form of the mortality function can have an effect on the expected relationship between the habitat characteristic and a species establishment probability. Figure 3.5 shows three different functions for the decrease in mortality as the habitat characteristic (Q_{hab}) increases from 0 to 1. The resulting relationship between establishment probability and the habitat characteristic is somewhat different for each mortality function. Note that even when mortality is a completely random variable (unrelated to Q_{hab}) there is still a slight signal of non-random pattern in the plot of $P(est)$ vs. Q_{hab} (Fig. 3.5D). This is simply the carry-over effect of habitat being the strongest environmental cue for the *intrinsic* settlement probability ($P(S_{int}) = 0.86$) and thus also having a relatively strong effect on the *extrinsic* settlement probability at each site.

As explained above, the DSE model can be used to investigate the effect of any environmental factor (cue) on the spatial distribution of a species, and compare among different factors. However, when a

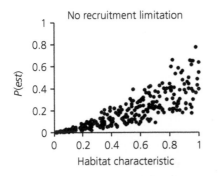

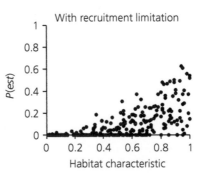

Figure 3.4 Relationship between species establishment probability, *P(est)*, and a given habitat characteristic when there is no recruitment limitation (left) and when recruitment limitation (*r*) is set as a random variable between 0 and 0.3 (right). Other model parameters are the same as in Fig. 3.3A.

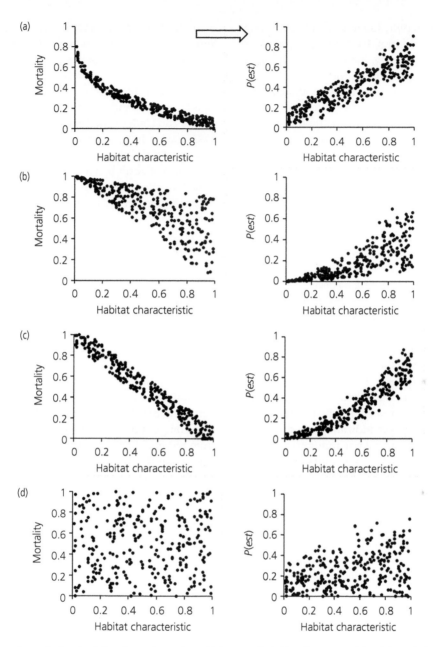

Figure 3.5 Effect of mortality function (m) on the resulting relationship between the habitat characteristic (Q) and the probability of successful establishment of the species at the location, $P(est)$. In panels A–C the mortality function is negatively correlated with the habitat characteristic (A) $m = 1 - [(0.95 \times Q^{0.3}) + \varepsilon(-0.05 \to 0.05)]$, (B) $m = 1 - [(Q \times \varepsilon(0.2 \to 1)]$, (C) $m = 1 - [(Q + \varepsilon(-0.1 \to 0.1)]$. In panel D, there is no relationship between m and Q. Stochasticity (ε) was added to each function as a random number between the specified bounds. In all panels, $N = 300$.

habitat characteristic is thought to be the most important cue then the DSE model becomes a way of testing what we could call the *habitat-cue hypothesis of species distribution*. This hypothesis is relevant to the process of habitat selection and the immense body of research focused on the behavioral aspects of that process and its ecological/evolutionary causes and consequences (Southwood 1977; Cody 1981,

1985; Rosenzweig 1981; Lima and Dill 1990; Huey 1991; Jones 2001; Morris 2003; Mayor et al. 2009; Boyce et al. 2016). Briefly, habitat selection can be defined as non-random (disproportionate) use of a particular habitat type(s) by an individual or species. It can also be defined (in a related way) as the active choice of a particular habitat type by an individual in response to the individual being able to detect and subsequently breed in the habitat and thereby increase its fitness. Both of these definitions implicitly suggest that habitat selection could ultimately determine the distribution of a species in nature. However, habitat selection *theory* is more about explaining the how and why that a given species selects (and uses) some habitat types rather than others, whereas the habitat-cue hypothesis is strictly focused on explaining a species' spatial distribution as a consequence of the DSE process in which habitat structural features are the strongest environmental cues of appropriate habitat. The habitat-cue hypothesis assumes that, in general, dispersing individuals settle into the type of habitat that maximizes fitness. The hypothesis simply suggests that habitat structural features are more important than food supply, density of antagonistic species, and other cues in indicating to a dispersing individual that it has arrived at appropriate habitat. Accordingly, habitat structural features are hypothesized to provide the first (lowest T_p) and most reliable (lowest E) cues to a dispersing individual. These cues then have a "lasting effect" in that the species' spatial distribution is the outcome of the continual DSE process that is based on the cues.

3.1 Habitat cueing in barnacles and reef fishes

Perhaps the most analogous historical precursor for the habitat-cue hypothesis of species distribution comes from marine ecology and particularly research on the life histories of barnacles (Connell 1985) and reef fishes (Sale et al. 1984). Although taxonomically they are very different types of organism, the common aspects of their life histories include a pelagic larval phase followed by directed dispersal to and eventual settlement on a hard substrate where individuals mature into adults within a relatively small territory (reef fishes) or become sessile (barnacles).

In particular, Connell (1985) examined settlement and recruitment rates of barnacles to address the question: *How does variation in either settlement or recruitment affect the abundance and distribution of adults living on the substratum as individuals or populations?* Connell and other marine ecologists (e.g., Denley and Underwood 1979; Grosberg 1982; Keough 1983; Caffey 1985; Raimondi 1991) were interested in knowing whether spatial patterns in the depths (zonation) of adult barnacle distribution were determined mostly by the supply of larvae to particular depths and locations, their rate of settling, or subsequent mortality among the settled recruits. The mortality could be due to various causes such as inter- and intraspecific competition, predation, and even desiccation. These marine ecologists essentially recognized four stages (larvae, settler, recruit, and adult) in the life cycle of barnacles and sought to determine which life stage most influenced the "final" pattern of spatial distribution of a barnacle species. Population-limiting processes were most likely to occur at which life stage?

Much of the interest in the structuring and spatial distribution of barnacle populations grew out of the "grand debate" about the importance of interspecific competition. This debate dominated ecology from the mid-1970s through the 1980s and still lingers today (Cody and Diamond 1975; Connor and Simberloff 1979; Diamond and Gilpin 1982; Strong et al. 1984; Brooker et al. 2005; Freckleton et al. 2009; Kikvidze and Brooker 2010; Brooker et al. 2013; Ricklefs 2015). In short, the debate was about whether deterministic competitive interactions regulated populations of competitors and their coexistence with one another or whether other biotic and abiotic factors (deterministic and stochastic) regulated populations and determined membership in an ecological community. A similar debate has played out in the past 20 years with regard to whether stochastic and dispersal-based or deterministic and niche-based processes are more important to species coexistence (Hubbell 2001; Chave 2004; Leibold and McPeek 2006; Siepielski and McPeek 2010; Clark 2012; Wennekes et al. 2012; Furniss et al. 2017). Before these debates, the classic field studies and theoretical papers of Robert MacArthur (e.g., MacArthur 1958, 1964, 1970; MacArthur and Levins 1967) strongly reinforced in many ecologists a view

THE PRIMACY OF HABITAT

of the overriding importance of interspecific competition in structuring nature, by affecting resource use by individuals, geographic distributions and abundances of species, composition of ecological communities, and evolution of species traits. Interspecific competition was seen as a grand unifying force, pervasive and ubiquitous in nature, and for good reason studies of barnacles and other sessile marine invertebrates had played a role in promoting the ecological grandeur of competition (Box 3.1).

In the 1960s and 1970s, the field studies of Joseph Connell, Robert Paine, A.J. Underwood, Paul Dayton, Bruce Menge, and others on marine intertidal communities indicated that interspecific competition and other species interactions could affect the distribution and abundance of various invertebrate species. With regard to barnacles, some ecologists eventually began to question whether interspecific competition was a complete or even necessary explanation. The doubters included Connell himself (1985) as well as Dayton and Oliver (1980), Keough (1983), Underwood and Denley (1984), and

---

Box 3.1 Competition for space among sessile marine invertebrates

Many marine invertebrates (e.g., barnacles and mussels) have a free-swimming or planktonic phase prior to settling on a hard substrate and becoming sessile as adults. Moreover, some species are restricted to tidal pools and other shallow areas where the surface area of hard substrate is limited. As such, organism densities can be quite high such that competition for space occurs. Tide pool communities formed an important study system early in the development of ecological theory. Ecologists such as Joseph Connell and others conducted experiments wherein different barnacle species were either removed or added to plots at different depths or points along the shore (e.g., Connell 1961a, b). The plots varied in the amount of time that they were above water and dry during the tidal cycle. Although the establishment and survival of the barnacles often depended on local edaphic conditions (some species were physiologically more tolerant of desiccation than others) and the density of predators, the barnacles also competed with one another for space to the extent that on some occasions one species could competitively exclude another from some plots.

Caffey (1985). These researchers obtained evidence that the early phases of the barnacle life cycle involving dispersal ("larval supply") and the active selection of particular types of microhabitat (depths and substrates) could be as or sometimes more influential than competition among adults. That is, the spatial distribution of adult barnacles often coincided very closely with the distribution of settled "juvenile" recruits, thus obviating the need to invoke any kind of strong population-thinning process occurring among the adults. In the context of Equation 3.4, this result suggests that mortality (m) and recruitment limitation (r) would be low enough for barnacles that a significant statistical relationship might exist between observed barnacle distribution ($P(est)$) and a habitat characteristic (Q_{hab}) such as depth that serves as the cue for larval settlement (as in Fig. 3.4). Moreover, if the habitat-cue hypothesis of species distribution is a valid explanation then the relationship between $P(est)$ and Q_{hab} would be stronger than that between $P(est)$ and any other possible environmental cue.

Perhaps surprisingly, Connell (1985) writes about dispersing larval barnacles searching for suitable attachment sites and choosing either to attach or not. This is a key component of my hypothesis. Organisms actively choose (select) certain habitat or microhabitat types based upon environmental cues. Settlement is non-random. It is interesting and noteworthy that even organisms as behaviorally simple as barnacles may still actively orient to and select a particular habitat and as a corollary actively avoid others. Intuitively, we might expect that intentional and non-random behavior in selecting a location for settlement is more likely to occur in organisms such as vertebrates that have well-developed perceptual abilities. Indeed, studies of habitat selection by reef fishes form the other set of literature most pertinent to the habitat-cue hypothesis of species distribution.

In the late 1960s and into the 1970s, Peter Sale initiated field studies on the coexistence and community structure of fishes inhabiting coral reefs, primarily in Australia. As with the studies on barnacles, attention was focused on the early life stages and the nature of habitat selection by recruits (Sale 1969a, b). Unlike the barnacle studies, the studies of reef fish community structure were not born out of the "competition is pervasive" paradigm. Perhaps for this reason, studies of recruitment limitation and habitat selection in coral reef fishes were not initially linked to or informed by the barnacle studies. Further, the early studies by Sale and colleagues (e.g., Sale and Dybdahl 1975; Sale and Douglas 1984; Sale et al. 1984) did not directly pose and test the question of whether spatial patterns in settlement and recruitment could have a carry-over effect in determining the spatial distribution of adult fishes (this was a question of primary interest to the barnacle ecologists). Nonetheless, they hinted at this possibility, from Sale et al. (1984), p. 91: "reef fish at the end of their larval phase [may] have specific behavioral responses which determine whether or not they will settle at a specific site" and "some species [of reef fish] show such narrow habitat distributions as adults that selection of these habitats at settlement is highly likely." Again, as with barnacles, the question was whether post-settlement and post-recruitment processes, such as interspecific competition among adults and predation on adults, could have such strong effects as to ameliorate any spatial pattern initially set by settlement and recruitment. Eventually, Wellington (1992), Tolimieri (1995), and Booth and Wellington (1998) directly examined the correspondence between the spatial distributions of newly recruited and adult damselfishes. They found that the latter coincided with the former fairly well—the distribution of adults at different depths, among different types of substratum, and on different reefs was largely explained by the selection of particular habitats or microhabitats of settling juveniles (Fig. 3.1). Returning to Equation 3.4, this means that mortality (m) and recruitment limitation (r) might be negligible for coral reef fishes, or at least, m is inversely proportional to the strength of the habitat cue.

The three studies cited above also linked recruitment limitation and habitat selection in reef fishes to the same processes in barnacles, by citing the conceptually similar studies by Connell and the other barnacle ecologists. This provided some early conceptual unification and taxonomic breadth to the idea that habitat selection by dispersing individuals could be a strong determinant of the distribution and abundance of a species. In addition, the reef fish ecologists were more deliberate than were the barnacle ecologists in describing how dispersing individuals actively find and choose a habitat

in response to an environmental cue. Perhaps this reflects the difference between reef fishes and barnacles in behavioral complexity.

The use of environmental cues by reef fishes when selecting habitat was finally studied directly and experimentally by the late 1990s. In particular, two studies, Gutiérrez (1998) and much more recently Majoris et al. (2018) serve well to thoroughly illustrate the DSE model and habitat-cue hypothesis of species distribution. Gutiérrez (1998) presented a study of habitat selection and spatial distribution of the dusky damselfish (*Stegastes adustus*, formerly *dorsopunicans*) and the three-spot damselfish (*S. planifrons*) inhabiting reefs along a small portion of the Caribbean coast of Panama. The reefs were relatively small (<10,000 m²) rock and coral outcroppings separated from one another by a relatively barren sand sea bottom. As with most damselfishes (Pomacentridae), females lay eggs on the substrate within the male's territory (Thresher et al. 1989). Upon hatching, the larval fish are free-swimming and pelagic for a duration of 23–32 days (Wellington and Robertson 2001). They then settle onto a reef and eventually become territorial and relatively sedentary as adults (Fig. 3.1). By observing the settlement rate of larval fish on natural and artificially created reefs, Gutiérrez documented that dusky damselfish typically settled on rock substrate on the reef crest whereas three-spot damselfish always selected the coral substrate on the reef slope and at a greater depth. By experimentally removing adult damselfish, she also discovered that presence of conspecifics did not affect settlement and recruitment (establishment) of dusky damselfish although it may have had a slight positive effect on three-spot damselfish. Rather than conspecific attraction, Gutiérrez suggested that both species use substrate type and possibly water depth as cues for settlement and thus habitat selection. Perhaps of most importance, the study revealed that local reef-level patterns in the spatial distribution of adult damselfish can be explained by larval fish selecting and settling in a particular type of habitat. These damselfish serve as an example of how dispersing individuals can cue in on specific structural features of a habitat and thus select the habitat—with the consequence that the distribution and abundance of the species is well explained by cue-based habitat selection.

An even better study of cue-based habitat selection in reef fish is that of Majoris et al. (2018). This study went a bit further than Gutiérrez (1998) in discovering the exact cue used to select habitat. As adults, neon gobies (specifically *Elacatinus lori*) occupy the inner tube of sponges—that's their habitat. The sponges grow on coral reefs in the Caribbean and the life cycle of neon gobies is roughly similar to that described for damselfishes. Majoris et al. (2018) proceeded in much the same way as Gutiérrez (1998), measuring rates of larval settlement and documenting patterns of adult distribution. They then set up an ingenious experiment to determine the exact cues that the larval fish use to settle in sponges. On a portion of sea bottom next to the reef and at 2 m depth, Majoris and colleagues inserted three 35-cm-high yellow tube sponges (*Aplysina fistularis*) alternating with gray PVC ("plastic") pipes of the same height arranged in a circular arena (Fig. 3.6). A diver then released a single neon goby (barely a centimeter long) from a glass jar into

the middle of the arena and then watched over the next few minutes to see whether the little fish swam to a sponge or a PVC pipe. The experiment was repeated a few hundred times (with a different fish each time) and with a different setup for the following five comparisons: (1) yellow sponge vs. PVC pipe, (2) yellow sponge vs. brown sponge (*Agelas conifera*) of same height, (3) long vs. short (10 cm) yellow sponges, (4) single-tube vs. multi-tube yellow sponges, and (5) yellow sponges with a resident neon goby vs. yellow sponges without a resident. This was an elegant and simple experiment to determine the habitat feature(s) that the neon gobies select. But, the researchers went even further.

To really get at the exact sensory cue used by the gobies, Majoris and colleagues also conducted experimental trials in which each structure (sponge or PVC pipe) was obscured either chemically or visually. For the former, they installed clear plastic sleeves around the structure; for the latter, they covered the structures with opaque mesh permeable

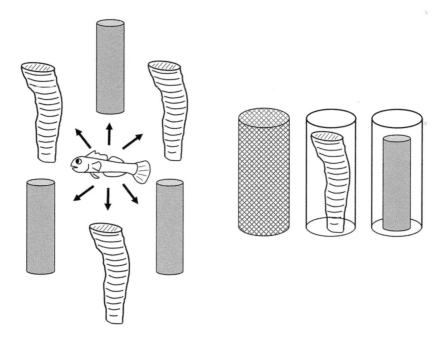

Figure 3.6 Diagram of the experiment used by Majoris et al. (2018) to study the environmental cues used by juvenile neon gobies (*Elacatinus lori*) to select habitat (yellow tube sponges). Each trial consisted of releasing a single fish in the middle of a 6-m diameter "arena" on the sea floor by the reef. The fish had a choice of selecting any of six structures; figure (left side) shows yellow tube sponges and PVC pipes being tested together. Majoris et al. (2018) further modified the experiment by placing an opaque mesh cover or transparent solid cover around the sponges and PVC pipes (right side of figure). All of the trials involved testing two types of structure and with either visual cues removed (opaque mesh cover) or olfactory cues removed (solid transparent cover). Figure is adapted from Majoris et al. (2018) and not drawn to scale.

to water flow and hence odors (Fig. 3.6). The results were interesting and informative. When the gobies had both chemical and visual cues they preferred the yellow sponges over the gray PVC pipes. Without chemical cues they still preferred sponges, but without visual cues they had no preference—they swam to the PVC pipes as often as they did to the yellow sponges. The gobies also seemed to prefer yellow over brown sponges but only when visually unobstructed such that they could see the color and shape of the sponges. They also preferred long over short sponges but again only when they could see the sponges. A single-tubed long sponge seemed to be a sufficient home as the gobies also did not preferentially swim to the multi-tubed sponges. Lastly, the gobies did not discriminate between sponge tubes occupied by a resident goby and those that were unoccupied—this makes sense as the occupant was always inside the tube and thus out of view of the approaching goby. Altogether these experimental results suggest that gobies primarily use vision to find long yellow sponges and it is the color or shape that is the sensory cue. Moreover, this habitat selection by neon gobies during the settlement phase subsequently determines the spatial distribution of adults on the reefs (Majoris et al. 2018). This is an excellent example of how a physical (color) and perhaps structural (long tube) feature can serve as a cue for a dispersing organism to select habitat.

3.2 Models of habitat selection

The DSE model as developed here is structurally similar to previous models of habitat selection. One of the first models was introduced by Levins (1968). In his model, a dispersing individual has a limited amount of time to find a suitable habitat within some searched area in which there are two types of habitat. One type of habitat provides greater fitness (survival and reproduction) and covers a proportion, p, of the searched area. Levins (1968) gave an equation for the probability that a dispersing individual finds or encounters the habitat as $P(encounter) = 1 - e^{-\lambda p t}$, where λ is a measure of the searching ability or mobility of the individual (or species) and t is a given amount of time (note that I'm using Levins' notation). This equation describes an exponential

decay over time such that the probability of finding the habitat within a given amount of time increases rapidly with time (especially when p and λ are large) and then gradually approaches but never reaches 1. That is, for a given p, the term $e^{-\lambda p t}$ quickly becomes very small as λ and t increase. Levins (1968) did not explicitly state why the process is modeled as an exponential decay. However, his equation also holds if t is replaced with $T - t$ where T is a set finite amount of time and $T - t$ is the amount of *time remaining*. In such a case, $P(encounter)$ initially decreases slowly but then rapidly approaches and reaches zero when $t = T$ and hence no time remains (a complete decay). In Levins' model, the higher-quality habitat is always used (selected) by the individual if it is encountered. However, it may not be encountered. Rather the lower-quality habitat might be encountered first. The individual must then "decide" whether to keep searching or use the lower-quality habitat. Levins (1968) used θ to symbolize the probability of using the lower-quality habitat. If the difference in fitness between the two habitat types is negligible, then θ should equal 1 and the individual uses whatever habitat is first encountered. If the difference in fitness is large, then θ should equal 0, particularly when the individual is really adept at finding the higher-quality habitat (large λ) or this habitat type is very common (large p). That is, the individual should continue to search for the higher-quality habitat even if it first encounters the lower-quality habitat because there's a very good chance that it will find the higher-quality habitat in the remaining time. Importantly, in Levins' model, the optimal θ is always either 0 or 1; it is determined by the difference in fitness between the two habitat types not by the amount of search time remaining (equation 2.9 in Levins 1968). That is, an individual should always only select the higher-quality habitat or should select whichever habitat is first encountered. To his great credit, Levins showed how this general model could also be applied to understand the distribution of alleles, genotypes, and phenotypes within a spatially and temporally changing environment, hence the title of his book, *Evolution in Changing Environments*. Except for a few pages here and there, his book has very little mention of habitat selection. He mainly used his model to explain the evolution of diapause in insects (Levins 1968) and then later to compare the fitness benefits of monophagy and polyphagy (Levins and MacArthur 1969).

Other early mathematical and conceptual models of habitat selection include those of Doyle (1975) and Ward (1987). The model of Doyle (1975) is set up like Levins (1968): a dispersing individual can encounter and settle in either of two habitat types (or substrates) and the goal of the model is to find the conditions that maximize fitness. However, Doyle explicitly included the probability that the dispersing individual dies if it does not choose either habitat. In Levins' model, mortality is implicit (not a modeled probability per se) as it presumably occurs by default when time expires. Doyle's model also incorporates the added complexity and realism of reproduction modeled as a probability specific to a given habitat type. Doyle's main goal was to develop a model of habitat selection to explain the settlement behavior of planktonic larvae of marine invertebrates, specifically the polychaete worm *Spirorbis borealis*. As an adult the worm secretes and lives in a calcareous tube attached to seaweeds such as *Ascophyllum* and *Fucus*.

Ward (1987) extended Levins' model by including an explicit term for mortality (as did Doyle) but more importantly he devised a way in which θ (recall this is the probability that a dispersing individual will select the lower-quality habitat) could initially be 0 but then become equal to 1 after a certain amount of time. That is, Ward (1987) modeled a situation in which there is a definite difference in fitness between two habitat types such that a dispersing individual always rejects the lower-quality habitat (θ = 0) during an initial "discrimination phase" but thereafter will accept the lower-quality habitat when it is encountered (θ = 1). As such, Ward's model directly reflected habitat preferences (or selectivities) in that the post-dispersal pattern of settlement of individuals among the two habitat types presumably arose due to the discrimination phase (Ward 1987). Duration of the discrimination phase depended on the fitness difference, availabilities (proportions) of the two habitat types, encounter rates, and the mortality rate. The model revealed the somewhat counterintuitive finding that the lower-quality habitat should be accepted at a lower rate as its availability increases; that is, a dispersing individual can afford to wait longer before accepting such habitat because it is relatively common and hence frequently encountered (Ward 1987).

Soon after it was published, the Levins model of habitat selection also influenced models of host-plant selection by phytophagous insects. In these models, a female insect searches for host plants on which to lay eggs, and she must choose among various types of host plant (Levins and MacArthur 1969; Jaenike 1978; Courtney 1982). Again, there is a time constraint and variation among host plants with regard to their "quality" or the fitness that can be realized by females laying eggs on plants best suited to the development of larvae. Some years later, Judy Stamps and colleagues used the general modeling framework of Levins (1968) and Ward (1987) to further develop habitat selection theory and explore the ecological and evolutionary consequences of the process (Stamps et al. 2005; Stamps and Krishnan 2005; Stamps et al. 2007).

A time constraint is a common feature of all the models so far discussed in this chapter. All of the models also involve some type of search process. That is, an individual organism moves through the environment searching for something (habitat, host plant, mate) required for it to realize some amount of reproductive fitness. The individual has limited time and either accepts or rejects encountered "items" based on an inherent decision rule that weighs the probability of finding a better item (in the sense of reproductive fitness) and the probability of dying in the time remaining. With regard to habitat selection, individuals either accept or reject habitats based on properties of the habitats, search costs, encounter rates, and time limitations (Ward 1987; Stamps et al. 2005). For a given set of conditions, the models are intended to find the optimal probability for accepting a particular habitat type wherein that optimal probability maximizes subsequent fitness of the individual. The models can also reveal the conditions under which the individual should in theory accept a wide range of habitat types or only one. However, none of the models attempts to directly explain the distribution and abundance of a species. The models are focused on explaining why a species might use more than one type of habitat and predicting which types of habitat are used based upon their fitness benefits and other characteristics (e.g., presence of conspecifics, predators). For example, the models could be used to generate a *list* of discrete habitat types that Species Y should, in theory, use given a set of prevailing conditions [X]. Such a list contains partial information on the spatial distribution of a species but it does not predict how the presence/absence of a species might relate to one or more continuous habitat variables (or other environmental cues affecting settlement) as in Fig. 3.3.

The DSE model that I am proposing also has a time constraint and a search process that involves acceptance and rejection of habitats. My model differs from the habitat selection models in focusing on how the strength of an environmental cue (e.g., a habitat structural feature) and an organism's ability to respond to it have a subsequent effect on the pattern of spatial distribution of the species. In general, the DSE model assumes that individuals end up in the "correct" habitat. Therefore the question of interest is not about the decision-making behavior that enables acceptance of some habitat types over others but rather what environmental factors are used by individuals as cues to settle at a particular location—and further whether those cues are good predictors of species distribution and abundance. Reproductive fitness is not explicitly included in the DSE model. Nonetheless, presumably natural selection acting on a species over evolutionary time induces matching of an individual's detection and response to an environmental cue with the habitat conditions that best promote survival and reproduction. The quality of the habitat may be directly proportional to the strength of the habitat cue.

Another early and quite different model of habitat selection is the ideal free distribution (IFD). This model describes how individuals of a population distribute themselves among pieces of habitat (i.e., a patch of habitat or a specific area of habitat). As originally framed by Fretwell and Lucas (1969), the driving force behind the IFD is territorial behavior and the assumption that individual organisms will select (or settle in) a habitat patch where their chance of success is highest—this is the so-called "ideality assumption." Fretwell and Lucas did not explicitly identify "chance of success" as fitness; in fact, they did not even use the word "fitness," but that's what they meant. Several times they did refer

to individuals being "ideally-adapted" particularly with regard to habitat selection behavior. They had birds in mind when they came up with the IFD, hence they wrote "the first assumption demands that the birds have habitat selection behavior which is ideal, . . . each bird selects the habitat best suited to its survival and reproduction." and "the second assumption demands . . . that the birds be free to enter any habitat" (Fretwell and Lucas 1969, p. 21).

The IFD assumes that each piece of habitat has a "basic suitability" and that the overall suitability of the location declines as more conspecific individuals arrive there. Moreover, at equilibrium when the distribution of individuals among habitats has become *ideal free*, then the suitabilities of all habitats are equal. The IFD is achieved when each individual selects the habitat of highest suitability at the given time. Careful reading of Fretwell and Lucas (1969) reveals that their original intent was to explain how individual birds in a population are spatially dispersed due to territorial behavior and implicit selection of habitat. As such, the IFD is a model of habitat selection but not one that attempts to describe selection as based on properties or characteristics of the habitat, except for conspecific density considered as a property of a patch of habitat. The IFD does not explicitly invoke any kind of search process or time constraint. However, Fretwell and Lucas (1969) did thoroughly discuss how birds might be using environmental cues to assess conspecific density and hence habitat suitability. They viewed the IFD as being temporally dynamic such that birds must have evolved ways of detecting and responding to cues associated with ever-changing densities of their neighbors. Fretwell and Lucas even suggested footprint abundance as a cue as well as overt territorial displays by resident birds. They went on further to state that the sensory reception (for environmental cues) of birds cannot be perfect and hence birds can sometimes be expected to misread cues—this is essentially the error, E, of Equation 3.1 of the DSE model. Fretwell and Lucas also recognized that the correlation between habitat cues and suitability is not perfect. In the context of the DSE model, this means that the strength of the cue (Q_{hab}) and subsequent post-settlement mortality in the habitat are not correlated in a perfectly inverse way.

Pulliam and Danielson (1991) relaxed the IFD assumption about equal suitabilities of habitats when they developed an extended habitat selection model that incorporated reproductive success and hence population growth—an important advancement. But as with Fretwell and Lucas their model did not explicitly include a search process or time constraint, unlike the model that I propose. Along with Levins (1968), the IFD was one of the first conceptual/mathematical models of habitat selection. It has been enormously influential in behavioral ecology. The IFD is a paragon of straightforward math and sound logic.

The label "habitat selection model" is often used in a very encompassing sense to refer to statistical models of species–habitat relationships, such as resource selection functions (see Section 11.1). That is, an identified association between a species presence or abundance and a particular habitat type or habitat variable is taken implicitly as individuals of the species selecting that type of habitat. However, "selection" used in this context does not necessarily invoke a process where the organism is literally moving about the landscape and choosing (or using) some habitat types over others. Also, "model" is being used in reference to a statistical model more so than a conceptual model of a behavioral process. Obviously, it is appropriate and acceptable for ecologists to use "habitat selection model" in strict reference to a statistical model; however, the reader should be aware of this particular usage in the literature.

In a very different context, habitat selection has also been invoked as a process (or step) potentially important in speciation. As mentioned in Chapter 1, Leavitt (1907) suggested that related species would always be found in adjacent but different habitats. However, in his verbal account, selection of habitat is an implied aspect of speciation; he did not describe an explicit role for habitat selection. However, later authors did, including Lack (1933), Thorp (1945), and Mayr (1947). They explained how part of a population might begin to select a different type of habitat and thus become physically separated from the remainder of the population. If gene flow between the two groups is precluded then speciation may proceed. None of these authors presented a model of habitat selection; instead in the course of discussing models of speciation they

described a process wherein dispersing individuals actively search for habitat. More recently, Tonnis et al. (2005) and Harvey et al. (2017) suggested a crucial role for habitat selection in the speciation of warbler-finches (*Certhidea*) in the Galapagos Islands and for various Amazonian birds. "Habitat selection" has wide and varied usage in the ecological and evolutionary literature. This is not surprising and it does not reflect misuse of the term (see Chapter 1). Rather, it reflects the fact that habitat selection by individuals and the habitat associations of species have broad implications for ecological/evolutionary processes ranging from the formation of new species (and hence biodiversity) to species adaptation and population persistence. Ultimately, these processes determine species distribution and abundance.

Lastly, this section was not intended as an exhaustive review of habitat selection models. The study of habitat selection is an expansive field of theory and empiricism within the larger disciplines of ecology and wildlife ecology. For brevity, I focused on presenting some of the first conceptual/mathematical models of habitat selection and those that explicitly incorporate a DSE process. Apart from those, I may have overlooked some important models; any reader specifically interested in habitat selection should look elsewhere for an introduction and review of this multi-faceted topic, such as Jones (2001) or Morris (2003).

3.3 Physical features and structure as habitat cues

The spatial distributions of many species seem to correlate with particular physical features of a habitat or its overall structure. For example, the amount of canopy cover can be important in defining the habitat of terrestrial species inhabiting forested and savanna landscapes (James 1971; Renken and Wiggers 1989; Diaz and Carrascal 1991; Mysterud and Østybe 1999; Ranius and Jansson 2000; Rockwell and Stephens 2018) as well as aquatic species that inhabit ponds and streams within such landscapes (Skelly et al. 1999; Compton et al. 2002). However, is canopy cover the actual cue that the dispersing individual uses in finding a place to settle (i.e., suitable habitat)? It makes sense evolutionarily that canopy cover and other characteristics of the physical structure of the habitat would be the environmental cues that guide individuals to appropriate habitat. That is, species could evolve morphological, physiological, and especially behavioral adaptations for finding and reacting to such cues. For a dispersing organism with limited time, particular structural features or characteristics of the habitat would be reliable in that they could be easy to detect (specifically using vision) and they are relatively permanent and stable over the course of days, weeks, or even a few months of a DSE process. Further, the structure of the habitat should be such that it promotes or at least does not hinder active foraging behavior. This has been demonstrated for birds (Robinson and Holmes 1982; Parrish 1995; Whelan 2001; Newmark and Stanley 2016), fish (Yeager and Hovel 2017), *Anolis* lizards (Moermond 1979), European hares (Mayer et al. 2018), African lions (Hopcraft et al. 2005), praying mantises (Hill et al. 2004), wolf spiders (Rypstra et al. 2007), barnacles (Larsson and Jonsson 2006), and likely many other types of organisms. Thus, habitat physical structure as a cue for settlement also directly relates to one of the first and on-going survival tasks of an individual: the ability to efficiently find food. Organisms should use habitat that facilitates the acquisition of food, which means that the habitat allows ease of movement and contains the food sources; although in some species, there may be a distinction between foraging and breeding habitat. Natural selection should select for a species to evolve sensory adaptations, a suite of behavioral, neurological, and physiological traits, such that individuals select a habitat or habitat type that *potentially* provides the type of food resource that the species is adapted to utilize. There should be a matching of food resource and habitat, but the presence of food need not be the actual cue that individuals use to find and settle into the habitat. Food supply is often ephemeral given that it is depleted by conspecifics or competitors and can sometimes depend on recent local weather conditions. Also a lack of food does not necessarily indicate poor habitat if the food supply can be replenished. Habitat structure does not get depleted. The habitat-cue hypothesis predicts that structural features of the habitat provide the main cues for settlement.

For many terrestrial bird species, the structure of the vegetation (e.g., canopy height, foliage density, stem density, grass height) appears to be the primary cue used in selecting habitat. For aquatic bird species, the amount of emergent vegetation and water depth may be important habitat features (Halse et al. 1993; Murkin et al. 1997; Lantz et al. 2010; Osborn et al. 2017). Avian ecologists long ago recognized the importance of vegetation structure in defining the habitat of forest and grassland birds (James 1971; Willson 1974; Rotenberry and Wiens 1980; Cody 1981). The physical structure of the vegetation influences foraging and breeding behavior (nest site selection), activities that obviously pertain to survival and reproduction, and hence the fitness of individuals and population persistence. Further, there is good evidence that birds cue in on vegetation structure because it is a reliable indicator of food resource abundance, particularly as an average expectation over multiple years if not in a single year (Seastedt and MacLean 1979; Marshall and Cooper 2004). The "structural cues" hypothesis (Smith and Shugart 1987) states that birds use structural features that correlate with prey abundance when selecting habitat and establishing breeding territories rather than assessing prey abundance directly (Newmark and Stanley 2016). Further, the hypothesis predicts that when a bird initially settles into and defends a territory it can "determine" the required territory size based on the structure of the habitat because the structure provides information on the foraging success that the bird can expect (Marshall and Cooper 2004). In addition, a bird without a territory can assess structure of the vegetation within a conspecific's territory and thereby acquire information on potential resource abundance and thus "decide" whether to contest the territory (Smith and Shugart 1987). Clearly, the structural cues hypothesis implies a critical role for physical structure or features as cues for settlement during a DSE process. Moreover, because the physical cues are also indicators of potential food abundance, the subsequent establishment phase of the DSE process is more likely to be successful and result in the bird (or other type of organism) reproducing in the selected habitat or territory. That is, the habitat itself (specifically its structure) serves as the cue for settlement and establishment such that the spatial distribution of the species is best predicted by habitat not by other environmental factors—this is the habitat-cue hypothesis of species distribution.

At its core, the habitat-cue hypothesis invokes an organism moving over the landscape (or seascape) either intentionally searching for or randomly encountering cues. The idea that organisms receive and respond to external cues or stimuli from either the environment or other individuals is deeply rooted in ethology and behavioral ecology. It dates all the way back to the pioneering work of the famous ethologists, Konrad Lorenz and Niko Tinbergen. However, neither wrote specifically about habitat selection. Highly influenced by Tinbergen (1948), Svärdson (1949) envisioned habitat selection in songbirds as occurring when males returning from migration settle into a breeding territory in response to stimuli or cues from physical features of the habitat. According to Svärdson, a male responds innately upon seeing particular habitat features; there is a neurological response that is due to stimulation of optical nerve cells and this results in the male becoming relatively stationary and singing at a higher rate. Svärdson (1949) referred to this process as the "releasing mechanism of habitat recognition" borrowing some of the releaser terminology of Tinbergen (1948). To further illustrate his idea, Svärdson presented some hypothetical data for four closely related bird species that occupy slightly different types of habitat: tree pipit (*Anthus trivialis*), meadow pipit (*Anthus pratensis*), tawny pipit (*Anthus campestris*) and water pipit (*Anthus spinoletta*) (Table 3.2). The data pertain to the influence that different physical cues of the habitat have on the settlement behavior of an individual bird. Svärdson viewed the habitat occupied by each species as a direct consequence of the environmental stimuli that males respond to when establishing a territory. His conceptual paper provides an early description of the innate recognition and response mechanism that could be involved when an organism selects habitat based upon physical characteristics or structural features of the habitat.

As emphasized throughout this chapter, there are several broad types of environmental cues that organisms could hypothetically use in finding appropriate locations to settle. These are habitat structural features, food supply, absence or low density of antagonistic species, and presence of conspecifics. It

Table 3.2 Hypothetical data presented in Svärdson (1949) to illustrate the contribution of different environmental cues to habitat selection and settlement behavior of four pipit species. Numbers in table correspond to what Svärdson called "reaction unities" and were intended to represent the amount of nerve cells dedicated to an individual bird's detection and response to each cue.

Habitat type occupied[1]
Tree pipit—forest edges and parkland
Meadow pipit—meadows with wet ground
Tawny pipit—dry sandy areas
Water pipit—mountainous areas or rocky areas along shore

Cue	Tree pipit	Meadow pipit	Tawny pipit	Water pipit
Open or semi-closed canopy; substantial light penetration	30	40	40	35
High branches (tall trees)	37	14	20	26
Green color	20	20	5	5
No vegetation on ground	5	5	30	15
Water nearby	2	15	1	15
Conspecific males nearby	4	4	2	2
Other external stimuli	2	2	2	2
Total	100	100	100	100

[1] Descriptions of occupied habitat types are as given by Svärdson (1949). Clearly, the numbers in the table were assigned so as to correspond with each species type of habitat.

seems difficult to envision a scenario where the *absence* of individuals of an antagonistic species would serve as the strongest cue for settling at a particular location. This is particularly true if the cue does not correspond to the ease with which an individual can forage in the habitat, which in turn depends on habitat physical structure and food availability. Such a scenario would also require that, over evolutionary time, natural selection favored dispersers that could *quickly* determine that heterospecific individuals of certain species were absent or at low densities prior to settling. Recall that the DSE process is time-constrained. A behavioral adaptation would need to evolve that gave a dispersing individual some level of "confidence" that after a certain amount of time assessing the local conditions, antagonistic species are absent. Natural selection has its limits. Perhaps such a trait could evolve in the context of a dispersing individual comparing multiple potential settlement sites and selecting the one with the lowest density of antagonistic individuals, but such a comparison itself would require a lot of valuable time. Compared with heterospecifics, conspecific individuals represent a fundamentally different type of cue, one in which presence elicits a positive response by the dispersing individual. The role of conspecific attraction in habitat selection has been widely researched (Stamps 1988; Donahue 2006; Fletcher 2006; Jeanson and Deneubourg 2007; Campomizzi et al. 2008; Bruinsma and Koper 2012). No doubt the presence of conspecifics, particularly conspecifics that are successfully breeding, can sometimes be a very reliable indicator of quality habitat (although see Van Horne 1983) and thus a strong cue enticing dispersing individuals to settle (Muller et al. 1997; Danchin et al. 1998; Serrano and Tella 2003; Betts et al. 2008; Robinson et al. 2011; Fobert and Swearer 2017; Folt et al. 2018). Contrary to this, some studies have revealed that increasing density of conspecifics makes it less likely that an individual will settle at a given location (D'Aloia et al 2011; Glorvigen et al. 2012; Brandl et al 2018). Regardless of whether conspecific attraction is an effective cue, it does not help explain how a species comes to occupy new localities (that do not already have the species) in the first place. The habitat-cue hypothesis of species distribution is intended to predict the spatial dispersion of a species, and this entails explaining how a species arrives at and establishes a population at new locales.

Therefore, if we discount the role of conspecific attraction and heterospecific avoidance in determining the spatial dispersion of a species, then we are left with two types of environmental cue that organisms might use to settle in a new location: structural features of the habitat and food supply. I am not aware of any studies that have simultaneously compared all four types of cue (habitat structure, food supply, antagonistic species, and conspecifics) on habitat selection of a given species. However, numerous studies have examined and compared the effects of habitat structure (often physical characteristics of the vegetation) and food

supply on the process of habitat selection for a wide variety of species. There is not a strong consensus as to which cue is more important. Several studies suggest that both are equally important (Robinson and Holmes 1982; Orians and Wittenberger 1991; Newmark and Stanley 2016; Zenzal et al. 2018), particularly when the physical structure of the habitat indicates (to the dispersing individual) the current level of food availability or the long-term expectation that food is typically available in such a habitat (Cody 1981; Smith and Shugart 1987; Marshall and Cooper 2004; Buler et al. 2007). For example, small forest-dwelling insectivorous passerines such as warblers (Parulidae) and vireos (Vireonidae) often establish breeding territories and select migratory stopover sites based more on foliage density rather than insect abundance, although the latter often correlates with the former (Marshall and Cooper 2004; Wolfe et al. 2014). Further, Wolfe et al. (2014) found that insectivorous species relied on the structure of forest vegetation to select migratory stopover sites rather than direct assessment of insect density, whereas frugivorous species selected stopover sites based on fruit biomass and even the sugar content of the fruit. These findings make a lot of sense. It is likely difficult or time-consuming for an insectivorous bird to detect and assess the abundance or even the presence of cryptic insect prey at a given location (Wolfe et al. 2014). In particular, such assessment might be nearly impossible to do from afar such as when flying over; however, visually assessing the amount of foliage and canopy cover is likely much easier. Detection of fruit on trees might be relatively easy for a frugivorous bird; hence food is a direct cue for stopover habitat for those species.

In Washington state (United States) female yellow-headed blackbirds (*Xanthocephalus xanthocephalus*) appear to select and settle in marsh habitat based upon density of dragonflies; however, the selection of a breeding territory and construction of a nest occurs in particular areas of the marsh that have moderately thick sedge and cattail vegetation so as to support nest construction (Orians and Wittenberger 1991). Nest site locations are also surrounded by 1-m-wide channels that presumably deter snake and mammal predators (Orians and Wittenberger 1991). Hence, habitat selection depends on structural and food cues, both of which relate to reproductive success. By contrast, conspecific attraction does not appear to be important in habitat selection by yellow-headed blackbirds (Ward et al. 2010).

For many types of species in various kinds of landscapes, habitat selection may be greatly influenced by the structural features and characteristics of the habitat. These serve as reliable cues to inform a dispersing individual that appropriate habitat has been found. Furthermore, physical structure may be the most important cue and literally determine the spatial distribution of the species in nature—this is the primacy of habitat. Physical structure largely defines habitat, exclusive of food, other species, and abiotic conditions. Physical structure may be the strongest cue. This means that, as cues, structural features or properties of the habitat should have relatively low perception time (T_p), minimal error (E), and substantial strength (Q) (see Equations 3.1 and 3.2). For example, a dispersing bird can quickly and easily assess specific structure of the vegetation in a forest, grassland, or marsh and "decide" whether to settle or not. A dispersing marine fish can quickly and easily assess specific structure of a coral reef, a stand of mangroves, or a bed of seagrass. A praying mantis can assess the rugosity and circumference of tree branches and thus settle on smooth small branches that facilitate faster running to capture prey (Hill et al. 2004). In an analogous way, African lions select habitats that provide extensive view-sheds and moderate woody or rocky cover; these habitats allow for easier prey detection and capture even though prey are more abundant in other habitats (Hopcraft et al. 2005).

3.4 Strength and efficacy of environmental cues

Returning to the mathematical formulation behind the DSE model, the strength of an environmental cue is simply (and strictly) an adjustment to a species-*intrinsic* settlement probability so as to get a *site-specific* settlement probability (see Equation 3.2). That is, a given environmental cue j at a given location i will have a strength (Q_{ij}) that essentially indicates the intensity or magnitude of the cue at the location. Cue strength is scaled from 0 to 1, with 0 representing the absence of the cue and 1 indicating the optimal amount of the cue. For example, if foliage density is the cue that a warbler responds to when settling in a patch of forest then locations with either maximal or optimal foliage density will have $Q_{ij} = 1$. For some cues, a species might best respond to an intermediate amount or level (rather than maximum) and hence $Q_{ij} = 1$ represents the optimum. Consider a hypothetical situation in which a species' strongest intrinsic response occurs when there is intermediate canopy cover. As such, locations with 50 percent canopy cover would have a cue strength $Q_{ij} = 1$. This is the way cue strength is quantified and incorporated into the DSE model.

However, we can also think about the strength of an environmental cue in a broader conceptual context. That is, a strong environmental cue is one that can attract a dispersing individual from a greater distance and retain the individual for a longer duration of time. As a brief aside, retention is important in the distinction between permanently settling at a location and habitat sampling or exploration; this is discussed in Section 3.6. Attraction or orientation of a dispersing individual toward the point location of a cue depends on the distance at which the individual can detect the cue. This distance is the perceptual range of the cue (or any feature on a landscape) and it depends on properties of the cue and sensory/behavioral traits of the species (Lima and Zollner 1996). As discussed in the previous section, it is reasonable to think that structural features or characteristics of a habitat often have a long perceptual range, perhaps longer than cues associated with food supply, conspecifics, and antagonistic species. This might be the case particularly if vision is the primary sense used by the dispersing individual to detect cues. On the other hand, if the primary sense is olfaction then food supply and conspecifics might also have relatively long perceptual ranges. Also, not all habitat structural features have a long perceptual range. Black-throated green warblers (*Dendroica virens*) use needle length to distinguish spruce trees when selecting habitat (Parrish 1995); presumably needle length can only be assessed when the warbler is very close to the tree.

In addition to strength of a cue, efficacy is another general property of an environmental cue that could be important in the DSE process. Broadly thinking, efficacy can be defined as the capacity of the cue to *always* attract or elicit a settling response from the dispersing individual, to function as a cue for settling *only* (not some other behavioral activity), and to be detectable via a species trait that itself is maintained by natural selection. As explained in the previous section, any hypothetical cue that is based on the absence or low "intensity" of a factor is likely to either not evolve or not be maintained over the long term. Detection of such cues is inefficient and error-prone. Therefore, effective cues are based on the presence of a factor for which there is high fidelity or matching between the cue and the individual's settlement response. Essentially, efficacy will usually entail that the cue has low error (Equation 3.1).

There is also an important distinction between negative and positive settlement cues (Morello and Yund 2016). Negative settlement cues cause an individual to move away from the spatial source of the cue as in repulsion or avoidance, whereas positive cues cause movement toward. In the DSE process of a dispersing organism actively moving about a landscape, positive cues are likely more important to *initial* settlement than are negative cues, although negative cues could be involved in causing the organism to leave the initial location and continue dispersing (see Section 3.6). Consider that if negative settlement cues alone were responsible for habitat selection then an individual would eventually end up settling in a location and selecting a habitat by default. This would be some small area of the landscape where the individual is the least repulsed from negative cues emanating elsewhere. Simply put, negative settlement cues are not effective.

Settlement cues as discussed here may not necessarily be exact indicators of habitat quality per se. Cues to habitat quality should be comprehensive, multi-faceted, and indicate potential reproductive success (Doligez et al. 2003; Sergio and Penteriani 2005; Forsman et al. 2008). Thus, cues for settlement and cues for habitat quality have fundamentally different functions. The former are intended only to elicit an immediate behavioral response, whereas the latter are intended to provide an organism with information about expected reproductive success in the near future. An individual's "assessment" of that information might be protracted, occur after initial settlement, and determine whether the individual is permanently retained at the settlement location (see Section 3.6). Also, some indicators of habitat quality might not function well as cues for settlement if they cannot be easily and quickly detected, given that the DSE process is time-limited. A settlement cue can be an indicator of habitat quality but not all indicators are effective cues for settlement. (Note that "cues" for habitat quality are probably best referred to as "indicators" given that they are not cues in the strict sense of the word, although the phrase "cue for habitat quality" is commonly used by authors.)

Some environmental factors might function as behavioral cues for immediate settlement as well as provide some information that the selected habitat is likely to have the resources and conditions necessary for survival and reproduction. This is habitat quality broadly defined. Further, in order for a species' recognition of a settlement cue to have evolved to begin with, that same cue is likely also to reliably indicate habitat quality. The cue then serves to place the individual in a habitat where it can reproduce and pass on the trait for cue recognition. Many of the barnacle and reef fish ecologists, mentioned earlier in the chapter, wrote about how the habitat selected by dispersing juveniles should correlate with fitness of the adults in the context of natural selection continually reinforcing adaptive behavioral traits for finding and responding to the cues.

The habitat-cue hypothesis of species distribution assumes that habitat structural cues have the greatest strength and efficacy. That is, when a dispersing individual senses a structural cue(s) it should always orient toward the cue and then engage in settling behavior such as defending a territory, digging a burrow, or attaching to a hard substrate upon arriving at the settlement location. If the physical structure of the selected habitat also facilitates foraging then the individual may eventually reproduce (assuming conspecifics also arrive) with subsequent establishment of a population.

3.5 Scaling of environmental cueing

The perceptual range of the environmental cue may be very important in determining whether the cue can potentially attract the individual from afar and hence enable the individual to quickly and efficiently find a settlement location of the appropriate type of habitat (Lima and Zollner 1996; Zollner and Lima 1997; Bowler and Benton 2005; Fagan et al. 2017). This is true whether the cue is a habitat structural feature, food supply, or conspecific individuals. Further, the perceptual range may depend on properties of the cue as well as species traits (Mech and Zollner 2002; Speigel et al. 2013), the dispersing individual's internal physiological state, and even the natal habitat or previous experience of the dispersing individual (Öckinger and Van Dyck 2012). Also, the fact that different cues have different perceptual ranges emphasizes that dispersal and habitat selection are processes playing out over a range of spatial scales. A juvenile individual leaves its natal habitat and begins to disperse. The first environmental cues that it responds to are likely to be those with the greatest perceptual range. Subsequently the individual may respond to cues with a shorter perceptual range particularly if these help "guide" it more precisely to the particular habitat type for which the species is best adapted. Structural features of the habitat may be useful cues during this entire process of homing in on a settlement location.

To briefly return to the structural cues hypothesis, pileated woodpeckers (*Dryocopus pileatus*) use downed and standing dead trees as cues to establish territories and possibly as indicators of their wood-boring invertebrate prey (Renken and Wiggers 1989). A woodpecker presumably can visually detect these physical features from some distance. Contrast this with black-throated green warblers using needle length to distinguish red spruce (*Picea rubens*) from white spruce (*Picea glauca*) when selecting habitat (Parrish 1995). The warblers prefer red spruce because it has shorter needles that are compressed against the stem making it easier to forage among the branches (Parrish 1995). However, needle length and arrangement are fine-scale cues that the warblers likely can detect only when they are very near to individual trees; that is, the habitat must first be "sampled" prior to the individual deciding to permanently settle. Again, some cues simply have a larger range of attraction than others, and a dispersing individual might use multiple types of cues during the DSE process to home in on an appropriate location to settle (Woodson et al.

2007; Rushing et al. 2015). Regardless, at some point in time, the individual "decides" to stop dispersing and thereby settle; this decision is likely most influenced by very proximate structural features of the habitat, rather than by presence of food or absence of competitors and predators.

3.6 Settling versus sampling of habitat

"Many habitats that are good enough to trigger initial exploration may not be good enough for settling" (Orians and Wittenberger 1991, p. S30), or put another way, an individual might leave a "settled" location prior to reproducing and being truly settled and established. There is clearly a distinction between an individual settling and remaining indefinitely at a location in a given area of habitat and the individual initially arriving at the location, lingering, and then leaving soon after (e.g., within days or weeks). That is, individuals, particularly juveniles, often try out multiple different locations prior to truly selecting one that becomes relatively permanent. Ecologists have referred to this trying-out behavior as "habitat sampling," "prospecting," and "exploratory forays." During this process, an individual is moving about the landscape or seascape receiving and responding to environmental cues. Given the particular sensory abilities of the individual, the cues have a perceptual range and hence might attract or orient the individual. Such cues could be structural features or characteristics of various habitat types, food supply, and presence of conspecifics, as discussed above. At some specific spatial location, the cues are such that they temporarily halt further dispersal. Importantly, although the cues may have lead the individual to halt further movement over the landscape, the cues may not be strong enough to cause the individual to permanently settle. That is, the cues are strong enough to attract the individual but not to retain it.

It is also possible that once a dispersing individual is sampling the habitat, other cues are detected or information assessed that cause the individual to leave the location. This may happen even if the initial positive cues that attracted the individual are still present at the location. Predation risk is extremely important in the overall process of habitat selection (Werner et al. 1983; Schlosser 1987;

Lima and Dill 1990; Martin 1993; Ahnesjö and Forsman 2006; Rypstra et al. 2007). Initially a dispersing prey individual may not detect the presence of a predator and thus the individual stops dispersing and tries out a given location. Given that absence of an antagonistic species is an ineffective settlement cue to begin with, prey species likely do not have adaptations to quickly determine predator absence at a potential settlement location when the location is first encountered. However, with more time and more habitat sampling at the location, the presence of a predator or any other negative settlement cue might become apparent enough to cause the prey individual to leave. Even without actual predation, a cue for leaving could also be manifested as interspecific aggression due to territorial behavior (Cody 1981). Based on this logic, negative settlement cues can override the positive settlement cues that initially attracted the individual to the location. The negative cues prevent the individual from being retained at the location. Negative settlement cues, such as intraspecific competition and aggression, could also act in a positive way to stimulate a young organism's initial attempt at dispersal away from its natal location (Clobert et al. 2009). More generally, negative cues are often involved in density-dependent dispersal such that an increase in either conspecifics or heterospecifics causes individuals to disperse or move away from the source of the cue (Hauzy et al. 2007).

The absence of an antagonistic species is not an effective (reliable) cue for settlement. Simply put, the absence of a negative cue does not translate into the existence of a positive cue. The *presence and detection* of an antagonistic species also would not necessarily be an effective cue for not settling or at least not sampling the habitat. Apart from a predator lurking in the bushes, the location might otherwise be very good habitat for the dispersing individual. Therefore, the presence/absence of a competing species or a predator is not a good cue for habitat recognition. It is best to view presence/absence of competitors and predators as indicators of whether the dispersing individual can actually use the location if it is appropriate habitat. Antagonistic species indicate if the location is available to the individual for use, rather than whether the location is suitable habitat. Density of an antagonistic species, as assessed during habitat sampling, can be

a reliable indicator about whether a given location of suitable habitat is usable. Habitat must be both suitable and immediately usable in order for a dispersing individual to occupy it.

With regard to the actual use of a habitat, an organism may "select" various parts of the habitat for different activities such as foraging, nesting, denning, burrowing, hiding from predators, and even thermoregulating. Ecologists use particular phrases to label the parts of a habitat used for each activity such as resource or food patches for foraging, nest or oviposition sites for egg-laying, refugia for hiding from predators. There is a rich literature for each of these subcategories of habitat features. Moreover, this fine-scale distinction recognizes that an individual organism uses its habitat in different ways that reflect spatial heterogeneity or patchiness of the habitat (Orians and Wittenberger 1991). During a period of habitat sampling, it is important for a dispersing individual to thoroughly assess the habitat at a potential settlement location. Once truly settled, an organism routinely engages in all these behavioral activities. Thus, the initial act of habitat selection is a choice that affects almost all of an organism's subsequent behavior and ultimately its fitness through survival and reproduction (Orians and Wittenberger 1991).

With regard to the DSE process and the habitat-cue hypothesis of species distribution, the cues that elicit the various behaviors (e.g., foraging and nesting) *within* the habitat do not matter. Likely, these cues have much shorter perceptual ranges than do the structural features or characteristics of a habitat that the dispersing individual uses to home in on appropriate habitat and make the initial choice of whether to settle or not. For example, as mentioned previously, warblers and other small insectivorous birds cue in on settlement locations that have a high density of foliage that they can presumably detect at some distance. However, when they select a particular foraging patch in a tree they seem to use insect-damaged leaves as the cue for the presence of food (Robinson and Holmes 1982) and perhaps such a subtle cue is only detected at short distances. The distinction between settling and sampling of habitat is also not that important to the DSE process or the habitat-cue hypothesis. The most important aspect of the habitat-cue hypothesis is that a dispersing individual uses habitat structure as a cue to permanently settle in a particular location regardless of any subsequent behavior by the individual. Further, the cues for settling are also cues for retaining the individual permanently, unless those cues are overridden by negative settlement cues causing the individual to depart the location. Habitat structural cues as retainers might be one reason why some juvenile individuals do not disperse from their natal habitat. In addition, consider that when the physical structure of a habitat is altered, through anthropogenic or natural disturbance, the individuals of many species disperse away. This may be because the retaining influence of the habitat structure has been lost.

3.7 From neon gobies to kangaroo rats: the generality of the DSE process

The neon goby example presented earlier in the chapter involves a dispersing organism using a very distinctive structural feature (tall, bright yellow tube sponges) to cue in on coral reef habitat. Further, the coral reefs are also very distinctive existing in a seascape matrix consisting of relatively barren sand flats. As such, during its pelagic phase a neon goby disperses through open water where the coral reef habitat is very apparent visually. The goby literally lives within the hollow sponge. Given this close and obligate relationship between organism and habitat cue, there has likely been strong natural selection for gobies that can efficiently find yellow tube sponges. Other scenarios of the DSE process and cueing on habitat structural features may not be as obvious, yet the process is presumably ubiquitous and proceeds in the same general way for a wide variety of species.

Some species inhabit landscapes that may not have conspicuous and distinct patches of habitat but rather consist of vast expanses of vegetation that appears on a broad scale to be relatively homogeneous. Picture a landscape scene of an unfragmented and intact area of forest, grassland, or desert. In such landscapes, a dispersing organism would seem to be immersed in habitat everywhere around its birth site such that the DSE process simply entails getting far enough away from the parents' territory. That is, perhaps the individual need not recognize and respond to any environmental cues. The cues for settlement might be

everywhere and yet also hard to detect against a background that provides little contrast. For example, consider a juvenile Merriam's kangaroo rat (*Dipodomys merriami*) that is dispersing through a vast expanse of creosote shrub vegetation, typical of parts of the southwestern United States where the species lives (Box 3.2). As it disperses away from its natal burrow, the juvenile kangaroo rat may be receiving various environmental cues that could be relatively subtle and yet strong enough to induce it to dig a burrow, likely one of the first tasks that the young kangaroo rat must do when settling in a new location. Such cues could be a light to moderate cover of shrub vegetation and a loose sandy or loamy substrate inasmuch as *D. merriami* and some other kangaroo rat species appear to have these habitat requirements (Box 3.2). The fact that such cues might be readily available without an individual having to disperse very far or search for very long does not preclude a DSE process from occurring.

Moreover, cues for settlement may not be apparent to the human observer, yet be very obvious to

---

Box 3.2 Kangaroo rats

Kangaroo rats (*Dipodomys* spp.) inhabit arid and semi-arid regions of western North America. Given their large hind feet, long tails, and bipedal hopping form of locomotion, these rodents superficially resemble kangaroos, hence the name (Fig. 3.7). Most of the 20 *Dipodomys* species are granivorous, a few species are also herbivorous. All kangaroo rats are strictly nocturnal, residing within burrows during the day. Many of the species have fairly strict habitat requirements related to vegetation and soil properties. Some species (e.g., *D. merriami, D. elator, D. compactus, D. deserti,* and *D. spectabilis*) require habitat that consists of a substantial amount of bare ground, thin sparse herbaceous vegetation, shrub cover that does not form a closed continuous canopy, and soil devoid of gravel and rocks (Reynolds 1958; Rosenzweig and Winakur 1969; Price 1978; Nelson et al. 2009; Stevens and Tello 2009; Bliss et al. 2019). These habitat features either facilitate their bipedal locomotion as when moving about on the ground surface to forage for seeds or ability to dig a burrow (Thompson 1982; Price and Heinz 1984; Brown and Harney 1993; Miller et al. 2003). Thus kangaroo rats select habitat based on the need to dig a burrow, forage efficiently, and avoid predation.

Figure 3.7 Merriam's kangaroo rat (*Dipodomys merriami*) and typical desert-scrub habitat. In this photo, the habitat is a mix of creosote (*Larrea tridentata*) and mesquite (*Prosopis glandulosa*) shrubs near Fort Stockton, Texas. Photo credits are unknown photographer (kangaroo rat) and author (habitat).

the dispersing organism. Additionally, at a landscape level, the cue(s) may have much more spatial heterogeneity than is apparent to the human eye. For example, *D. elator* exists in a small region of north-central Texas where natural land cover within the landscape is substantially fragmented. The fragmentation of grassland habitat is due to anthropogenic factors such as cropland farming and roads. However, there is also substantial spatial heterogeneity of soil type and of course such fragmentation is natural (humans do not alter soil type on a landscape scale). Heterogeneity in soil type matters to *D. elator* because it requires clay-based soils, rather than sandy, loamy, silty, or gravelly soils (Roberts and Packard 1973; Martin and Matocha 1991; Ott et al. 2019). The species also tends to avoid areas of dense grassy vegetation and with a moderate to high percentage cover of shrubs (Ott et al. 2019 and references within). Although no one has documented the exact environmental cues that a dispersing *D. elator* individual uses to recognize appropriate habitat for settlement, it is easy to imagine that soil texture and density of the vegetation at ground level are likely two important cues. Kangaroo rats are constantly "sampling" the soil as their main seed harvesting behavior is to sift seeds from the soil (Eisenberg 1963; Morgan and Price 1992). When encountered, clay soil and sparse grassy vegetation would be very apparent to a kangaroo rat hopping along the ground even if these

cues do not have a long perceptual range. Thus, although kangaroo rats and neon gobies are very different types of organism living in very different environments, the selection of habitat and a settlement location might proceed similarly—that is, a DSE process occurs in which the dispersing individual settles at a particular location based upon the cues that it receives.

3.8 Heuristic predictions

The DSE model applies basic probability to describe a multi-phase process that leads to a dispersing organism arriving and permanently settling at a particular location on a landscape. Further, the model entails that the dispersing organism is receiving and responding to environmental cues during the time-limited search process. But the cues can be anything, not just structural features of the habitat. As such, the DSE model is intended to be omnibus, flexible in application, and able to generate testable predictions. When a habitat feature is presumed to be the most important cue then the DSE model becomes a model of habitat selection (broadly defined) and a way of testing the habitat-cue hypothesis of species distribution. This hypothesis extends the DSE process in predicting that the spatial distribution of adults of a species will be best explained by the habitat feature that was the initial cue for dispersing juvenile individuals to settle.

Alternatively, the most important cue may not be a habitat structural feature but rather some other aspect of habitat (e.g., availability of a food resource or density of conspecifics within the habitat) in which case the habitat-cue hypothesis of species distribution is falsified. Furthermore, individuals might respond to an environmental cue that is physically external to the habitat (strictly defined) such that the DSE process does not initially result in the species occupying the habitat *sensu stricto*. For example, a dispersing individual might encounter a rich food resource and hence linger at the resource for some days even though the resource is not confined within the actual habitat (or the area where the individual will eventually spend most of its time). Also, a dispersing individual might be attracted to a group of conspecifics even though they are not literally in the habitat. These scenarios represent intervening events and do not preclude an individual from eventually settling in the habitat. They also point out that for some species that range widely on a daily basis (e.g., birds of prey, sea birds) the exact habitat may be difficult to delimit in a physical sense and some activities (e.g., nesting and foraging) may occur in widely separated areas that do not necessarily appear to be the same habitat (see Chapter 6).

To the extent that organisms use a DSE process to find and occupy habitat, the habitat-cue hypothesis should generate heuristic predictions or "spin-off" hypotheses that are testable and thus falsifiable. Here I briefly propose seven such hypotheses: (1) When given a choice among various types of cues, organisms should respond most quickly and consistently to cues representing a physical structure or feature. Keeping in mind that the DSE process is time-constrained, natural selection in the past and present should favor individuals that can quickly find and respond to structural cues if habitat structure is the primary cue leading to settlement. This prediction could be tested in either a natural or laboratory setting by providing cues of various types (physical structure, food, conspecifics, antagonistic species) and testing the response of individuals to those cues. These types of studies are relatively common, although sometimes for purposes other than studying habitat selection. They often involve measuring how quickly and consist-

ently an individual moves toward the source of a cue, the distance at which the cue can be detected, and the amount of time the individual remains in close proximity to the cue (e.g., Casterlin and Reynolds 1977; Woodson et al. 2007; Forero-Medina and Vieira 2009; Dahms et al. 2017; Fobert and Swearer 2017).

(2) Related to the above, if it is most important for an individual to detect structural cues, rather than non-structural cues such as food and conspecifics, then the sensory and behavioral adaptations used to detect physical structure should be the best developed. During the DSE process, vision and the ability to quickly move to the source of the cue (i.e., physical structure detected at some distance) could be the best developed set of sensory/behavioral adaptations for many species. This hypothesis would be supported by evidence showing that mass or neurological connections to eyes and the visual cortex of the brain are better developed than for olfactory organs in those species that cue in on habitat structure. Interestingly, the olfactory organ and gustatory cells (taste buds) in neon gobies do not appear to be specialized or highly advanced (Hu et al. 2019); this makes sense if the gobies are primarily using vision to find and detect the yellow tube sponges. As a corollary to this hypothesis, species that have better developed olfactory senses than visual senses should be less likely to use structural cues in selecting a settlement location, particularly if structural cues are primarily detected through vision. However, perhaps structural cues can be detected in ways other than vision? For example, free-swimming oyster (*Crassostrea virginica*) larvae actually orient toward the sound emanating from oyster reefs (i.e., the congregated shells of adults) and thus settle on the hard substrate provided by the reef (Lillis et al. 2013). Marine fish are capable of using a wide array of sensory cues to detect the location and direction of physical structure provided by coral reefs (Montgomery et al. 2001). In terrestrial environments, perhaps olfaction can be used to smell and distinguish between different structural habitat types based upon the plant species present.

Continuing with the list of hypotheses: (3) Artificial physical structure that mimics naturally occurring structure should be effective in attracting dispersing individuals. This may be particularly true for species that use vision to locate cues. For example, a wide range of corals, invertebrates, and fish will colonize and permanently settle on artificial reefs such as those that develop on shipwrecks. Similarly, some fish species that require a rocky substrate are known to aggregate to rock and cobble piles when such materials and other refuse is deposited or discarded by humans into the sea (e.g., Fabrizio et al. 2013). Non-native plants may sometimes closely mimic the structure of native vegetation and thus attract some species of fish (Figueiredo et al. 2015)—this is fairly solid evidence that the fish are responding to a cue that is literally just the physical structure or arrangement of the vegetation rather than some other property that only the native plants and habitat would have. In an ingenious experiment, Paxton and Smith (2018) discovered that the density of reef fish can be increased by 30–50 percent by adding mirrors to an artificial reef. They attributed the increase to an enhancement of the visual cues that the fish were receiving even though the physical structure representing the cues was obviously only an illusion. In fact, regardless of demonstrating that organisms respond to artificial structural cues, the ability of some species to reside in completely artificial habitat (e.g., fish on real but artificial reefs) is indirect support for the hypothesis that such structure stimulates settlement.

(4) If individuals select habitat based on structural cues, then when the cues are removed through either a natural or anthropogenic event, the same individuals should leave the habitat. This may be the case particularly if the cues are also important in retaining the individuals during some relatively prolonged period (e.g., days or weeks) of habitat sampling. A researcher could test this hypothesis by removing the cue and then seeing if individuals depart the habitat. However, it is important to note that the lack of departure would not necessarily mean that the individuals did not initially use structural cues in finding the habitat. Also, there is an element of time underlying this hypothesis and its testing. Presumably, removal of the attracting structural cues becomes less likely to cause individuals to depart as more time elapses from the initial selection of the habitat. Indeed, if the individuals have already passed out of an initial habitat sampling phase then removal of structural cues might not cause departure.

(5) Slow-moving species or those inhabiting landscapes and seascapes with widely separated and infrequent patches of habitat should be particularly adept at detecting and responding to habitat structural cues in that the DSE process is time-constrained. Such individuals should also be able to respond even when the cue is rather subtle. Given that individuals may not encounter many cues, they must be very adept at detecting the cue from a substantial distance or as their searching movement brings them near to the cue. They do not have many opportunities to make a mistake—overlook a cue that indicates an appropriate settlement location. Evidence in favor of this hypothesis would exist as slow-moving organisms or those inhabiting naturally fragmented landscapes having longer perceptual ranges for structural cues, whether these are detected through vision or olfaction.

(6) Species with time-consuming and energetically costly foraging (e.g., web-building spiders) or long-duration foraging (e.g., sit-and-wait predators) should use structural features, rather than food or conspecifics, as cues for appropriate habitat. In these species, habitat sampling to assess prey availability would require too much time and energy. The presence of conspecifics might indicate food availability and hence appropriate habitat, but conspecifics also would be competitors for a possibly limited food source, and hence settlement should also not be based on conspecific density except perhaps as a negative settlement cue. McNett and Rypstra (1997, 2000) found that female orb-weaving spiders, *Argiope trifasciata*, were more likely to build webs where vegetational complexity, measured as the number of potential contact points for web strands, was greatest. Further, web-building was unrelated to prey availability, although spiders did tend to abandon webs and relocate elsewhere if prey capture did not exceed a minimum threshold.

(7) In its strongest form, the habitat-cue hypothesis predicts that dispersing individuals will not settle at a given location if it has the appropriate food source but is lacking the structural cue(s) for settlement. The daily or routine acquisition of food

is a major survival task for most species. Therefore, if a dispersing individual foregoes settling at a location even though it has an obvious and detectable food source but the wrong type of habitat structure, then indirect but strong support for the habitat-cue hypothesis is obtained. This scenario could be examined in a natural setting by providing a food source at locations that differ in one or more presumed structural habitat cues, but are otherwise similar and close enough to one another that a dispersing individual could potentially encounter each in the time it has available. The prediction is that the organism will select and settle at the location that has food and the correct structural cue. An even bolder prediction would be that the organism settles at a location that has the correct structural cue but no food rather than those locations that have food but no structural cue(s). Obtaining falsifying data (results) for any one of these seven hypotheses is evidence against the habitat-cue hypothesis of species distribution. Alternatively, finding support for any of these hypotheses does not necessarily indicate the validity of the habitat-cue hypothesis. Ultimately, the strongest evidence for the habitat-cue hypothesis is documenting a pattern revealing a strong association between the species presence, and perhaps abundance, and a habitat structural feature as in Fig. 3.4.

Lastly, the process of habitat selection may sometimes begin very early in the life of a young organism, even before it has dispersed. *Natal habitat preference induction* (NHPI) describes a condition or scenario in which a juvenile organism's familiarity with the habitat of its birthplace has a carry-over effect in influencing its habitat preference at a later life stage such as when dispersing (Jaenike 1983; Davis and Stamps 2004). Essentially, NHPI is a form of habitat imprinting. It is relevant to the DSE process in that imprinting suggests that an organism might form search images or have other neurological-behavioral mechanisms by which to recognize appropriate habitat when dispersing. Further, these mechanisms could involve the organism recognizing and responding to extrinsic cues such as structural features or characteristics of the habitat (Merrick and Koprowski 2016). Therefore, studies of NHPI and habitat imprinting in general could include indirect tests of the habitat-cue hypothesis. These tests would essentially involve the researcher attempting to determine the specific cues that an individual organism imprints on rather than just demonstrating that the habitat selected and occupied as an adult is very similar to the habitat that the individual originates from. Although a DSE process could be involved in NHPI, the two are not the same. The DSE process and the habitat-cue hypothesis are about an organism finding and settling at a particular location and the consequences of that for patterns of species distribution and abundance. Unlike NHPI, the habitat-cue hypothesis does not seek to explain why a species has certain habitat preferences.

3.9 Summary

While ecologists have long recognized that various environmental factors can restrict the spatial distribution of species, until now there has been no conceptual/analytical framework for explaining how a species' response to environmental cues for settlement can subsequently determine the spatial distribution of the species. This is the intent of the habitat-cue hypothesis of species distribution. The hypothesis invokes active selection of habitat by individual organisms as occurs when an individual disperses, settles, and establishes itself at a particular location (i.e., the DSE process). This process can be quantified by a series of event probabilities that combine to give the probability that the species exists at a particular location given the environmental conditions at the location. Currently, some of the best empirical support for the habitat-cue hypothesis comes from reef fish and barnacles. These organisms appear to actively select habitat based upon the structural features of coral reefs that they recognize as cues for settlement. Further, the settlement behavior of juvenile individuals establishes the pattern of adult distribution. As such, the species' distribution can be predicted by the distribution of habitat. The habitat-cue hypothesis proposes that habitat structural features or properties are more important than food, conspecific density, and absence of antagonistic species as cues for settlement. This is the primacy of habitat in explaining the distribution and abundance of species.

References

Ahnesjö, J. and Forsman, A. (2006). Differential habitat selection by pygmy grasshopper color morphs; interactive effects of temperature and predator avoidance. *Evolutionary Ecology*, 20, 235–57.

Betts, M.G., Hadley, A.S., Rodenhouse, N., and Nocera, J.J. (2008). Social information trumps vegetation structure in breeding-site selection by a migrant songbird. *Proceedings of the Royal Society B*, 275, 2257–63.

Binckley, C.A. and Resetarits, W.J. (2005). Habitat selection determines abundance, richness and species composition of beetles in aquatic communities. *Biology Letters*, 1, 370–74.

Bliss, L.M., Veech, J.A., Castro-Arellano, I., and Simpson, T.R. (2019). GIS-based habitat mapping and population estimation for the Gulf Coast kangaroo rat (*Dipodomys compactus*) in the Carrizo Sands Region of Texas, USA. *Mammalian Biology*, 98, 17–27.

Booth, D.J. and Wellington, G. (1998). Settlement preferences in coral-reef fishes: effects on patterns of adult and juvenile distributions, individual fitness and population structure. *Australian Journal of Ecology*, 23, 274–79.

Bowler, D.E. and Benton, T.G. (2005). Causes and consequences of animal dispersal strategies: relating individual behaviour to spatial dynamics. *Biological Reviews*, 80, 205–25.

Boyce, M.S., Johnson, C.J., Merrill, E.H., Nielsen, S.E., Solberg, E.J., and van Moorter, B. (2016). Can habitat selection predict abundance? *Journal of Animal Ecology*, 85, 11–20.

Brandl, H.B., Griffith, S.C., and Schuett, W. (2018). Wild zebra finches do not use social information from conspecific reproductive success for nest site choice and clutch size decisions. *Behavioral Ecology and Sociobiology*, 72, e114.

Brooker, R., Kikvidze, Z., Pugnaire, F.I., et al. (2005). The importance of importance. *Oikos*, 109, 63–70.

Brooker, R., Kikvidze, Z., Kunstler, G., Liancourt, P., and Seifan, M. (2013). The concept and measurement of importance: a comment on Rees et al. 2012. *Journal of Ecology*, 101, 1369–78.

Brown, J.H. and Harney, B.A. (1993). Population and community ecology of heteromyid rodents in temperate habitats. Pages 618–51 in *Biology of the Heteromyidae*, Brown, J.H. and Genoways, H.H. (editors), Brigham Young University Press, Provo, Utah.

Bruinsma, D.R.W. and Koper, N. (2012). Review of conspecific attraction and area sensitivity of grassland birds. *Great Plains Research*, 22, 187–94.

Buler, J.J., Moore, F.R., and Woltmann, S. (2007). A multi-scale examination of stopover habitat use by birds. *Ecology*, 88, 1789–02.

Caffey, H.M. (1985). Spatial and temporal variation in settlement and recruitment of intertidal barnacles. *Ecological Monographs*, 55, 313–32.

Campomizzi, A.J., Butcher, J.A., Farrell, S.L., et al. (2008). Conspecific attraction is a missing component in wildlife habitat modeling. *Journal of Wildlife Management*, 72, 331–36.

Casterlin, M.E. and Reynolds, W.W. (1977). Aspects of habitat selection in the mosquitofish *Gambusia affinis*. *Hydrobiologia*, 55, 125–27.

Chave, J. (2004). Neutral theory and community ecology. *Ecology Letters*, 7, 241–53.

Chesson, P. (1998). Recruitment limitation: a theoretical perspective. *Australian Journal of Ecology*, 23, 234–40.

Clark, J.S. (2012). The coherence problem with the unified neutral theory of biodiversity. *Trends in Ecology and Evolution*, 27, 198–02.

Clobert, J., Le Galliard, J.-F., Cote, J., Meylan, S., and Massot, M. (2009). Informed dispersal, heterogeneity in animal dispersal syndromes and the dynamics of spatially structured populations. *Ecology Letters*, 12, 197–09.

Cody, M.L. (1981). Habitat selection in birds: the roles of vegetation structure, competitors, and productivity. *Bioscience*, 31, 107–13.

Cody, M.L. (1985). *Habitat Selection in Birds*. Academic Press, Orlando, Fla.

Cody, M.L. and Diamond, J.M. (1975). *Ecology and Evolution of Communities*. Harvard University Press, Cambridge, Mass.

Compton, B.W., Rhymer, J.M., and McCollough, M. (2002). Habitat selection by wood turtles (*Clemmys insculpta*): an application of paired logistic regression. *Ecology*, 83, 833–43.

Connell, J.H. (1961a). Effects of competition, predation by *Thais lapillus*, and other factors on natural populations of the barnacle *Balanus balanoides*. *Ecological Monographs*, 31, 61–104.

Connell, J.H. (1961b). The influence of interspecific competition and other factors on the distribution of the barnacle *Chthamalus stellatus*. *Ecology*, 42, 710–23.

Connell, J.H. (1985). The consequences of variation in initial settlement vs. post-settlement mortality in rocky intertidal communities. *Journal of Experimental Marine Biology and Ecology*, 93, 11–45.

Connor, E.F. and Simberloff, D. (1979). The assembly of species communities: chance or competition? *Ecology*, 60, 1132–40.

Cornwell, W.K., Schwilk, D.W., and Ackerly, D.D. (2006). A trait-based test for habitat filtering: convex hull volume. *Ecology*, 87, 1465–71.

Courtney, S.P. (1982). Coevolution of pierid butterflies and their cruciferous foodplants. V. Habitat selection, community structure, and speciation. *Oecologia*, 54, 101–107.

D'Aloia, C.C., Majoris, J.E., and Buston, P.M. (2011). Predictors of the distribution and abundance of a tube sponge and its resident goby. *Coral Reefs*, 30, 777–86.

Dahms, H.-U., Tseng, L.-C., and Hwang, J.-S. (2017). Are vent crab behavioral preferences adaptations for habitat choice? *PLoS ONE*, 12, e0182649.

Danchin, E., Boulinier, T., and Massot, M. (1998). Conspecific reproductive success and breeding habitat selection: implications for the study of coloniality. *Ecology*, 79, 2415–28.

Davis, J.M. and Stamps, J.A. (2004). The effect of natal experience on habitat preferences. *Trends in Ecology and Evolution*, 19, 411–16.

Dayton, P.K. and Oliver, J.S. (1980). An evaluation of experimental analyses of population and community patterns in benthic marine environments. Pages 93–120 in *Marine Benthic Dynamics*, Tenore, K.R. and Coull, B.C. (editors). University of South Carolina Press, Columbia, SC.

Denley, E.J. and Underwood, A.J. (1979). Experiments on factors influencing settlement, survival, and growth of two species of barnacles in New South Wales. *Journal of Experimental Marine Biology and Ecology*, 36, 269–93.

Diamond, J.M. and Gilpin, M.E. (1982). Examination of the "null" model of Connor and Simberloff for species co-occurrences on islands. *Oecologia*, 52, 64–74.

Diaz, J.A. and Carrascal, L.M. (1991). Regional distribution of a Mediterranean lizard: influence of habitat cues and prey abundance. *Journal of Biogeography*, 18, 291–97.

Doerr, V.A.J., Doerr, E.D., and Jenkins, S.H. (2006). Habitat selection in two Australasian treecreepers: what cues should they use? *Emu*, 106, 93–03.

Doligez, B., Cadet, C., Danchin, E., and Boulinier, T. (2003). When to use public information for breeding habitat selection? The role of environmental predictability and density dependence. *Animal Behaviour*, 66, 973–88.

Donahue, M.J. (2006). Allee effects and conspecific cueing jointly lead to conspecific attraction. *Oecologia*, 149, 33–43.

Dorazio, R.M., Connor, E.F. and Askins, R.A. (2015). Estimating the effects of habitat and biological interactions in an avian community. *PLoS ONE*, 10, e0135987.

Doyle, R.W. (1975). Settlement of planktonic larvae: a theory of habitat selection in varying environments. *American Naturalist*, 109, 113–26.

Eisenberg, J.F. (1963). *The Behavior of Heteromyid Rodents*. University of California Press, Berkeley, Calif.

Fabrizio, M.C., Manderson, J.P., and Pessutti, J.P. (2013). Habitat associations and dispersal of black sea bass from a mid-Atlantic Bight reef. *Marine Ecology Progress Series*, 482, 241–53.

Fagan, W.F., Gurarie, E., Bewick, S., Howard, A., Cantrell, R.S., and Cosner, C. (2017). Perceptual ranges, information gathering, and foraging success in dynamic landscapes. *American Naturalist*, 189, 474–89.

Fichaux, M., Béchade, B., Donald, J., et al. (2019). Habitats shape taxonomic and functional composition of neotropical ant assemblages. *Oecologia*, 189, 501–13.

Figueiredo, B.R.S., Mormul, R.P., and Thomaz, S.M. (2015). Swimming and hiding regardless of the habitat: prey fish do not choose between a native and a non-native macrophyte species as a refuge. *Hydrobiologia*, 746, 285–90.

Fletcher, R.J. (2006). Emergent properties of conspecific attraction in fragmented landscapes. *American Naturalist*, 168, 207–19.

Freckleton, R.P., Watkinson, A.R., and Rees, M. (2009). Measuring the importance of competition in plant communities. *Journal of Ecology*, 97, 379–84.

Fretwell, S.D. and Lucas, H.L. (1969). On territorial behavior and other factors influencing habitat distribution in birds. *Acta Biotheoretica*, 19, 45–52.

Fobert, E.K. and Swearer, S.E. (2017). The nose knows: linking sensory cue use, settlement decisions, and post-settlement survival in a temperate reef fish. *Oecologia*, 183, 1041–51.

Folt, B., Donnelly, M.A., and Guyer, C. (2018). Spatial patterns of the frog *Oophaga pumilio* in a plantation system are consistent with conspecific attraction. *Ecology and Evolution*, 8, 2880–89.

Forero-Medina, G. and Vieira, M.V. (2009). Perception of a fragmented landscape by neotropical marsupials: effects of body mass and environmental variables. *Journal of Tropical Ecology*, 25, 53–62.

Forsman, J.T., Hjernquist, M.B., Taipale, J., and Gustafsson, L. (2008). Competitor density cues for habitat quality facilitating habitat selection and investment decisions. *Behavioral Ecology*, 19, 539–45.

Furniss, T.J., Larson, A.J., and Lutz, J.A. (2017). Reconciling niches and neutrality in a subalpine temperate forest. *Ecosphere*, 8, e01847.

Glorvigen, P., Bjørnstad, O.N., Andreassen, H.P., and Ims, R.A. (2012). Settlement in empty versus occupied habitats: an experimental study on bank voles. *Population Ecology*, 54, 55–63.

Grosberg, R.K. (1982). Intertidal zonation of barnacles: the influence of planktonic zonation of larvae on vertical distribution of adults. *Ecology*, 63, 894–99.

Grosberg, R.K. and Levitan, D.R. (1992). For adults only? Supply-side ecology and the history of larval biology. *Trends in Ecology and Evolution*, 7, 130–33.

Gu, W. and Swihart, R.K. (2004). Absent or undetected? Effects of non-detection of species occurrence on wildlife-habitat models. *Biological Conservation*, 116, 195–03.

Gutiérrez, L. (1998). Habitat selection by recruits establishes local patterns of adult distribution in two species of damselfishes: *Stegustes dorsopunicans* and *S. planifrons*. *Oecologia*, 115, 268–77.

Halse, S.A., Williams, M.R., Jaensch, R.P., and Lane, J.A.K. (1993). Wetland characteristics and waterbird use of wetlands in southwestern Australia. *Wildlife Research*, 20, 103–26.

Harvey, M.G., Aleixo, A., Ribas, C.C., and Brumfield, R.T. (2017). Habitat association predicts genetic diversity and population divergence in Amazonian birds. *American Naturalist*, 190, 631–48.

Hauzy, C., Hulot, F.D., Gins, A., and Loreau, M. (2007). Intra- and interspecific density-dependent dispersal in an aquatic prey-predator system. *Journal of Animal Ecology*, 76, 552–58.

Hill, P.J.B., Holwell, G.I., Göth, A., and Nerberstein, M.E. (2004). Preference for habitats with low structural complexity in the praying mantid *Ciulfina* sp. (Mantidae). *Acta Oecologia*, 26, 1–7.

Hopcraft, J.G.C., Sinclair, A.R.E., and Packer, C. (2005). Planning for success: Serengeti lions seek prey accessibility rather than abundance. *Journal of Animal Ecology*, 74, 559–66.

Hu, Y., Majoris, J.E., Buston, P.M., and Webb, J.F. (2019). Potential roles of smell and taste in the orientation behaviour of coral-reef fish larvae: insights from morphology. *Journal of Fish Biology*, 95, 311–23.

Hubbell, S.P. (2001). *The Unified Neutral Theory of Biodiversity and Biogeography*. Princeton University Press, Princeton, NJ.

Huey, R.B. (1991). Physiological consequences of habitat selection. *American Naturalist*, 137, S91–S115.

Jaenike, J. (1978). On optimal oviposition behavior in phytophagous insects. *Theoretical Population Biology*, 14, 350–56.

Jaenike, J. (1983). Induction of host preference in *Drosophila melanogaster*. *Oecologia*, 58, 320–25.

James, F.C. (1971). Ordinations of habitat relationships among breeding birds. *Wilson Bulletin*, 83, 215–36.

Jeanson, R. and Deneubourg, J.-L. (2007). Conspecific attraction and shelter selection in gregarious insects. *American Naturalist*, 170, 47–58.

Jenkins, S.R. (2005). Larval habitat selection, not larval supply, determines settlement patterns and adult distribution in two chthamalid barnacles. *Journal of Animal Ecology*, 74, 893–04.

Jones, J. (2001). Habitat selection studies in avian ecology: a critical review. *The Auk*, 118, 557–62.

Keough, M.J. (1983). Patterns of recruitment of sessile invertebrates in two subtidal habitats. *Journal of Experimental Marine Biology and Ecology*, 66, 213–45.

Kikvidze, Z. and Brooker, R. (2010). Toward a more exact definition of the importance of competition—a reply to Freckleton et al. (2009). *Journal of Ecology*, 98, 719–24.

Krebs, C.J. (2009). *Ecology: the Experimental Analysis of Distribution and Abundance*. Sixth edition. Pearson Publishing, London.

Lack, D. (1933). Habitat selection in birds, with special reference to the effects of afforestation on the Breckland avifauna. *Journal of Animal Ecology*, 2, 239–62.

Lantz, S.M., Gawlik, D.E., and Cook, M.I. (2010). The effects of water depth and submerged aquatic vegetation on the selection of foraging habitat and foraging success of wading birds. *Condor*, 112, 460–69.

Larsson, A.I. and Jonsson, P.R. (2006). Barnacle larvae actively select flow environments supporting post-settlement growth and survival. *Ecology*, 87, 1960–1966.

Leavitt, R.G. (1907). The distribution of closely related species. *American Naturalist*, 41, 207–40.

Leibold, M.A. and McPeek, M.A. (2006). Coexistence of the niche and neutral perspectives in community ecology. *Ecology*, 87, 1399–10.

Levins, R. (1968). *Evolution in Changing Environments*. Princeton University Press, Princeton, NJ.

Levins, R. and MacArthur, R.H. (1969). An hypothesis to explain the incidence of monophagy. *Ecology*, 50, 910–11.

Lillis, A., Eggleston, D.B., and Bohnenstiel, D.R. (2013). Oyster larvae settle in response to habitat-associated underwater sounds. *PLoS ONE*, 8, e79337.

Lima, S.L. and Dill, L.M. (1990). Behavioral decisions made under the risk of predation: a review and prospectus. *Canadian Journal of Zoology*, 68, 619–40.

Lima, S.L. and Zollner, P.A. (1996). Towards a behavioral ecology of ecological landscapes. *Trends in Ecology and Evolution*, 11, 131–35.

MacArthur, R.H. (1958). Population ecology of some warblers of northeastern coniferous forests. *Ecology*, 39, 599–19.

MacArthur, R.H. (1964). Environmental factors affecting bird species diversity. *American Naturalist*, 98, 387–97.

MacArthur, R.H. (1970). Species packing and competitive equilibrium for many species. *Theoretical Population Biology*, 1, 1–11.

MacArthur, R.H. and Levins, R. (1967). The limiting similarity, convergence, and divergence of coexisting species. *American Naturalist*, 101, 377–85.

MacKenzie, D.I., Nichols, J.D., Lachman, G.B., Droege, S., Royle, J.A., and Langtimm, C.A. (2002). Estimating site occupancy rates when detection probabilities are less than one. *Ecology*, 83, 2248–55.

Majoris, J.E., D'Aloia, C.C., Francis, R.K., and Buston, P.M. (2018). Differential persistence favors habitat preferences that determine the distribution of a reef fish. *Behavioral Ecology*, 29, 429–39.

Mannan, R.W. and R.J. Steidl (2013). Habitat. Pages 229–45 in *Wildlife Management and Conservation: Contemporary Principles and Practices*. Krausman, P.R.

and Cain, J.W. (editors), Johns Hopkins University Press, Baltimore, MD.

Marshall, M.R. and Cooper, R.J. (2004). Territory size of a migratory songbird in response to caterpillar density and foliage structure. *Ecology*, 85, 432–45.

Martin, T.E. (1993). Nest predation and nest sites. *Bioscience*, 43, 523–32.

Martin, R.E. and Matocha, K.G. (1991). The Texas kangaroo rat, *Dipodomys elator*, from Motley Co., Texas, with notes on habitat attributes. *Southwestern Naturalist*, 36, 354–56.

Mayer, M., Ullmann, W., Sunde, P., Fischer, C., and Blaum, N. (2018). Habitat selection by the European hare in arable landscapes: the importance of small-scale habitat structure for conservation. *Ecology and Evolution*, 8, 11619–33.

Mayor, S.J., Schneider, D.C., Schaefer, J.A., and Mahoney, S.P. (2009). Habitat selection at multiple scales. *Ecoscience*, 16, 238–47.

Mayr, E. (1947). Ecological factors in speciation. *Evolution*, 1, 263–88.

McNett, B.J. and Rypstra, A.L. (1997). Effects of prey supplementation on survival and web site tenacity of *Argiope trifasciata* (Araneae, Areneidae): a field experiment. *Journal of Arachnology*, 25, 352–60.

McNett, B.J. and Rypstra, A.L. (2000). Habitat selection in a large orb-weaving spider: vegetational complexity determines site selection and distribution. *Ecological Entomology*, 25, 423–32.

Mech, S.G. and Zollner, P.A. (2002). Using body size to predict perceptual range. *Oikos*, 98, 47–52.

Merrick, M.J. and Koprowski, J.L. (2016). Evidence of natal habitat preference induction within one habitat type. *Proceedings of the Royal Society B*, 283, e20162106.

Moermond, T.C. (1979). The influence of habitat structure on *Anolis* foraging behavior. *Behaviour*, 70, 147–67.

Montgomery, J.C., Tolimieri, N., and Haine, O.S. (2001). Active habitat selection by pre-settlement reef fishes. *Fish and Fisheries*, 2, 261–77.

Morello, S.L. and Yund, P.O. (2016). Response of competent blue mussel (*Mytilus edulis*) larvae to positive and negative settlement cues. *Journal of Experimental Marine Biology and Ecology*, 480, 8–16.

Morgan, K.R. and Price, M.V. (1992). Foraging in heteromyid rodents: the energy costs of scratch-digging. *Ecology*, 73, 2260–72.

Morris, D.W. (2003). Toward an ecological synthesis: a case for habitat selection. *Oecologia*, 136, 1–13.

Miller, M.S., Wilson, K.R., and Andersen, D.C. (2003). Ord's kangaroo rats living in floodplain habitats: factors contributing to habitat attraction. *Southwestern Naturalist*, 48, 411–18.

Muller, K.L., Stamps, J.A., Krishnan, V.V., and Willits, N.H. (1997). The effects of conspecific attraction and habitat quality on habitat selection in territorial birds (*Troglodytes aedon*). *American Naturalist*, 150, 650–61.

Murkin, H.R., Murkin, E.J., and Ball, J.P. (1997). Avian habitat selection and prairie wetland dynamics: a 10-year experiment. *Ecological Applications*, 7, 1144–59.

Munday, P.L. (2000). Interactions between habitat use and patterns of abundance in coral-dwelling fishes of the genus *Gobiodon*. *Environmental Biology of Fishes*, 58, 355–69.

Mysterud, A. and Østybe, E. (1999). Cover as a habitat element for temperate ungulates: effects on habitat selection and demography. *Wildlife Society Bulletin*, 27, 385–94.

Nelson, A.D., Goetze, J.R., Watson, E., and Nelson, M. (2009). Changes in vegetation patterns and their effect on Texas kangaroo rats (*Dipodomys elator*). *Texas Journal of Science*, 61, 119–30.

Newmark, W.D. and Stanley, T.R. (2016). The influence of food abundance, food dispersion and habitat structure on territory selection and size of an Afrotropical terrestrial insectivore. *Ostrich*, 87, 199–07.

Öckinger, E. and Van Dyck, H. (2012). Landscape structure shapes habitat finding ability in a butterfly. *PLoS ONE*, 7, e41517.

Orians, G.H. and Wittenberger, J.F. (1991). Spatial and temporal scales in habitat selection. *American Naturalist*, 137, S29–S49.

Osborn, J.M., Hagy, H.M., McClanahan, M.D., Davis, J.B., and Gray, M.J. (2017). Habitat selection and activities of dabbling ducks during non-breeding periods. *Journal of Wildlife Management*, 81, 1482–93.

Ott, S.L., Veech, J.A., Simpson, T.R., Castro-Arellano, I., and Evans, J. (2019). Mapping potential habitat and range-wide surveying for the Texas kangaroo rat. *Journal of Fish and Wildlife Management*, 10, 619–30.

Parrish, J.D. (1995). Effects of needle architecture on warbler habitat selection in a coastal spruce forest. *Ecology*, 76, 1813–20.

Paxton, A.B. and Smith, D. (2018). Visual cues from an underwater illusion increase relative abundance of highly reef-associated fish on an artificial reef. *Marine and Freshwater Research*, 69, 614–19.

Price, M.V. (1978). The role of microhabitat in structuring desert rodent communities. *Ecology*, 59, 910–21.

Price, M.V. and Heinz, K.M. (1984). Effects of body size, seed density, and soil characteristics on rates of seed harvest by heteromyid rodents. *Oecologia*, 61, 420–25.

Pulliam, H.R. and Danielson, B.J. (1991). Sources, sinks, and habitat selection: a landscape perspective on population dynamics. *American Naturalist*, 137, S50–66.

Raimondi. P.T. (1991). Settlement behavior of *Chthamalus anisopoma* larvae largely determines the adult distribution. *Oecologia*, 85, 349–60.

Ranius, T. and Jansson, N. (2000). The influence of forest regrowth, original canopy cover and tree size on saproxylic beetles associated with old oaks. *Biological Conservation*, 95, 85–94.

Renken, R.B. and Wiggers, E.P. (1989). Forest characteristics related to pileated woodpecker territory size in Missouri. *Condor*, 91, 642–52.

Reynolds, H.G. (1958). The ecology of the Merriam kangaroo rat (*Dipodomys merriami* Mearns) on the grazing lands of southern Arizona. *Ecological Monographs*, 28, 111–27.

Ricklefs, R.E. (2015). Intrinsic dynamics of the regional community. *Ecology Letters*, 18, 497–03.

Roberts, J.D. and Packard, R.L. (1973). Comments on movements, home range and ecology of the Texas kangaroo rat, *Dipodomys elator* Merriam. *Journal of Mammalogy*, 54, 957–62.

Robinson, S.K. and Holmes, R.T. (1982). Foraging behavior of forest birds: the relationships among search tactics, diet, and habitat structure. *Ecology*, 63, 1918–1931.

Robinson, B.G., Larsen, K.W., and Kerr, H.J. (2011). Natal experience and conspecifics influence the settling behaviour of the juvenile terrestrial isopod *Armadillidium vulgare*. *Canadian Journal of Zoology*, 89, 661–67.

Rockwell, S.M. and Stephens, J.L. (2018). Habitat selection of riparian birds at restoration sites along the Trinity River, California. *Restoration Ecology*, 26, 767–77.

Rosenzweig, M.L. (1981). A theory of habitat selection. *Ecology*, 62, 327–35.

Rosenzweig, M.L. and Winakur, J. (1969). Population ecology of desert rodent communities: habitats and environmental complexity. *Ecology*, 50, 558–72.

Rotenberry, J.T. and Wiens, J.A. (1980). Habitat structure, patchiness, and avian communities in North American steppe vegetation: a multivariate analysis. *Ecology*, 61, 1228–50.

Rushing, C.S., Dudash, M.R., and Marra, P.P. (2015). Habitat features and long-distance dispersal modify the use of social information by a long-distance migratory bird. *Journal of Animal Ecology*, 84, 1469–79.

Rypstra, A.L., Schmidt, J.M., Reif, B.D., DeVito, J., and Persons, M.H. (2007). Tradeoffs involved in site selection and foraging in a wolf spider: effects of substrate structure and predation risk. *Oikos*, 116, 853–63.

Sale, P.F. (1969a). A suggested mechanism for habitat selection by the juvenile Manini (*Acanthrus triostegus sandvicensis* Streets). *Behaviour*, 35, 27–44.

Sale, P.F. (1969b). Pertinent stimuli for habitat selection by the juvenile Manini, *Acanthrus triostegus sandvicensis*. *Ecology*, 50, 616–23.

Sale, P.F. and Douglas, W.A. (1984). Temporal variability in the community structure of fish on coral patch reefs and the relation of community structure to reef structure. *Ecology*, 65, 409–22.

Sale, P.F. and Dybdahl, R. (1975). Determinants of community structure for coral reef fishes in an experimental habitat. *Ecology*, 56, 1343–55.

Sale, P.F., Douglas, W.A., and Doherty, P.J. (1984). Choice of microhabitats by coral reef fishes at settlement. *Coral Reefs*, 3, 91–99.

Schlosser, I.J. (1987). The role of predation in age- and size-related habitat use by stream fishes. *Ecology*, 68, 651–59.

Seastedt, T.R. and MacLean, S.F. (1979). Territory size and composition in relation to resource abundance in Lapland Longspurs breeding in Arctic Alaska. *The Auk*, 96, 131–42.

Sergio, F. and Penteriani, V. (2005). Public information and territory establishment in a loosely colonial raptor. *Ecology*, 86, 340–46.

Serrano, D. and Tella, J.L. (2003). Dispersal within a spatially structured population of lesser kestrels: the role of spatial isolation and conspecific attraction. *Journal of Animal Ecology*, 72, 400–10.

Siepielski, A.M. and McPeek, M.A. (2010). On the evidence for species coexistence: a critique of the coexistence program. *Ecology*, 91, 3153–64.

Southwood, T.R.E. (1977). Habitat, the template for ecological strategies? *Journal of Animal Ecology*, 46, 336–65.

Skelly, D.K., Werner, E.E., and Cortwright, S.A. (1999). Long-term distributional dynamics of a Michigan amphibian assemblage. *Ecology*, 80, 2326–37

Smith, T.M. and Shugart, H.H. (1987). Territory size variation in the ovenbird: the role of habitat structure. *Ecology*, 68, 695–04.

Speigel, O., Getz, W.M., and Nathan, R. (2013). Factors influencing foraging search efficiency: why do scarce lappet-faced vultures outperform ubiquitous white-backed vultures? *American Naturalist*, 181, E102–15.

Stamps, J.A. (1988). Conspecific attraction and aggregation in territorial species. *American Naturalist*, 131, 329–47.

Stamps, J.A. and Krishnan, V.V. (2005). Nonintuitive cue use in habitat selection. *Ecology*, 86, 2860–67.

Stamps, J.A., Krishnan, V.V., and Reid, M.L. (2005). Search costs and habitat selection by dispersers. *Ecology*, 86, 510–18.

Stamps, J.A., Davis, J.M., Blozis, S.A., and Boundy-Mills, K,L. (2007). Genotypic variation in refractory periods and habitat selection by natal dispersers. *Animal Behaviour*, 74, 599–10.

Stevens, R.D. and Tello, J.S. (2009). Micro- and macrohabitat associations in Mojave Desert rodent communities. *Journal of Mammalogy*, 90, 388–03.

Strong, D.R., Simberloff, D., Abele, L.G., and Thistle, A.B. (1984). *Ecological Communities: Conceptual Issues and the Evidence*. Princeton University Press, Princeton, NJ.

Svärdson, G. (1949). Competition and habitat selection in birds. *Oikos*, 1, 157–74.

Thompson, S.D. (1982). Structure and species composition of desert heteromyid rodent species assemblages: effects of a simple habitat manipulation. *Ecology*, 63, 1313–21.

Thorp, W.H. (1945). The evolutionary significance of habitat selection. *Journal of Animal Ecology*, 14, 67–70.

Thresher, R.E., Colin, P.L., and Bell, L.J. (1989). Planktonic duration, distribution, and population structure of western and central Pacific damselfishes (Pomacentridae). *Copeia*, 1989, 420–34.

Tilman, D. (1997). Community invasibility, recruitment limitation, and grassland biodiversity. *Ecology*, 78, 81–92.

Tinbergen, N. (1948). Social releasers and the experimental method required for their study. *Wilson Bulletin*, 60, 6–51.

Tolimieri, N. (1995). Effects of microhabitat characteristics on the settlement and recruitment of a coral reef fish at two spatial scales. *Oecologia*, 102, 52–63.

Tonnis, B., Grant, P.R., Grant, B.R., and Petren, K. (2005). Habitat selection and ecological speciation in Galápagos warbler finches (*Certhidea olivacea* and *Certhidea fusca*). *Proceedings of the Royal Society B*, 272, 819–26.

Ulrich, W., Hajdamowicz, I., Zalewski, M., Stańska, M., Ciurzycki, W., and Tykarski, P. (2010). Species assortment or habitat filtering: a case study of spider communities on lake islands. *Ecological Research*, 25, 375–81.

Underwood, A.J. and Denley, E.J. (1984). Paradigms, explanations, and generalizations in models for the structure of intertidal communities on rocky shores. Pages 181–95 in *Ecological Communities: Conceptual Issues and the Evidence*. Strong, D.R., Simberloff, D., Abele, L.G., and Thistle, A.B. (editors), Princeton University Press, Princeton, NJ.

Vallès, H., Kramer, D.L., and Hunte, W. (2008). Temporal and spatial patterns in the recruitment of coral-reef fishes in Barbados. *Marine Ecology Progress Series*, 363, 257–72.

Van Horne, B. (1983). Density as a misleading indicator of habitat quality. *Journal of Wildlife Management*, 47, 893–01.

Ward, S.A. (1987). Optimal habitat selection in time-limited dispersers. *American Naturalist*, 129, 568–79.

Ward, M.P., Benson, T.J., Semel, B., and Herkert, J.R. (2010). The use of social cues in habitat selection by wetland birds. *Condor*, 112, 245–51.

Wellington, G.M. (1992). Habitat selection and juvenile persistence control the distribution of two closely related Caribbean damselfishes. *Oecologia*, 90, 500–508.

Wellington, G.M. and Robertson, D.R. (2001). Variation in larval life-history traits among reef fishes across the Isthmus of Panama. *Marine Biology*, 138, 11–22.

Wennekes, P.L., Rosindell, J., and Etienne, R.S. (2012). The neutral-niche debate: a philosophical perspective. *Acta Biotheoretica*, 60, 257–71.

Werner, E.E., Gilliam, J.F., Hall, D.J., and Mittelbach, G.G. (1983). An experimental test of the effects of predation risk on habitat use in fish. *Ecology*, 64, 1540–48.

Whelan, C.J. (2001). Foliage structure influences foraging of insectivorous forest birds: an experimental study. *Ecology*, 82, 219–31.

Willson, M.F. (1974). Avian community organization and habitat structure. *Ecology*, 55, 1017–29.

Wolfe, J.D., Johnson, M.D., and Ralph, C.J. (2014). Do birds select habitat or food resources? Nearctic-Neotropic migrants in northeastern Costa Rica. *PLoS ONE*, 9, e86221.

Woodson, C.B., Webster, D.R., Weissburg, M.J., and Yen, J. (2007). Cue hierarchy and foraging in calanoid copepods: ecological implications of oceanographic structure. *Marine Ecology Progress Series*, 330, 163–77.

Wright, S.J. (2002). Plant diversity in tropical forests: a review of mechanisms of species coexistence. *Oecologia*, 130, 1–14.

Yeager, M.E. and Hovel, K.A. (2017). Structural complexity and fish body size interactively affect habitat optimality. *Oecologia*, 185, 257–67.

Zenzal, T.J., Smith, R.J., Ewert, D.N., Diehl, R.H., and Buler, J.J. (2018). Fine-scale heterogeneity drives forest use by spring migrant landbirds across a broad, contiguous forest matrix. *Condor*, 120, 166–84.

Zollner, P.A. and Lima, S.L. (1997). Landscape-level perceptual abilities in white-footed mice: perceptual range and the detection of forest habitat. *Oikos*, 80, 51–60.

CHAPTER 4

Goals of Characterizing a Species Habitat

There are many reasons for analyzing species–habitat associations and ideally achieving a complete characterization of a species' habitat, or at least identifying the habitat characteristics that seem to be most important to the species. The end goal may be a practical one such as when better knowledge is needed for the management, manipulation, or restoration of a given species' habitat—examples include species of particular conservation concern or species that are harvested recreationally (i.e., hunted). For this task we need to know the habitat requirements of the species just as well as we know the resource (food) requirements. A different, but related, goal might be to obtain detailed knowledge of habitat associations in order to facilitate investigations into habitat preference and selection as processes that occur on ecological time scales and as outcomes of evolution occurring over long periods of time. The study of these processes is a huge area of research within evolutionary ecology and behavioral ecology. In this chapter, I briefly discuss five reasons for analyzing the habitat associations of a species: (1) to achieve proper habitat management, manipulation, and restoration; (2) to obtain a better understanding of carrying capacity and population dynamics; (3) to obtain a better understanding of a species potential and realized interactions with other species occurring in the same habitat; (4) to study habitat preference and selection as ecological and evolutionary processes; and (5) to predict where a species might occur by constructing and using species distribution models.

4.1 Habitat management

Habitat management is often a crucial conservation strategy for rare and threatened species and those in which some degree of human intervention is necessary to facilitate population growth and persistence, regardless of endangerment. For most terrestrial animal species and some aquatic ones, vegetation is habitat. The amount of woody vegetation in the form of stem density, canopy cover, tree or shrub height, and density of the foliage can be important determinants of habitat suitability for terrestrial organisms. The amount of herbaceous vegetation (grasses and forbs at the ground level) may also be important particularly with regard to the thickness or density and the percentage coverage. Apart from the physical habitat structure created by vegetation, the presence or absence of certain plant species might be a habitat requirement. In aquatic ecosystems, the flow, depth, and temperature of the water often defines habitat as suitable or not. The amount of vegetation, underwater vertical structure, and type of substrate are also important habitat factors.

In both terrestrial and aquatic ecosystems, the physical manipulation or modification of vegetation is often necessary in the management of habitat, including the restoration of habitat from a severely degraded condition (e.g., completely revegetating an area). In aquatic habitats, management often consists of restoring natural flow regimes and the spatial heterogeneity of flow and depth.

Habitat Ecology and Analysis. Joseph A. Veech, Oxford University Press (2021). © Joseph A. Veech. DOI: 10.1093/oso/9780198829287.003.0004

Presumably prior to any habitat modification or restoration, the conservation practitioner would have a certain outcome in mind that represents good habitat for the species and ideally that target would have been informed by knowledge of the habitat requirements of the species (i.e., a habitat analysis). Even apart from any actual habitat modification or restoration, a species conservation plan might involve reintroduction and establishment of "new" populations. In that context, the conservation practitioner would first want to identify locations that have the most suitable habitat. Such knowledge is best obtained by careful design and analysis of studies aimed at determining the most appropriate habitat conditions for the species and perhaps identifying habitat features or characteristics to be avoided.

To the extent that vegetation is habitat, the natural process of ecological succession can change an area from having suitable habitat to being at a successional stage that does not represent appropriate habitat. This natural change may occur very gradually and have relatively little influence (positive or negative) on species that are habitat generalists. However, habitat specialists may require a certain successional stage. Therefore, species–habitat relationships must be viewed as somewhat dynamic such that proper habitat management does not strictly entail a single static idealized state for the habitat but rather the management allows for natural cyclical transitions. This perspective suggests that analysis of species–habitat associations need not always uncover strong patterns of central tendency such as species abundance being highly correlated to habitat factor X, but rather analysis could also be geared toward delimiting the wide range of habitat conditions that the species can tolerate. Either way, effective habitat management requires obtaining as much knowledge of a species' habitat ecology as possible.

4.2 Carrying capacity and population dynamics

Carrying capacity is the maximum number of individuals that an environment can support given limitations in resource availability and renewal rates. Although the limiting resource is often considered to be food, particularly when carrying capacity is defined and measured based on daily energy requirements of individuals (Petit and Pors 1996; Nolet et al. 2006; Aryal et al. 2014; Osborn et al. 2017), the physical habitat requirements of a species can also play a role in determining carrying capacity. This may be particularly true if limiting "resources" are considered to be aspects of the vegetation (or other physical structures) that are important to foraging and reproductive behavior. An example is the limited availability of tree cavities for cavity-nesting bird species (Newton 1994; Sanchez et al. 2007; Cockle et al. 2010)—cavities are a distinct habitat requirement and the density of cavities in a given area may in part determine carrying capacity (Catry et al. 2013). In such a case, cavities may be a limiting resource as well as a habitat requirement. In a statistical analysis we would expect to find a relationship between cavity density and species presence–absence or abundance of breeding pairs. Understanding the importance of cavities as a limiting resource and a required habitat feature would enable a better understanding of carrying capacity and hence population dynamics of the species.

Analyzing species–habitat associations also informs us on the importance of carrying capacity in the following way. For a given species, different habitats (as specific spatial locations) might vary in resource quality and availability; hence populations in the different habitats have different carrying capacities. However, some of the difference in carrying capacity among the habitats might be due to differences in non-resource characteristics of the habitats. Returning to the example of cavity-nesting birds, the density and species of trees in the forest are not resources per se but both might affect the availability of the cavities if some tree species are more likely to have decaying centers and if the number of suitable cavities is a function (linear or not) of tree density (Martin 2014), or forest successional stage (Katayama et al. 2017). Moreover, different forest types may differ in availability of cavities because they differ in tree species composition, density, and local climatic factors that cause fungal heart-rot in trees (Remm and Lohmus 2011). Presumably, carrying capacity will differ between the different forest types (habitats). Obtaining a deeper understanding of this difference would require knowledge of how tree species composition

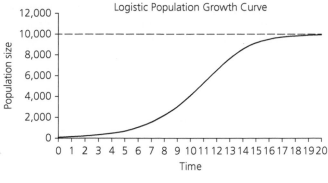

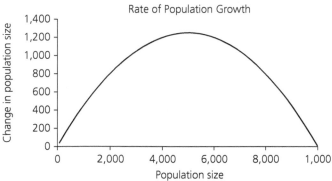

Figure 4.1 Logistic population growth over time for a hypothetical population (top panel) and rate of population growth as it depends on population size N at time t (bottom panel). Population size at time t is calculated as $N_t = [rN_{t-1}(1 - (N_{t-1}/K))] + N_{t-1}$ and change in population size is calculated as $dN/dt = rN_{t-1}[(K - N_{t-1})/K]$ where K = carrying capacity and r = intrinsic rate of increase (essentially the difference between per-capita birth rate and death rate). In this example, $K = 10,000$, $r = 0.5$, and $N_0 = 100$. Note that the population reaches the carrying capacity fairly rapidly (dashed line in top panel), within 20 time steps. The rate of population growth is greatest and equal to $rK/4$ (or 1,250) when the population is at $K/2$ (or 5,000); the rate then declines toward zero as the population approaches the carrying capacity. The logistic population growth model is one of the classic discoveries of ecology that dates back well over a century.

and density are factors important in defining and characterizing the habitat of the species.

The link between carrying capacity and population dynamics is embodied in the classic logistic population growth model and variants of it. Population growth slows as the population size nears the carrying capacity (Fig. 4.1). Regardless of whether natural populations adhere to logistic population dynamics or not, carrying capacities are real and intuitively obvious. With few exceptions, all natural populations have a carrying capacity, although some populations may typically be so far below carrying capacity that resource limitation does not restrict growth. In such populations, biotic and abiotic factors (other than intraspecific resource competition) regulate population growth and hence the size of the population. Some of these factors may be physical habitat characteristics and indicate the general suitability of the habitat for the species. Therefore, analyzing species–habitat associations can inform us on population dynamics apart from whether a population is practically restricted by a carrying capacity. Indeed, with a well-designed habitat analysis it should be possible in practice to identify habitat factors that are more important than resource availability in determining population density.

4.3 Species interactions

Every species interacts to some extent with other species, be it competition, predation, parasitism, or mutualism. As such, in studying species interactions and the structuring of ecological communities, there is a need to understand the extent to which the interacting species actually and potentially co-occur. The co-occurrence (or lack thereof) consequentially depends on habitat requirements and preferences. Therefore, it would seem that any study of species interactions (and these continue to be super-common and worthwhile in ecology) could benefit from a characterization of the habitat of the focal species. Analyzing each species–habitat association could provide an assessment of the extent to which they can potentially co-occur in the same habitat(s), apart from the specific goal(s) of studying the interaction itself. My recommendation here is not that every study of species interactions (and their resultant effects) needs to formally and explicitly incorporate a habitat analysis, but rather that researchers be aware of the added benefit of having thorough knowledge of habitat requirements when designing their studies and in interpreting the results.

Knowledge of species interactions due to co-occurrence in shared habitat can also be useful in species conservation. Although habitat management is often conceived and planned in the context of a single target species, no species exists in isolation from all others. Further, more likely than not, the exact habitat requirements of two or more species will not completely overlap. Inevitably, certain management actions designed to benefit the target species can either harm or benefit other species, even if unintentional. These unintended consequences might not be fully understood if knowledge of the habitat ecology of those other species is lacking. Therefore, the hidden implications of single-species habitat management can best be revealed by knowing the habitat requirements of all the other species that co-occur with the target species. Indeed, species-centric habitat management is gradually being replaced by management that is focused on communities and ecosystems.

4.4 Habitat preference and selection

The study of habitat preference and selection continues to be an active arena of investigation in behavioral and evolutionary ecology (Lack 1933; Thorp 1945; Wecker 1964; Rosenzweig 1981, 1991; Morris 1987, 1990; Jaenike and Holt 1991; Resetarits 2005; Ahnesjö and Forsman 2006; Grossman 2014; Matthiopoulos et al. 2015; Acker et al. 2017; Webber and Vander Wal 2017). Studies typically examine the *causes* or mechanisms leading to habitat preferences and selection or they focus on the *consequences* for single-species adaptation, niche divergence, foraging behavior, population dynamics, and community structure (relative abundance and species diversity). It is nearly impossible to overstate the importance of habitat selection to many other different processes in ecology and evolutionary biology—there are literally well over 10,000 published studies that involve some form of reference to "habitat selection." The ecological-evolutionary context of

habitat preference and selection is also becoming more and more incorporated into the conservation and management of species (Morris 2003; Battin 2004; Fisher and Davis 2010; Grossman 2014; Gibson et al. 2016).

Studies of habitat selection are diverse in purpose and approach. The need to know more about individual dispersal is a crucial conceptual and empirical aspect of some studies of habitat selection (Morris et al. 2004; Bowler and Benton 2005; Stamps et al. 2005; Acker et al. 2017). Individuals, usually juveniles, disperse to find appropriate habitat as an integral step in the overall process of habitat selection. Some studies examine the differences in fitness among individuals occurring in different habitats (Morris 1987, 2011; Martin 1998; Brown et al. 2002; Beck et al. 2004; Haugen et al. 2006; Matthiopoulos et al. 2015). Related to this, researchers sometimes study habitat selection to learn more about resource partitioning and species coexistence (Ben-David et al. 1996; Arlettaz 1999; Parra 2006; Stewart et al. 2010; Legault et al. 2012). The domain of these studies is further extended when researchers examine the role of habitat selection in producing and maintaining patterns of species diversity (MacArthur et al. 1966; Binckley and Resetarits 2005; Resetarits 2005; Kilpatrick et al. 2006; Veech and Crist 2007).

For all these various types of studies of habitat selection, there is value in conducting a formal habitat analysis in which important habitat associations are identified for each species being studied. That is, in contemporary ecological time representing the study duration and period of data collection (perhaps one to several field seasons), the study species likely respond to certain habitat factors. Identification of the factors and knowledge of the quantitative response (presence/absence or abundance) would be helpful in interpreting the results pertaining to habitat selection even if the inference from those results is being made to a much longer ecological or evolutionary time scale. For example, if a given habitat factor is transient or highly variable over relatively short time periods, then even if a species' abundance is positively and strongly associated with the factor there may not be much role for the factor in explaining habitat selection as an adaptive trait evolved over a much longer period of time. Alternatively, if the positive habitat factor is persistent and steady over time, then it may well be important in habitat selection. Even more fundamentally, a habitat analysis conducted prior to the habitat selection study can inform the researcher as to whether the "habitats" to be included in the study (i.e., to be sampled) truly are suitable habitat for the species.

4.5 Species distribution models

Species distribution models (SDMs) are a popular technique for modeling the potential spatial distribution of a species over a fairly large geographic extent (e.g., an area 10,000 km^2 or greater and often much larger) (Guisan and Thuiller 2005; Phillips et al. 2006; Elith and Leathwick 2009; Franklin 2010; Guisan et al. 2017) (Section 11.5). SDMs are typically presented as probability surfaces or gridded maps wherein each grid cell has a calculated "probability" of containing the species. This probability is based upon the predictor variables included in the SDM and the particular values of those predictors in the grid cell. The predictors often include habitat factors such as variables related to vegetation, soil, and topography. Other variables not directly assessing habitat, such as climatic factors, are often included in SDMs. When an analyst undertakes construction of an SDM, he or she usually has in mind a list of factors thought to affect the distribution of the species. The modeling task then involves finding the particular set of factors that seems to best predict presence and absence of the species. As such, a well-designed and thorough habitat analysis could be an important initial phase of research prior to doing species distribution modeling. Indeed, an analyst building an SDM might often be relying on knowledge of habitat associations that was built up by other researchers many years previously.

Herein I am not considering species distribution modeling as a technique for directly analyzing species–habitat associations given that the ultimate goal of an SDM is to serve as an accurate predictor of spatial distribution rather than a direct test of habitat associations. However, in a round-about way SDMs could be used to learn about habitat associations. That is, an SDM is built with one or more habitat factors (with or without non-habitat predictors such as climate variables) and then the predictive performance of the model is assessed in any of the standard ways (Section 10.3). If the model has acceptable and minimal rates of error (low rates of predicting false presence and false absence) *and* the habitat variables have high importance values, then presumably those habitat variables represent factors important to the species, that is, habitat requirements. If not, then the habitat factors included in the SDM are not important to the species and the analyst can look for other habitat variables to test by including additional SDMs—in this way SDMs become a way to hunt for important habitat variables. The efficiency of this approach for identifying important habitat factors is a decision for the analyst to make. However, the typical goal of species distribution modeling is to obtain a predictive (trustworthy) model without hunting for influential variables. The analyst has good knowledge a priori as to the most important habitat and non-habitat factors to include in the SDM and the ultimate goal is to understand spatial patterns of distribution, rather than to exhaustively hunt for important habitat factors. A more detailed treatment of SDMs is provided in Section 11.5.

4.6 Conclusion

In this chapter I have presented five goals or general reasons for characterizing a species' habitat—there are likely additional ones that I have missed. Indeed the most fundamental reason might be simply to better understand the distribution and abundance of a species—and the primary role that habitat plays in determining where a species exists—the topic of Chapter 3. In the remainder of Part II, I describe the types of data and the framework that is common to all methods of habitat analysis (Chapter 5). I then discuss some statistical and sampling issues (Chapters 6 and 7) that an analyst should consider prior to engaging in an analysis of habitat associations. This is then followed by a separate treatment of each method in which I point out strengths and weaknesses and demonstrate the application of the methods to data (Chapters 8 and 9). Part II ends with my philosophical view on the study of habitat and some practical advice (Chapter 12).

References

Acker, P., Besnard, A., Monnat, J.-Y., and Cam, E. (2017). Breeding habitat selection across spatial scales: is grass always greener on the other side? *Ecology*, 98, 2684–97.

Ahnesjö, J. and Forsman, A. (2006). Differential habitat selection by pygmy grasshopper color morphs; interactive effects of temperature and predator avoidance. *Evolutionary Ecology*, 20, 235–57.

Arlettaz, R. (1999). Habitat selection as a major resource partitioning mechanism between the two sympatric sibling bat species *Myotis myotis* and *Myotis blythii*. *Journal of Animal Ecology*, 68, 460–71.

Aryal, A., Brunton, D., Ji, W., and Raubenheimer, D. (2014). Blue sheep in the Annapurna Conservation Area, Nepal: habitat use, population biomass and their contribution to the carrying capacity of snow leopards. *Integrative Zoology*, 9, 34–45.

Battin, J. (2004). When good animals love bad habitats: ecological traps and the conservation of animal populations. *Conservation Biology*, 18, 1482–91.

Beck, H., Gaines, M.S., Hines, J.E., and Nichols, J.D. (2004). Comparative dynamics of small mammal populations in treefall gaps and surrounding understorey within Amazonian rainforest. *Oikos*, 106, 27–38.

Ben-David, M., Bowyer, R.T., and Faro, J.B. (1996). Niche separation by mink and river otters: coexistence in a marine environment. *Oikos*, 75, 41–48.

Binckley, C.A. and Resetarits, W.J. (2005). Habitat selection determines abundance, richness and species composition of beetles in aquatic communities. *Biology Letters*, 1, 370–74.

Bowler, D.E. and Benton, T.G. (2005). Causes and consequences of animal dispersal strategies: relating individual behaviour to spatial dynamics. *Biological Reviews*, 80, 205–25.

Brown, D.R., Strong, C.M., and Stouffer, P.C. (2002). Demographic effects of habitat selection by hermit thrushes wintering in a pine plantation landscape. *Journal of Wildlife Management*, 66, 407–16.

Catry, I., Franco, A.M.A., Rocha, P., et al. (2013). Foraging habitat quality constrains effectiveness of artificial nest-site provisioning in reversing population declines in a colonial cavity nester. *PLoS ONE*, 8, e58320.

Cockle, K.L., Martin, K., and Drever, M.C. (2010). Supply of tree-holes limits nest density of cavity-nesting birds in primary and logged subtropical Atlantic forest. *Biological Conservation*, 143, 2851–57.

Elith, J. and Leathwick, J.R. (2009). Species distribution models: ecological explanation and prediction across space and time. *Annual Review of Ecology, Evolution, and Systematics*, 40, 677–97.

Fisher, R.J. and Davis, S.K. (2010). From Wiens to Robel: a review of grassland-bird habitat selection. *Journal of Wildlife Management*, 74, 265–73.

Franklin, J. (2010). *Mapping Species Distributions: Spatial Inference and Prediction*. Cambridge University Press, Cambridge.

Gibson, D., Blomberg, E.J., Atamian, M.T., and Sedinger, J.S. (2016). Nesting habitat selection influences nest and early offspring survival in greater sage-grouse. *Condor*, 118, 689–702.

Grossman, G.D. (2014). Not all drift feeders are trout: a short review of fitness-based habitat selection models for fishes. *Environmental Biology of Fishes*, 97, 465–73.

Guisan, A. and Thuiller, W. (2005). Predicting species distribution: offering more than simple habitat models. *Ecology Letters*, 8, 993–1009.

Guisan, A., Thuiller, W., and Zimmerman, N.E. (2017). *Habitat Suitability and Distribution Models*. Cambridge University Press, Cambridge.

Haugen, T.O., Winfield, I.J., Vøllestad, L.A., Fletcher, J.M., James, J., and Stenseth, N.C. (2006). The ideal free pike: 50 years of fitness-maximizing dispersal in Windermere. *Proceedings of the Royal Society B*, 273, 2917–24.

Jaenike, J. and Holt, R.D. (1991). Genetic variation for habitat preference: evidence and explanations. *American Naturalist*, 137, S67–90.

Katayama, M.V., Zima, P.V.Q., Perrella, D.F., and Francisco, M.R. (2017). Successional stage effect on the availability of tree cavities for cavity-nesting birds in an Atlantic Forest park from the state of São Paulo, Brazil. *Biota Neotropica*, 17, e20170391.

Kilpatrick, M., Mitchell, W.A., Porter, W.P., and Currie, D.J. (2006). Testing a mechanistic explanation for the latitudinal gradient in mammalian species richness across North America. *Evolutionary Ecology Research*, 8, 333–44.

Lack, D. (1933). Habitat selection in birds, with special reference to the effects of afforestation on the Breckland avifauna. *Journal of Animal Ecology*, 2, 239–62.

Legault, A., Theuerkauf, J., Rouys, S., Chartendrault, V., and Barré, N. (2012). Temporal variation in flock size and habitat use of parrots in New Caledonia. *Condor*, 114, 552–63.

MacArthur, R., Recher, H., and Cody, M. (1966). On the relation between habitat selection and species diversity. *American Naturalist*, 100, 319–32.

Martin, T.E. (1998). Are microhabitat preferences of coexisting species under selection and adaptive? *Ecology*, 79, 656–70.

Martin, T.E. (2014). Consequences of habitat change and resource selection specialization for population limitation in cavity-nesting birds. *Journal of Applied Ecology*, 52, 475–85.

Matthiopoulos, J., Fieberg, J., Aarts, G., Beyer, H.L., Morales, J.M., and Haydon, D.T. (2015). Establishing the link between habitat selection and animal population dynamics. *Ecological Monographs*, 85, 413–36.

Morris, D.W. (1987). Ecological scale and habitat use. *Ecology*, 68, 362–69.

Morris, D.W. (1990). Temporal variation, habitat selection and community structure. *Oikos*, 59, 303–12.

Morris, D.W. (2003). Toward an ecological synthesis: a case for habitat selection. *Oecologia*, 136, 1–13.

Morris, D.W. (2011). Adaptation and habitat selection in the eco-evolutionary process. *Proceedings of the Royal Society B*, 278, 2401–11.

Morris, D.W., Diffendorfer, J.E., and Lundberg, P. (2004). Dispersal among habitats varying in fitness: reciprocating migration through ideal habitat selection. *Oikos*, 107, 559–75.

Newton, I. (1994). The role of nest sites in limiting the numbers of hole-nesting birds: a review. *Biological Conservation*, 70, 265–76.

Nolet, B.A., Gyimesi, A., and Klaassen, R.H.G. (2006). Prediction of bird-day carrying capacity on a staging site: a test of depletion models. *Journal of Animal Ecology*, 75, 1285–92.

Osborn, J.M., Hagy, H.M., McClanahan, M.D., Davis, J.B., and Gray, M.J. (2017). Habitat selection and activities of dabbling ducks during non-breeding periods. *Journal of Wildlife Management*, 81, 1482–93.

Parra, G.J. (2006). Resource partitioning in sympatric delphinids: space use and habitat preferences of Australian snubfin and Indo-Pacific humpback dolphins. *Journal of Animal Ecology*, 75, 862–74.

Petit, S. and Pors, L. (1996). Survey of columnar cacti and carrying capacity for nectar-feeding bats on Curacao. *Conservation Biology*, 10, 769–75.

Phillips, S.J., Anderson, R.P., and Schapire, R.E. (2006). Maximum entropy modeling of species geographic distributions. *Ecological Modelling*, 190, 231–59.

Remm, J. and Lõhmus, A. (2011). Tree cavities in forests—the broad distribution pattern of a keystone structure for biodiversity. *Forest Ecology and Management*, 262, 579–85.

Resetarits, W.J. (2005). Habitat selection behaviour links local and regional scales in aquatic systems. *Ecology Letters*, 8, 480–86.

Rosenzweig, M.L. (1981). A theory of habitat selection. *Ecology*, 62, 327–35.

Rosenzweig, M.L. (1991). Habitat selection and population interactions: the search for mechanism. *American Naturalist*, 137, S5–S28.

Sanchez, S., Cuervo, J.J., and Moreno, E. (2007). Suitable cavities as a scarce resource for both cavity and non-cavity nesting birds in managed temperate forests, a case study in the Iberian Peninsula. *Ardeola*, 54, 261–74.

Stamps, J.A., Krishnan, V.V., and Reid, M.L. (2005). Search costs and habitat selection by dispersers. *Ecology*, 86, 510–18.

Stewart, K.M., Bowyer, R.T., Kie, J.G., and Hurley, M.A. (2010). Spatial distributions of mule deer and North American elk: resource partitioning in a sage-steppe environment. *American Midland Naturalist*, 163, 400–12.

Thorp, W.H. (1945). The evolutionary significance of habitat selection. *Journal of Animal Ecology*, 14, 67–70.

Veech, J.A. and Crist, T.O. (2007). Habitat and climate heterogeneity maintain beta-diversity of birds among landscapes within ecoregions. *Global Ecology and Biogeography*, 16, 650–56.

Webber, Q.M.R. and Vander Wal, E. (2017). An evolutionary framework outlining the integration of individual social and spatial ecology. *Journal of Animal Ecology*, 87, 113–27.

Wecker, S.C. (1964). Habitat selection. *Scientific American*, 211, 109–17.

PART II

Statistical Methods of Analysis

CHAPTER 5

Types of Data and the General Framework of Analysis

The previous chapters hopefully demonstrate to you that the analysis of habitat (in some form or another) is involved in a wide variety of ecological studies. Moreover, researchers collect and use various types of environmental data when studying the habitat associations of a species. These data may span a wide range of spatial scales, from landscapes on a scale of a few to hundreds of square kilometers down to so-called microhabitats that are typically conceived of as an area of a few square meters immediately surrounding an animal's resting location, nesting site, or other spatial point where an animal engages in some activity while remaining relatively stationary. Of course, the relevant scale for studying habitat associations depends on the organism's size and capacity for movement as well as the particular goals of the researcher (as discussed in Chapters 6 and 4, respectively). At the scale of landscapes, researchers often examine land cover composition surrounding recorded locations of the species (examples 1–3 in Table 5.1). Such analysis typically involves the use of GIS (Geographic Information Systems) software and remote sensing imagery to derive the percentage cover of various habitat types (e.g., forest, grassland, wetland) within the landscape. In the parlance of GIS, the landscape is a "buffer," often a circle of a given radius (e.g., several hundred meters to 10 km or so) centered on a location of the species. Alternatively, the landscape may be a large area that one or more individuals of the species are known to use, and the area is depicted as a polygon in GIS. Going down in spatial scale, researchers often collect and analyze data on the vegetation surrounding locations of

individual animals (examples 4–7 in Table 5.1). This requires being in the field and literally measuring the vegetation and hence these data are usually collected on spatial scales ranging from a few square meters to 100 m of linear distance (as in the use of the line–transect–intercept method to measure percentage cover of vegetation).

The actual lack of vegetation is a habitat requirement for some species. Bare ground is a habitat feature that can be measured as a percentage cover at various spatial scales (examples 8 and 9 in Table 5.1). Bare ground could also be thought of as a characteristic of the substrate. Indeed, characteristics of the substrate are often analyzed in habitat studies, including the type of soil (soils are much more complex and interesting than most ecologists realize, just ask a soil scientist or geologist), amount of rock in or on the substrate, particle size, and porosity. Of note, substrate characteristics can matter to organisms (i.e., be important in defining habitat) whether the substrate is vegetated or not.

The presence or number of refugia ("hiding places") is often measured and analyzed (example 10 in Table 5.1). Of course, what counts as a refugium depends on the size, physiology, and behavior of the species. Refugia at ground level could be thought of as part of the substrate (e.g., rock crevices) whereas those off the ground such as formed within the vegetation (e.g., tree hollows) clearly are not.

Properties of the terrain such as aspect, slope, and elevation may be considered potential habitat characteristics in some cases (examples 11–12 in Table 5.1). Aspect is often analyzed as a potentially

Habitat Ecology and Analysis. Joseph A. Veech, Oxford University Press (2021). © Joseph A. Veech. DOI: 10.1093/oso/9780198829287.003.0005

Table 5.1 Examples of the types of environmental variables that have been measured and used in analyzing the habitat associations of particular species.[1]

Environmental variable	Species	Location	Source
1 Amount of juniper–oak woodland at landscape scales (3–200 ha)	Golden-cheeked warbler (*Dendroica chrysoparia*)	Texas, USA	Magness et al. 2006
2 Amount of pine–hardwood mixed forest in landscape	Eastern box turtle (*Terrapene carolina*)	Georgia, USA	Greenspan et al. 2015
3 Percentage forest canopy cover	Brown-throated sloth (*Bradypus variegatus*)	Costa Rica	Neam and Lacher 2018
4 Stem density of tree saplings, shrubs, and vines	Swainson's warbler (*Limnothlypis swainsonii*)	South-eastern USA	Graves 2002
5 Tree height	Indiana bat (*Myotis sodalis*)	Indiana, USA	Bergeson et al. 2018
6 Plant species richness	Scimitar-horned oryx (*Oryx dammah*)	Tunisia	Cooke et al. 2016
7 Vegetation height and soil moisture	Various beetle and spider species	Greenland	Hansen et al. 2016
8 Proportion of bare ground	Woodlark (*Lullula arborea*)	Dorset, UK	Mallord et al. 2007
9 Proportion of bare ground	Yellow meadow ant (*Lasius flavus*)	Germany	Streitberger and Fartmann 2016
10 Presence of soil crevices	Chevron skink (*Oligosoma homalonotum*)	Great Barrier Island, New Zealand	Neilson et al. 2006
11 South-facing aspect	Sharp-tailed snake (*Contia tenuis*)	British Columbia, Canada	Wilkinson et al. 2007
12 Slope of the terrain	Newell's shearwater (*Puffinus newellii*)	Kauai, Hawaiian Islands	Troy et al. 2014
13 Total length of rocky substrate in stream segment	Booroolong frog (*Litoria booroolongensis*)	New South Wales, Australia	Hunter and Smith 2013
14 Salinity of water	Reticulated flagtail (*Kuhlia sandvicensis*) and strange-tailed flagtail (*Kuhlia xenura*)	Hawaii, USA	McRae et al. 2011
15 Depth of the surface mixed layer	Striped marlin (*Kajikia audax*)	Pacific Ocean	Lam et al. 2015
16 Percentage canopy cover over stream	Various fish species	French Guiana, South America	Mérigoux et al. 1998

[1] In each of the examples the presence, abundance, or amount of activity of the species was significantly ($P < 0.05$) and positively associated with the measured variable. Examples have been intentionally selected to represent a wide variety of species. All of the studies analyzed other variables in addition to the one listed; for brevity I have listed the ones with strongest effects.

important environmental variable defining habitat. The type and amount of vegetation may be influenced by aspect (and hence aspect has an indirect effect in forming habitat) or aspect might be important in that it influences the availability of direct solar radiation and exposure to wind both of which might be critical to some organisms. Similarly, slope and elevation can influence vegetation and define the climatic conditions of a location and thus determine whether the location is suitable habitat for a given species.

So far, all of these environmental variables have been discussed implicitly in the context of habitat for terrestrial organisms. However, each variable

may more or less have an aquatic equivalent (example 13 in Table 5.1) although there are also environmental variables that would seem to be exclusively and obviously aquatic such as water temperature and salinity (examples 14 and 15). Although there are definite differences in the way in which some aquatic and terrestrial environmental variables are measured and data collected, the statistical analysis of the variables is blind to whether they are aquatic or terrestrial.

In this book I do not generally consider climatic factors and especially weather conditions as components of habitat. Obviously, weather and climate are important to species inasmuch as species must be adapted to local climatic conditions and be able to withstand weather extremes for individual survival and population persistence. My exclusion of climate and weather (as components of habitat) is simply to allow for a cohesive and focused perspective on the importance of physical structure in defining the habitat of a species (Chapters 2 and 3). Certainly, the effect of climate on species occurrences (e.g., range limits) has always been a very active area of ecological research. Related to the goals of this book, species distribution modeling (SDM, Sections 4.5 and 11.5) often uses climate variables to predict the probability of occurrence of a species over a relatively large region. As such, SDMs make use of *species–climate* relationships to map species distributions. There is some general similarity here in that one purpose of this book is to describe methods to analyze *species–habitat* relationships. Therefore, many of the techniques discussed in Chapter 9 could be applied with actual climate variables rather than the types of environmental variables (meant to represent habitat in a strict sense) listed in Table 5.1.

Further, exclusion of climate and weather variables does not necessarily mean that climate and weather are of no consequence to a species' habitat requirements. For example, soil moisture could be an important environmental variable that defines habitat for some species and it is partly due to climate as well as weather. Also, water temperature in streams and lakes can be very important to some aquatic species and it also is due indirectly to weather and climate, such as spring snowmelt affecting the temperature of streams and rivers. Here is another example. The *oceanic surface mixed layer* describes a layer of water that is homogeneous throughout (temperature and salinity) due to the action of surface winds, heat exchange, evaporation, and water currents creating turbulence and mixing the water. Thus, climate and weather affect the depth of the surface mixed layer (and incidentally it in turn can affect climate patterns on a regional and global scale) and this depth can be important in determining the bathymetric occurrence of some marine organisms (example 15 in Table 5.1). Finally, the environmental variables that are important in defining habitat for a given species might sometimes seem unexpected and somewhat

out of place physically. The canopy formed by vegetation overhanging streams can create shading and effectively moderate water temperature and thus be a critical habitat component for some stream fishes (example 16 in Table 5.1). In effect, trees are part of the habitat of small stream-dwelling organisms. This example also represents the situation where a structural component of the habitat works to lessen the effect of a weather variable. This situation occurs over and over with both aquatic and terrestrial organisms that must use shaded and sunny places to thermoregulate. The life history, diet, behavior, and physiology of a species will say a lot about the environmental variables to examine when conducting a habitat analysis. Researchers should be encouraged to think creatively and thoroughly about the environmental variables that might be functioning as crucial habitat requirements for their species of interest.

In order to be thorough in conducting a habitat analysis and thinking about the importance of habitat for your species of interest, you may be faced with analyzing a wide variety of environmental variables. Fortunately, we can conceptualize all the disparate types of data and environmental variables in a relatively simple and useful framework. All forms of habitat analysis share the common foundation in which a researcher has obtained data (measurements) on a set of environmental characteristics or properties at particular locations during a specific time period. These are measured or categorized variables, which we can represent as $X_1, X_2, X_3, \ldots X_N$ and denote collectively as [X]. In addition to the environmental data, the researcher would have also collected species data at the same locations and ideally at roughly the same time that the environmental data were collected (although simultaneous collection of environmental and species data is not always possible or strictly necessary). The species data can take several forms: they can be presence/absence data, abundance or density data, reproductive output, time spent at the locations and/or number of visits by individuals to the locations, and even data on population growth rates. In essence, the species data are demographic or behavioral data measuring something about the species presence, use, or persistence at each location. We can denote these data as [Y]. Thus, a typical habitat analysis consists of quantifying the strength of the relationship between [Y] and any combination of [X] and also testing the statistical significance of the relationship. This is the analytical task that is common to all the methods of analysis to be described in Chapter 9. For the purpose of identifying meaningful habitat associations the relevant question is: *How do we determine the extent to which each environmental variable singly and in combination with others is most closely associated with (or best predicts) the presence, abundance, activity, or persistence of the species?*

To further illustrate the type of data used in characterizing a species' habitat, I will discuss the study of Neam and Lacher (2018) that examined the habitat of brown-throated sloths (*Bradypus variegatus*) in Costa Rica. The species is found from Central America southward into most of northern South America (Moraes-Barros et al. 2014). As with all sloth species, *B. variegatus* is highly arboreal. With regard to diet, brown-throated sloths are herbivorous, primarily consuming the new leaves of various tree species (Hayssen 2010). Therefore, forest is the general type of habitat for this species. However, beyond this habitat association, Neam and Lacher (2018) sought to determine more specific features or characteristics that might define the habitat of brown-throated sloths, and in particular within a landscape altered by human activities such as crop farming and livestock grazing. The 16-km² landscape

comprising the study area in this region of Costa Rica (on the Caribbean slope of the Cordillera de Tilarán) is highly fragmented due in part to these human land use activities (Neam and Lacher 2018).

Neam and Lacher (2018) surveyed for sloths by slowly walking along transects while visually scanning the forest canopy. When a sloth was seen, it was photographed so as to distinguish among individuals. Although transects were walked multiple times on separate occasions resulting in a count of sloth sightings, Neam and Lacher collapsed the count data into 17 confirmed presence locations of a unique individual sloth. The data for sloth absence consisted of 15 point locations (100 m² area) that were surveyed extensively without a sloth ever being recorded. Moreover, these locations were within 126 m (typical radius of a sloth home range) of a presence location. This ensured that the absence location could potentially have been occupied by a sloth (it was not too distant from a known individual) and was in forest habitat.

Neam and Lacher (2018) measured environmental variables at a local scale (100-m² plots centered on the presence and absence locations) and at landscape scales of 0.5, 2, and 5 ha surrounding each location. At the local scale, the variables assessed structural complexity of the habitat at the exact locations of sloths and in the 100-m² plots where sloths were recorded as absent (Table 5.2). At

Table 5.2 Environmental variables measured by Neam and Lacher (2018) in their study of the habitat associations of brown-throated sloths at local and landscape scales.[1]

Scale	Variable	Mean of presence locations	Mean of absence locations	Significant difference?[2]
Local	Mean tree height (m)	16.5	17.4	No
	Tree density	6.65	6.73	No
	Canopy cover (%)	79.2	68.9	Yes
	Total basal area of tree trunks (cm²)	4486	3499	No
	Standard deviation of basal area	718.7	302.9	Yes
	Standard deviation of height	6.75	4.86	Yes
Landscape	Percentage secondary forest (0.5 ha)	27.8	12.3	No
	Percentage secondary forest (2 ha)	31.4	17.2	No
	Percentage secondary forest (5 ha)	36.4	21.0	Yes

[1] Local scale was defined as 100-m² plots centered on the location. Landscape scale was plots of 0.5, 2, and 5 ha centered on the location.

[2] Significant difference ($P < 0.05$) refers to a *t*-test applied to test for a difference between the mean of the presence locations and the mean of the absence locations.

the landscape scales, Neam and Lacher used a variety of metrics quantifying the configuration and composition of land cover types. They had high-resolution aerial imagery of the study site in Costa Rica and were able to identify 11 different land use and land cover types within the 16-km² landscape: old growth forest, secondary forest, riparian forest, tree plantations, cropland, pasture, mixed-use areas, urban areas, unpaved roads, bare ground, and water bodies. Using ArcGIS software, the imagery was converted to pixelated maps that could then be used to derive metrics such as area:edge ratio, shape, aggregation, and diversity of patches as well as mean distance of certain types of patches (e.g., secondary forest) to landscape features such as roads or rivers. These various metrics assess landscape fragmentation and configuration (spatial arrangement of land use or land cover types), and as such, the metrics are not environmental variables characterizing habitat. Rather, these metrics assess the spatial dispersion, amount, and location of areas (patches) of potential habitat and thus quantify the availability of habitat rather than measure properties or characteristics of the habitat. This is an important distinction. Using ArcGIS, Neam and Lacher (2018) also measured the percentage of land in the 0.5-, 2-, and 5-km landscapes that was composed of secondary forest (Table 5.2). Although this is a measure of habitat availability (assuming that secondary forest is the main vegetation type serving as habitat for brown-throated sloths), it is often useful to examine whether a given species associates positively or negatively with a given land cover type (examples 1 and 2 in Table 5.1). This then allows one to determine if the land cover type is habitat for the species.

Brown-throated sloths in this particular area of Costa Rica tend to occur within landscapes that have a greater percentage cover of secondary forests compared with landscapes where sloths were apparently absent. Somewhat surprisingly, this difference was most pronounced for larger landscapes (Table 5.2) indicating that sloths might respond to the amount of forest habitat at landscapes as large as 5 ha even though daily movement and use of habitat typically occurs within an area of 0.5 ha (Neam and Lacher 2018). At a local scale, sloths were more likely to occur at locations that had greater canopy coverage and greater heterogeneity in tree height and girth (Table 5.2). Presumably this habitat preference relates to their use of tree canopies for horizontal and vertical movement while foraging (Neam and Lacher 2018). A well-developed canopy with variation in tree height and girth would entail a wide variety of climbing structures to facilitate a sloth's ability to move almost anywhere within its territory.

This example from the study of Neam and Lacher 2018 illustrates that the environmental variables used in an analysis of habitat need not be very complex. Further, the main response variable can be a measurement as simple as presence/absence. Their actual habitat analysis was substantially more sophisticated and detailed than what I've presented here. Nonetheless, they essentially had a dataset of environmental variables [X] recorded at locations or within landscapes where they also had a variable Y representing species presence/absence. Hence, the foundation of their analyses was then to test for statistical relationships between Y and [X]. This general framework of linking Y to [X] should help researchers think clearly about the environmental variables to be measured and the habitat analysis to be conducted.

References

Bergeson, S.M., O'Keefe, J.M., and Haulton, G.S. (2018). Managed forests provide roosting opportunities for Indiana bats in south-central Indiana. *Forest Ecology and Management*, 427, 305–16.

Cooke, R.S.C., Woodfine, T., Petretto, M., and Ezard, T.H.G. (2016). Resource partitioning between ungulate populations in arid environments. *Ecology and Evolution*, 6, 6354–65.

Graves, G.R. (2002). Habitat characteristics in the core breeding range of the Swainson's warbler. *Wilson Bulletin*, 114, 210–20.

Greenspan, S.E., Condon, E.P., and Smith, L.L. (2015). Home range and habitat selection in the eastern box turtle (*Terrapene carolina carolina*) in a longleaf pine (*Pinus palustris*) reserve. *Herpetological Conservation and Biology*, 10, 99–111.

Hansen, R.R., Hansen, O.L.P., Bowden, J.J., Treier, U.A., Normand, S., and Høye, T. (2016). Meter scale variation in shrub dominance and soil moisture structure Arctic arthropod communities. *PeerJ*, 4, e2224.

Hayssen, V. (2010). *Bradypus variegatus* (Pilosa: Bradypodidae). *Mammalian Species*, 42, 19–32.

Hunter, D. and Smith, M.J. (2013). Multiscale habitat assessment for the endangered booroolong frog (*Litoria booroolongensis*): implications for threatened species management in the rural landscape of southeastern Australia. *Herpetological Conservation and Biology*, 8, 122–30.

Lam, C.H., Kiefer, D.A., and Domeier, M.L. (2015). Habitat characterization for striped marlin in the Pacific Ocean. *Fisheries Research*, 166, 80–91.

Magness, D.R., Wilkins, R.N., and Hejl, S.J. (2006). Quantitative relationships among golden-cheeked warbler occurrence and landscape size, composition, and structure. *Wildlife Society Bulletin*, 34, 473–79.

Mallord, J.W., Dolman, P.M., Brown, A.F., and Sutherland, W.J. (2007). Linking recreational disturbance to population size in a ground-nesting passerine. *Journal of Applied Ecology*, 44, 185–95.

Moraes-Barros, N., Chiarello, A., and Plese, T. (2014). *Bradypus variegatus*, The IUCN Red List of Threatened Species, eT3038A47437046.

McRae, M.G., McRae, L.B., and Fitzsimons, J.M. (2011). Habitats used by juvenile flagtails (*Kuhlia* spp.; Perciformes: *Kuhliidae*) on the Island of Hawai'i. *Pacific Science*, 65, 441–50.

Merigoux, S., Ponton, D., and de Mérona, B. (1998). Fish richness and species-habitat relationships in two coastal streams of French Guiana, South America. *Environmental Biology of Fishes*, 51, 25–39.

Neam, K.D. and Lacher, T.E. (2018). Multi-scale effects of habitat structure and landscape context on a vertebrate with limited dispersal ability (the brown-throated sloth, *Bradypus variegatus*). *Biotropica*, 50, 684–93.

Neilson, K., Curran, J.M., Towns, D.R., and Jamieson, H. (2006). Habitat use by chevron skinks (*Oligosoma homalonotum*) (Sauria: Scincidae) on Great Barrier Island, New Zealand. *New Zealand Journal of Ecology*, 30, 345–56.

Streitberger, M. and Fartmann, T. (2016). Vegetation heterogeneity caused by an ecosystem engineer drives oviposition-site selection of a threatened grassland insect. *Arthropod-Plant Interactions*, 10, 545–55.

Troy, J.R., Holmes, N.D., Veech, J.A., Raine, A.F., and Green, M.C. (2014). Habitat suitability modeling for the Newell's shearwater on Kauai. *Journal of Fish and Wildlife Management*, 5, 315–29.

Wilkinson, S.F., Gregory, P.T., Engelstoft, C., and Nelson, K.J. (2007). The Rumsfeld Paradigm: knowns and unknowns in characterizing habitats used by the endangered sharp-tailed snake, *Contia tenuis*, in southwestern British Columbia. *Canadian Field-Naturalist*, 121, 142–49.

CHAPTER 6

Issues Based on Species Life History and Behavior

The general framework of a habitat analysis, as presented in Chapter 5, can be applied to any type of species. Nonetheless, the particular environmental variables that are examined, how the data are collected, and the subsequence inferences from the results of the analysis will vary among species. Species differ tremendously in their life histories and behavior and this influences appropriate study design for collecting presence/absence or abundance data and measurement of environmental variables. Although this book is not formally about the design of habitat studies or field methods for data collection, there are several issues related to life history and behavior that the researcher should be aware of before commencing a study. Some of these considerations are as follows: differences between breeding and foraging habitat, individuals at different life stages (or ages) using different habitats, seasonality of habitat use, and individual movement and the scale of habitat use. These factors could be decisive in designing the sampling/surveying aspects of a study and in determining which statistical method to use for analysis. In addition, proper inference from the results is more likely when the researcher has a thorough understanding of these issues.

6.1 Difference between breeding and foraging habitat

Some species differ in the habitat that is used for breeding and foraging. Many sea bird species nest in a habitat (or at a location) that is different and somewhat distant from where most foraging activity occurs out at sea (Furness and Monaghan 1987;

Fasola and Bogliani 1990; Hamer et al. 2000; Fritz et al. 2003). Other more sedentary bird species will often have a breeding territory centered around the nest and yet also forage in areas that are not strictly defended. This distinction between the breeding and foraging habitats of birds is sometimes recognized in studies of habitat associations (e.g., Steele 1993; Matsuoka et al. 1997; Heldbjerg et al. 2017). Terrestrial amphibians provide another example of a distinct difference between breeding and foraging habitat. Many species require a water body of some type for reproduction, which includes chorusing by males, egg-laying, and development of the tadpoles. Outside the breeding season, adults may be primarily terrestrial and exist some distance from the nearest water body. The distinction between aquatic and terrestrial habitats is sometimes incorporated into studies of the habitat requirements of various amphibian species (e.g., Porej et al. 2004; Pearl et al. 2005; Simon et al. 2009).

In a much more general and inclusive context, there are many types of animals that nest, burrow, or otherwise have a repeatedly used place of shelter somewhere within the larger overall territory or home range that they occupy. Moreover, this resting place might also be used for rearing offspring. As such, the environmental conditions immediately surrounding it could be considered as representing breeding habitat (or microhabitat). For example, tree cavities of a certain size or height above ground might be a specific habitat requirement for some cavity-nesting birds that otherwise use (for foraging) a much larger area of the landscape surrounding the nest cavity. Another example is the use of burrows as dens for birthing and raising of a litter by some mammal species (e.g., Lesmeister et al. 2008; Zielinski et al. 2010; Punjabi et al. 2013; Sprayberry and Edelman 2018; Symes et al. 2019). Many fish species use rock crevices or depressions in the substrate for egg-laying. On a much larger spatial scale, anadromous fish species (e.g., Salmonidae) clearly use different types of habitat for reproduction and for non-reproductive existence. We can even extend this idea of breeding vs. foraging habitat to organisms as behaviorally simple as insects if egg-laying occurs in a type of microhabitat (e.g., host plant or particular set of soil conditions) that is different and spatially distinct from where the adults otherwise spend most of their time. In all these scenarios, it may be useful to distinguish between breeding and foraging habitat (or microhabitat) to better understand the intricacies of a species' habitat associations.

Whether the distinction is useful or not will often depend on the spatial scale and extent of the study. Some studies simply subsume the breeding location (or microhabitat) as part of the greater overall type of habitat occupied by the individual and species. Thus, for a given area to be appropriate habitat (and identified as such) it must include potential breeding sites or resources (e.g., Johnson et al. 2008). If the distinction between breeding and foraging habitats is regarded as important based upon the goals of the study, then two separate analyses of habitat associations may be warranted. However, even if the actual habitat analysis does not formally separate the environmental variables corresponding to breeding from those for foraging, it still may be beneficial to recognize the distinction when communicating the results of the study.

6.2 Use of different types of habitat during different life stages

Similar to differences between breeding and foraging habitat, many species use different types of habitat at different life stages or ages. These differences may be even stronger than those corresponding to breeding and foraging habitat. Differences in habitat use based on life stage often involve organisms in which the life stages are fundamentally

different from one another with regard to resource (food) use, behavior, physiology, and sometimes morphology. Amphibians with indirect larval development are a good example; the tadpoles of a given species are very different from the adults. Similarly, some insect species (e.g., dragonflies and damselflies, Odonata) have a larval or juvenile life stage that develops in water whereas the adults are free flying. Other examples include fish species with anadromous life cycles and some fish species that are relatively sedentary as adults but have earlier life stages represented by pelagic larvae. Age- or life stage-based differences in habitat use have also been revealed for species in which the life stages are not greatly different from one another. Young et al. (2019) discovered differences in the habitat used by adult Henslow's sparrows (*Centronyx henslowii*) and that used by fledglings, and they further suggested that these differences need to be considered in managing habitat for the species. On a much larger geographic scale, recently fledged black-footed albatrosses (*Phoebastria nigripes*) use foraging habitat in an area of the Pacific Ocean that is separate from the area used by adults (Gotowsky et al. 2014).

When the individuals of a species are very distinct from one another as based on life stage or age, then a study of habitat associations should proceed almost as though the different life stages or age classes represent different species. Indeed, the study design and protocol for collecting data on the species response variable and the environmental predictor variables may be very different among the different life stages. As such, separate analyses of habitat associations are typically necessary for species with distinctly different life stages. For amphibians, examples of this approach include Roznick et al. (2009) and Grant et al. (2018). For fish, examples include Santos et al. (2011), Compton et al. (2012), Dance and Rooker (2016), and Dunn and Angermeier (2016). Any species that undergoes an ontogenetic habitat shift or age-based habitat partitioning (Stamps 1983; Werner and Hall 1988; Macpherson 1998; Dahlgren and Eggleston 2000; Blouin-Demers et al. 2007; Hisam et al. 2020) will require that habitat associations are analyzed separately for the different life stages or age classes.

6.3 Seasonality of habitat use

Individuals of a species might also shift their habitat use with a change in season. This could be due to physiological tolerances with individuals moving to more abiotically benign types of habitat (e.g., Hofmann and Fischer 2002; Menéndez and Gutiérrez 2004; Kraus and Morse 2005; Croak et al. 2013) or perhaps due to resource availability with individuals moving into a type of habitat that provides greater food resources (e.g., Prins and Ydenberg 1985; Adler et al. 1999; Luiselli et al. 2005; Copeland et al. 2007; Gilroy et al. 2010; Yoganand and Owen-Smith 2014; Yohannes et al. 2014; Simard et al. 2015) or both physiology and resources (e.g., Ager et al. 2003; Rettig and Brewer 2011). Seasonality of habitat use might also be more common in migratory species in that the winter and summer habitats are often substantially different (Nicholson et al. 1997; Pilliod et al. 2002; Nikula et al. 2004; Lamb et al. 2019; Lerche-Jørgensen et al. 2019). Importantly, seasonality in habitat use is not change in habitat conditions at a given location based on season (although this certainly can occur due to weather and climate), but rather individuals relocating to use different habitats. Sometimes the seasonal shift in habitat use can be quite dramatic and occur over a relatively small spatial scale. For example, around Narragansett Bay (Rhode Island, United States) the wolf spider *Pardosa lapidicina* occupies tidal cobble beaches during most of the year, but in autumn, individuals move into adjacent forest habitat to overwinter in the leaf litter (Kraus and Morse 2005). In other instances, the seasonal change in habitat use might simply represent a relatively minor change such as fish species in all types of water bodies relocating to either deeper or shallower water depending on seasonal changes in water temperature. Seasonal habitat use can also come about, by necessity, when the species occupies an ephemeral habitat (e.g., vernal pools) during at least part of the year. Once the habitat (or certain environmental conditions) disappears, then individuals must relocate to a different type of habitat.

Prior to initiating a study of habitat associations, the researcher might not know whether the focal species has a seasonal shift in habitat use. Indeed, the goal of the study might be to determine this important aspect of life history. As such, it might be necessary to conduct surveying at predetermined locations throughout an entire year (all seasons). Even better, given that the habitat shift likely involves movement (relocation) by individuals of the focal species, then some form of tracking might be necessary, or at least establishment of new survey plots in areas where the species is presumed to have relocated to. In either case, a wide variety of environmental variables will likely need to be measured and some of these could vary seasonally (as well as spatially). The basic plan here is to conduct the study on a broad enough spatial scale and for a sufficiently long duration to encompass all the seasonal habitat types and environmental variables that the species might encounter and use.

It may also be worthwhile to carefully consider whether the study species is truly using two or more distinctly different habitat types or whether it is more realistic to conceive of the species as using *different parts* of the same basic type of habitat. For example, Rettig and Brewer (2011) discuss how three small benthic stream fishes shift between using deep pools and riffles on a seasonal basis. However, the pools and riffles are often in very close proximity and are simply different parts of the same overall stream habitat. The fish are primarily responding to water depth and current velocity as the two most important habitat variables (Rettig and Brewer 2011). There is still a seasonal component to the species' habitat associations, but the seasonality is not so profound that individuals shift to using a completely different type of habitat.

6.4 Individual movement and the scale of habitat use

One of the main challenges with regard to studying and analyzing the habitat of animals is that they move around. In fact, the other issues discussed above all arise at least in part because most animals move about either during their daily activities or during some phase of their life cycle. This also means that habitat selection and use can occur at multiple spatial scales and are played out as hierarchical processes (Johnson 1980; Orians and Wittenberger 1991; Jones 2001; Mayor et al. 2009; Matthiopoulos et al. 2015; McGarigal et al. 2016).

For example, a female yellow-headed blackbird (*Xanthocephalus xanthocephalus*) first selects a marsh that has moderately dense vegetation and extensive water channeling (Orians and Wittenberger 1991). These are habitat associations on a broad spatial scale of hectares. Within the larger marsh, a female will then establish a territory based on finer-scale characteristics of the vegetation that relate to prey density and represent habitat associations on a relatively small spatial scale (Orians and Wittenberger 1991). As another example, coral gobies in the genus *Gobiodon* first select a particular coral reef, then the zone (reef crest, slope, or flat), and lastly establish a territory typically in association with a particular species of coral (Munday 2000).

The relevance of multi-scale habitat selection to a habitat analysis is that the species response variable (presence/absence, abundance, level of activity) and environmental variables should be measured at several spatial extents in order to get a full and complete assessment of habitat associations. This could entail having spatial sampling plots of various sizes embedded within one another whether the study design is based on surveying of pre-established plots or a posteriori establishment of plots at locations of tracked individuals (Section 7.6). Also, different variables might be measured at different spatial extents. With regard to identifying habitat associations, a separate habitat analysis (i.e., the general framework explained in Chapter 5) might need to be conducted at each spatial extent if the data for the species response variable fundamentally change across spatial scale. Alternatively, one single habitat analysis could be conducted simultaneously including all the environmental variables regardless of the spatial scale at which they were measured. There are many ways of including variable spatial scales in a habitat analysis, both in the data (predictor variables) that are used and the way they are analyzed. Also, there may be value in determining the optimal scale for each variable, defined as that spatial scale at which the species responds most strongly to the variable (McGarigal et al. 2016). Lastly, even when the spatial scale of habitat use has not been directly incorporated into the study design and analysis, it is important to recognize that documented habitat associations always depend to some extent on spatial scale.

References

Adler, G.H., Mangan, S.A., and Suntsov, V. (1999). Richness, abundance, and habitat relations of rodents in the Lang Bian Mountains of southern Viet Nam. *Journal of Mammalogy*, 80, 891–98.

Ager, A.A., Johnson, B.K., Kern, J.W., and Kie, J.G. (2003). Daily and seasonal movements and habitat use by female Rocky Mountain elk and mule deer. *Journal of Mammalogy*, 84, 1076–88.

Blouin-Demers, G., Bjorgan, L.P.G., and Weatherhead, P.J. (2007). Changes in habitat use and movement patterns with body size in black ratsnakes (*Elaphe obsoleta*). *Herpetologica*, 63, 421–29.

Compton, T.J., Morrison, M.A., Leathwick, J.R., and Carbines, G.D. (2012). Ontogenetic habitat associations of a demersal fish species, *Pagrus auratus*, identified using boosted regression trees. *Marine Ecology Progress Series*, 462, 219–30.

Copeland, J.P., Peek, J.M., Groves, C.R., et al. (2007). Seasonal habitat associations of the wolverine in central Idaho. *Journal of Wildlife Management*, 71, 2201–12.

Croak, B.M., Crowther, M.S., Webb, J.K., and Shine, R. (2013). Movements and habitat use of an endangered snake, *Hoplocephalus bungaroides* (Elapidae): implications for conservation. *PLoS ONE*, 8, e61711.

Dahlgren, C.P. and Eggleston, D.B. (2000). Ecological processes underlying ontogenetic habitat shifts in a coral reef fish. *Ecology*, 81, 2227–40.

Dance, M.A. and Rooker, J.R. (2016). Stage-specific variability in habitat associations of juvenile red drum across a latitudinal gradient. *Marine Ecology Progress Series*, 557, 221–35.

Dunn, C.G. and Angermeier, P.L. (2016). Development of habitat suitability indices for the candy darter, with cross-scale validation across representative populations. *Transactions of the American Fisheries Society*, 145, 1266–81.

Fasola, M. and Bogliani, G. (1990). Foraging ranges of an assemblage of Mediterranean seabirds. *Colonial Waterbirds*, 13, 72–74.

Fritz, H., Said, S., and Weimerskirch, H. (2003). Scale-dependent hierarchical adjustments of movement patterns in a long-range foraging seabird. *Proceedings Royal Society London B*, 270, 1143–48.

Furness, R.W. and Monaghan, P. (1987). *Seabird Ecology*. Chapman and Hall, New York.

Gilroy, J.J., Anderson, G.Q.A., Grice, P.V., Vickery, J.A., and Sutherland, W.J. (2010). Mid-season shifts in the habitat associations of yellow wagtails (*Motacilla flava*) breeding in arable farmland. *Ibis*, 152, 90–104.

Gotowsky, S.E., Tremblay, Y., Kappes, M.A., et al. (2014). Divergent post-breeding distribution and habitat associations of fledgling and adult black-footed albatrosses (*Phoebastria nigripes*) in the North Pacific. *Ibis*, 156, 60–72.

Grant, T.J., Otis, D.L., and Koford, R.R. (2018). Comparison between anuran call only and multiple life-stage occupancy designs in Missouri River floodplain wetlands following a catastrophic flood. *Journal of Herpetology*, 52, 371–80.

Hamer, K.C., Phillips, R.A., Wanless, S., Harris, M.P., and Wood, A.G. (2000). Foraging ranges, diets and feeding locations of gannets (*Morus bassanus*) in the North Sea: evidence from satellite telemetry. *Marine Ecology Progress Series*, 200, 257–64.

Heldbjerg, H., Fox, A.D., Thellesen, P.V., Dalby, L., and Sunde, P. (2017). Common starlings (*Sturnus vulgaris*) increasingly select for grazed areas with increasing distance-to-nest. *PLoS ONE*, 12, e0182504.

Hisam, F., Hajisamae, S., Ikhwanuddin, M., and Pradit, S. (2020). Distribution pattern and habitat shifts during ontogeny of the blue swimming crab, *Portunus pelagicus* (Linnaeus, 1758) (Brachyura, Portunidae). *Crustaceana*, 93, 17–32.

Hofmann, N. and Fischer, P. (2002). Temperature preferences and critical thermal limits of burbot: implications for habitat selection and ontogenetic habitat shift. *Transactions of the American Fisheries Society*, 131, 1164–72.

Johnson, D.H. (1980). The comparison of usage and availability measurements for evaluating resource preference. *Ecology*, 6, 65–71.

Johnson, J.R., Mahan, R.D., and Semlitsch, R.D. (2008). Seasonal terrestrial microhabitat use by gray treefrogs (*Hyla versicolor*) in Missouri oak-hickory forests. *Herpetologica*, 64, 259–69.

Jones, J. (2001). Habitat selection studies in avian ecology: a critical review. *Auk*, 118, 557–62.

Kraus, J.M. and Morse, D.H. (2005). Seasonal habitat shift in an intertidal wolf spider: proximal cues associated with migration and substrate preference. *Journal of Arachnology*, 33, 110–23.

Lamb, J.S., Satgé, Y.G., and Jodice, P.G.R. (2019). Seasonal variation in environmental and behavioural drivers of annual-cycle habitat selection in a nearshore seabird. *Diversity and Distributions*, 26, 254–66.

Lerche-Jørgensen, M., Mallord, J.W., Willemoes, M., et al. (2019). Spatial behavior and habitat use in widely separated breeding and wintering distributions across three species of long-distance migrant *Phylloscopus* warblers. *Ecology and Evolution*, 9, 6492–00.

Lesmeister, D.B., Gompper, M.E., and Millspaugh, J.J. (2008). Summer resting and den site selection by eastern spotted skunks (*Spilogale putorius*) in Arkansas. *Journal of Mammalogy*, 89, 1512–20.

Luiselli, L., Akani, G.C., Angelici, F.M., Ude, L., and Wariboko, S.M. (2005). Seasonal variation in habitat use in sympatric Afrotropical semi-aquatic snakes, *Grayia smythii* and *Afronatrix anoscopus* (Colubridae). *Amphibia-Reptilia*, 26, 372–76.

Macpherson, E. (1998). Ontogenetic shifts in habitat use and aggregation in juvenile sparid fishes. *Journal of Experimental Marine Biology and Ecology*, 220, 127–50.

Matsuoka, S.M., Handel, C.M., Roby, D.D., and Thomas, D.L. (1997). The relative importance of nesting and foraging sites in selection of breeding territories by Townsend's warblers. *Auk*, 114, 657–67.

Matthiopoulos, J., Fieberg, J., Aarts, G., Beyer, H.L., Morales, J.M., and Haydon, D.T. (2015). Establishing the link between habitat selection and animal population dynamics. *Ecological Monographs*, 85, 413–36.

Mayor, S.J., Schneider, D.C., Schaefer, J.A., and Mahoney, S.P. (2009). Habitat selection at multiple scales. *Ecoscience*, 16, 238–47.

McGarigal, K., Wan, H.Y., Zeller, K.A., Timm, B.C., and Cushman, S.A. (2016). Multi-scale habitat selection modeling: a review and outlook. *Landscape Ecology*, 31, 1161–75.

Menéndez, R. and Gutiérrez, D. (2004). Shifts in habitat associations of dung beetles in northern Spain: climate change implications. *Ecoscience*, 11, 329–37.

Munday, P.L. (2000). Interactions between habitat use and patterns of abundance in coral-dwelling fishes of the genus *Gobiodon*. *Environmental Biology of Fishes*, 58, 355–69.

Nicholson, M.C., Bowyer, R.T., and Kie, J.G. (1997). Habitat selection and survival of mule deer: tradeoffs associated with migration. *Journal of Mammalogy*, 78, 483–04.

Nikula, A., Heikkinen, S., and Helle, E. (2004). Habitat selection of adult moose *Alces alces* at two spatial scales in central Finland. *Wildlife Biology*, 10, 121–35.

Orians, G.H. and Wittenberger, J.F. (1991). Spatial and temporal scales in habitat selection. *American Naturalist*, 137, S29–S49.

Pearl, C.A., Adams, M.J., Leuthold, N., and Bury, R.B. (2005). Amphibian occurrence and aquatic invaders in a changing landscape: implications for wetland mitigation in the Willamette Valley, Oregon, USA. *Wetlands*, 25, 76–88.

Pilliod, D., Peterson, C.R., and Ritson, P.I. (2002). Seasonal migration of Columbia spotted frogs (*Rana luteiventris*) among complementary resources in a high mountain basin. *Canadian Journal of Zoology*, 80, 1849–62.

Porej, D., Micacchion, M., and Hetherington, T.E. (2004). Core terrestrial habitat for conservation of local populations of salamanders and wood frogs in agricultural landscapes. *Biological Conservation*, 120, 399–09.

Prins, H.H.T. and Ydenberg, R.C. (1985). Vegetation growth and a seasonal habitat shift of the barnacle goose (*Branta leucopsis*). *Oecologia*, 66, 122–25.

Punjabi, G.A., Chellam, R., and Vanak, A.T. (2013). Importance of native grassland habitat for den-site selection of Indian foxes in a fragmented landscape. *PLoS ONE*, 8, e76410.

Rettig, A.V. and Brewer, S.K. (2011). Seasonal habitat shifts by benthic fishes in headwater streams. *Proceedings Annual Conference Southeast Association Fish and Wildlife Agencies*, 65, 105–11.

Roznick, E.A., Johnson, S.A., Greenberg, C.H., and Tanner, G.W. (2009). Terrestrial movements and habitat use of gopher frogs in longleaf pine forests: a comparative study of juveniles and adults. *Forest Ecology and Management*, 259, 187–94.

Santos, J.M., Reino, L., Porto, M., et al. (2011). Complex size-dependent habitat associations in potamodromous fish species. *Aquatic Sciences*, 73, 233–45.

Simard, P., Wall, C.C., Allen, J.B., et al. (2015). Dolphin distribution on the West Florida Shelf using visual surveys and passive acoustic monitoring. *Aquatic Mammals*, 41, 167–87.

Simon, J.A., Snodgrass, J.W., Casey, R.E., and Sparling, D.W. (2009). Spatial correlates of amphibian use of constructed wetlands in an urban landscape. *Landscape Ecology*, 24, 361–73.

Sprayberry, T.R. and Edelman, A.J. (2018). Den-site selection of eastern spotted skunks in the southern Appalachian Mountains. *Journal of Mammalogy*, 99, 242–51.

Stamps, J.A. (1983). The relationship between ontogenetic habitat shifts, competition and predator avoidance in a juvenile lizard (*Anolis aeneus*). *Behavioral Ecology and Sociobiology*, 12, 19–33.

Steele, B.B. (1993). Selection of foraging and nesting sites by black-throated blue warblers: their relative influence on habitat choice. *Condor*, 95, 568–89.

Symes, S.A., Klafki, R., Packham, R., and Larsen, K.W. (2019). Discriminating different-purpose burrows of the North American badger *Taxidea taxus*. *Wildlife Biology*, 2019, 1–9.

Werner, E.E. and Hall, D.J. (1988). Ontogenetic habitat shifts in bluegill: the foraging rate-predation risk trade-off. *Ecology*, 69, 1352–66.

Yoganand, K. and Owen-Smith, N. (2014). Restricted habitat use by an African savanna herbivore through the seasonal cycle: key resources concept expanded. *Ecography*, 37, 969–82.

Yohannes, E., Arnaud, A., and Béchet, A. (2014). Tracking variations in wetland use by breeding flamingos using stable isotope signatures of feather and blood. *Estuarine, Coastal and Shelf Science*, 136, 11–18.

Young, A.C., Cox, W.A., McCarty, J.P., and Wolfenbarger, L.L. (2019). Postfledging habitat selection and survival of Henslow's sparrow: management implications for a critical life stage. *Avian Conservation and Ecology*, 14, 10.

Zielinski, W.J., Hunter, J.E., Hamlin, R., Slauson, K.M., and Mazurek, M.J. (2010). Habitat characteristics at den sites of the Point Arena mountain beaver (*Aplodontia rufa nigra*). *Northwest Science*, 84, 119–30.

CHAPTER 7

Design and Statistical Issues Related to Habitat Analysis

Although this book is not intended to be a comprehensive guide to study design or the collection of ecological/environmental data in the field, discussion of some pertinent issues is warranted. These issues essentially pertain to (1) factors or aspects of the study that need to be considered prior to collecting any data, or (2) readying the data (predictor and response variables) prior to the actual habitat analysis based on any of the six methods discussed in Chapter 9. Some of these issues are decisions under the full control of the researcher (e.g., number of predictor variables to include—Section 7.12) whereas others represent conditions of the study that likely are not under the control of the researcher (e.g., background density of the species and how this affects sampling—Section 7.7). Some of the issues might require direct corrective action by the researcher, whereas others simply represent a condition of the study that perhaps cannot be controlled but nonetheless can be acknowledged and dealt with when making inferences from the results. If neglected, all of these issues can lead to potential flaws, weaknesses, limitations, and errors of omission in a study, all or some of which a reviewer of the research might notice. Hence, it is important to be aware of and generally knowledgeable about these issues. Ultimately, these issues relate to proper statistical practice and sound inferential science—and they are relevant to all methods of habitat analysis.

7.1 Area to include in the study

One practical limitation of the analysis of any species' habitat requirements is the area to include in the study. The researcher must decide the extent of the overall area to survey, an entire region or part thereof, the entirety of a landscape or some smaller portion of it, or perhaps an even smaller spatial extent of a few hectares? This decision will be based on the behavior and ecology of the focal species as well as guided by logistical constraints (time, effort, and money). For example, the area of study for a large-bodied mobile species will be much larger than the area of study for a small-bodied sedentary species. The area of study is essentially the area contained within the mostly widely separated spatial sampling or survey units, thus the spatial spread of these units (which often is under the control of the researcher) determines the spatial extent of the study. The researcher must decide how many units to use, how far apart, and what type of deployment or spatial arrangement (completely random, semi-random, stratified). Much of this chapter is predicated on a study design in which the researcher has established or *identified* a priori spatially distinct survey units (plot, transect, or trap at a point location) that will be searched, trapped, surveilled, or otherwise examined for the presence of the species. Additionally, within or near to these survey units, a set of environmental variables will be measured (Chapter 5). The physical size of the survey units also matters, and it should be determined by the traits and behavior of the focal species. Ideally, the spatial sampling units are far enough apart that the data collected in each are statistically independent (although see Section 7.10).

The spatial deployment of the survey units can matter greatly, and this is an issue that must be given careful consideration and planning prior to initiating data collection. Typically, a researcher will

Habitat Ecology and Analysis. Joseph A. Veech, Oxford University Press (2021). © Joseph A. Veech. DOI: 10.1093/oso/9780198829287.003.0007

have some basic knowledge about the general type of habitat where the species can be found, and this knowledge can inform the placement of survey units. Units can be placed in areas or at locations that may not be the most appropriate habitat but that otherwise are physically accessible by individuals of the species. In such units, we might expect to record many instances of species absence or relatively low abundance. This is permissible and indeed it may even be desirable to end up with a dataset in which there are numerous zero- or low-abundance survey units (Sections 7.2 and 7.4). Such datasets can provide greater scope for identifying the environmental variables (and particular values of the variables) that a species most strongly associates with, as long as there are also numerous enough survey units in which the species was recorded as present or at relatively high abundance. It is inappropriate and nothing is to be gained by establishing survey units in areas that clearly could not be habitat for the species and hence where the species never occurs.

Other considerations are the duration of search or survey effort for each unit. Obviously, the size of survey units and amount of time surveying should be standardized when possible. The actual method to use in searching (or surveying) is beyond the scope of this book given that such a task is very species specific. However, one form of surveying, trapping using either a real trap or a camera trap, presents the researcher with the additional challenge of trying to figure out the amount of surrounding area that the trap is actually sampling. This may not be very evident, and if the trap is somehow baited or has a lure, then it is even less clear what area is being sampled. Presumably, if the trap is deployed (active or open) for x amount of time, then it is sampling an area in which an individual y distance from the trap could encounter (randomly or by attraction) the trap during time period x. It may be difficult to have exact knowledge of y, although if values of y (and x) do not vary among traps then there likely is no bias introduced in the trapping data. With regard to environmental data, the researcher must measure the variables at the exact location of the trap or near enough to it such that those variables can reasonably be thought (assumed) to affect trap success

and hence presence of the species. Most of the design considerations pertaining to survey area and other spatial aspects depend very much on the particular species of interest; hence it is not possible to provide precise recommendations in the context of this book.

7.2 How to handle absence data

As mentioned above, absence data are not bad. The exception is when a dataset consists of an inordinate number of observations (survey units) of species absence, perhaps 90 percent or more. In such a case, it may not be clear if the species is very rare and truly absent from survey units at a high rate, or if the detection probability (for the given survey protocol and survey units) of the species is really low, or both. A high percentage of survey units with the species recorded as absent (zero-abundance) really limits the types of statistical analyses that can be performed even if the zeros are trustworthy (Chapter 9). One reason for this is that a variable with many observations of the same value (zero or any particular value) will not have a normal distribution and there is no data transformation that can remedy that. Occupancy modeling is a type of habitat analysis that can handle a relatively large proportion of zeros in the species response data (Section 9.6). However, it is a technique that requires very careful study design a priori, including repeated surveying of the same survey units. Hence, a dataset that turns out to have many zeros cannot be retrofitted for use in an occupancy model. When a dataset ends up having a high percentage of zero-abundance survey units the best approach may simply be to redesign the survey protocol (search or trapping procedure) and perhaps also reconfigure the spatial placement of the survey units. Also, the researcher should always keep in mind that no species saturates all available habitat (Chapter 1), hence some zero-abundance survey units might actually represent true absences (see next section) and occur at locations with appropriate habitat conditions.

7.3 Imperfect detection probability

For any survey protocol applied to any species, detection probability is always less than one, or imperfect. Detection probability is the probability that at least one individual of the focal species will be detected (i.e., visually revealed, captured, or otherwise recorded as present) within the survey unit during the survey or observation period. Detection probability is determined by traits of the species (e.g., behavior, crypsis) as well as by attributes of the survey protocol. The importance of recognizing imperfect detection and estimating a detection probability is that it is an explicit acknowledgment that survey units where the species was not detected might actually have the species (at least one individual). Such units represent false negatives. Detection probability can also be considered at the level of the individual, although it is much more difficult to estimate detection probabilities for individuals. Nonetheless, recognizing that even individuals have detection probabilities means that abundance can be underestimated. Thus, imperfect detection is an issue for count data as well as for presence–absence data. Some methods of habitat analysis, for example, occupancy modeling (Section 9.6), estimate species detection probability and take it into account when determining the actual pattern of species occurrence (true positives and negatives among the survey units).

Ecologists have known about imperfect detection and the practicality of estimating detection probability for nearly 100 years. Estimating a detection, recapture, or resighting probability as a way of correcting count data is an inherent part of statistical techniques for estimating population size and individual survival rates (see reviews in Seber 1982; Pollock 1991; Nichols 1992; White 2005; Johnson 2008). Nonetheless, imperfect detection may not be too problematic for the study of habitat associations of a species if there is no bias in detection probability among survey units and in particular no bias based on the environmental factors (presumed habitat characteristics) within the survey units. In such a situation, the presence–absence and abundance (count) data will have some amount of measurement error, but the error will not depend on and is not correlated with any of the measured environmental variables. On the other hand, if detection probability does depend on an environmental variable (e.g., amount of ground vegetation hindering easy visual detection of a small-bodied species) or a design variable (e.g., time of day that surveying is conducted), then the method of habitat analysis likely needs to directly estimate detection probability and allow for it to vary among survey units (see Section 9.6 on occupancy modeling). No matter what method of habitat analysis is used, it is important to remember that all count data have errors due to imperfect detection; the only exception is when one is intending to do and has achieved a complete census. Given this fact of life, one need not become overly concerned about imperfect detection. Johnson (2008) provides a very lucid and thorough discussion of the pros and cons of estimating detection probability. Further, to really account for imperfect detection in a statistically appropriate way, typically one would need to know a priori what factors could cause it and design the study accordingly. Such intricate information on the detection process is not always available.

7.4 Presence, absence, and availability

Some types of habitat analysis require the comparison of survey units wherein each unit is categorized as presence, absence, or available. Presence defines units where the species was found (at least one individual recorded) during the survey period and absence defines units where the species was not found (and hence presumed to be absent) during the survey period. Importantly, such units represent a "true absence" in the sense that they were actually surveyed, albeit such a unit could also represent a false negative particularly if detection probability is low. "Pseudo-absence" is sometimes used to refer to a unit or location that was not surveyed but where the species is assumed to not be present. This label can be somewhat gray and imprecise with regard to how confident the researcher is about the absence of the species. It could be that there has previously been substantial survey effort near to the unit (that is considered to represent pseudo-absence) without the species being detected and hence the researcher is fairly confident in concluding absence. Alternatively, pseudo-absence might represent a random unsurveyed location pinpointed on a map of the species geographic range and perhaps the presumed veracity of this label is simply due to the species being rare and so it is *probably* absent at the particular location(s). "Pseudo-absence" spans an entire continuum of confidence/certainty between these two scenarios. In most situations, it is much more appropriate to think of unsurveyed locations or units as representing "availability" rather than absence or pseudo-absence. That is, the environmental conditions at such locations have been measured and they are taken to represent what is *available* to the species (to select) whether the species is actually present at the location or not. Of course, availability also entails or assumes that the location is accessible to individuals of the species. Thus, the set of observations (locations) composing the "available" category includes locations where the species is present as well as locations where it is truly absent, although the researcher does not know which is which because the locations have not been surveyed for the species (Keating and Cherry 2004; Johnson et al. 2006).

For many studies of habitat associations, the distinction between "absence" and "availability" may not matter too much. After all, one can never prove a negative and hence even the "absence" survey units are most cautiously thought of as available. In many statistical analyses (Chapter 9), absence and available would both be coded as 0 (as the response variable) and compared with presence coded as 1. However, stronger inference can be made from a comparison of presence with absence than from presence with available (Manly et al. 2002). Also, the empirical and conceptual differences between the three labels are important to some analytical methods related to habitat analysis (see Section 11.1 on Resource Selection Functions).

7.5 Non-count data indicating habitat use or activity

The species response variable for a habitat analysis need not be limited to presence–absence or count data. Indeed, any variable indicating the amount of time in, level of activity within, or number of visits to the survey unit is potentially indicative of habitat use and hence can be used to identify meaningful habitat associations (Humes et al. 1999; Zollner et al. 2000; Evans et al. 2005; Gjerdrum et al. 2008; Abba et al. 2016; Law et al. 2018). Importantly, such

variables may be most effective for mobile organisms that are not expected to always be completely confined to a relatively small survey unit or even a larger survey area (e.g., brown hyenas, Thorn et al. 2011). These data could be collected by a human observer or more likely by automated devices such as motion- or heat-activated cameras and acoustic recorders. These devices can be effective for surveying species that move in and out of the survey plots and species that are really cryptic and thus not easy to detect in the limited time period that a human observer would have for searching a plot. For example, Suarez-Rubio et al. (2018) deployed ultrasonic recording devices to measure the nightly level of bat visitation and activity, detected by the low-frequency sound emissions of the bats, at 180 sampling points dispersed throughout Vienna, Austria. In their study, median duration of the calls (vocalizations) recorded for a given species during a nightly sampling period was taken as an indicator of habitat use and used in subsequent statistical analysis of habitat associations. They also reported that median call duration was highly correlated with the number of calls, thus they could have alternatively used number of calls as the response variable. Automated recording devices have also been used to measure activity level of frog species with subsequent analysis of habitat associations (Liner et al. 2008; Hilje and Aide 2012; Iwai et al. 2018). Of course, automated recorders can also be used to obtain presence–absence data as the species response variable.

Species response variables based on the amount or rate of use of a survey unit are potentially better than presence–absence or count data. They can provide a more complete and richer, more detailed account of the extent that individuals of the focal species are influenced behaviorally by the environmental conditions (habitat characteristics) within the survey unit. Presumably, an individual would make more back-and-forth trips (visits) to and/or spend more time in survey units that are nearer to the ideal habitat conditions for the species. Presence–absence is too crude to capture any level of a behavioral response. Indeed, the study of Suarez-Rubio et al. (2018) found that differences in bat activity among survey locations was partly explained by differences in the structural complexity of vegetation, which in turn affected the maneuverability and echolocation of the bats and hence their foraging success. Thus, species response variables that are somehow based on behavior (even if indirectly) get a bit closer to measuring actual habitat use and the importance of habitat to survival and reproduction (Chapter 3). Indeed, studies of habitat use, preference, and selection typically require a response variable that goes beyond just presence–absence, whereas the latter is sufficient for simply identifying and quantifying the habitat associations of a species (Section 7.14).

7.6 Habitat sampling plots established a posteriori

Sometimes the spatial units for measuring environmental variables (habitat characteristics) are established after the species data have been compiled or collected. This form of a posteriori habitat sampling can be linked to species surveying in several ways: (1) The researcher might be using previously collected and archived data on species occurrence. These data could emanate from large-scale government-funded monitoring programs (e.g., Swiss Breeding Bird Survey, UK Breeding Bird Survey, North American Breeding Bird Survey, Amphibian Research and Monitoring Initiative), citizen-science initiatives (e.g., eBird, Aussie Backyard Bird Count, iNaturalist), and museum records (e.g., VertNet), or any other survey effort in which species occurrence at spatially referenced (latitude and longitude) locations has been recorded. The researcher then collects or acquires environmental (habitat) data from "plots" centered on the locations of the species. (2) The researcher might conduct a survey in which some large area is searched for individuals of the focal species such that searching is not strictly confined to a circumscribed spatial unit. Upon finding an individual or evidence of an individual such as a nest or burrow, the location is recorded, and the researcher collects environmental data from a plot centered on the location. This approach is widely used in the study of nesting habitat of birds (e.g., Li and Martin 1991; Blakesley et al. 2005; Plumb et al. 2005; Kirkpatrick and Conway 2010) and the habitat where animals of various species place their burrows (e.g., Heemeyer

et al. 2012; Punjabi et al. 2013; Troy et al. 2016; Koshkina et al. 2020). Some studies simply survey for individual organisms. Winchell et al. (2017) visually surveyed for *Anolis* lizards, recorded their perching locations, and then measured characteristics of the habitat immediately surrounding a lizard. (3) The researcher might have conducted a study that involved tracking the movement of individuals of a mobile species (e.g., through radio-telemetry or GPS tracking). When a given individual "stops" for a sufficiently long period of time, then the location can be considered as used or selected by the individual. Environmental or habitat data are then collected from the location. This approach has been used to study the habitat associations of species ranging from gray wolves (*Canis lupus*) in Scandinavia (Sanz-Pérez et al. 2018) to leatherback turtles (*Dermochelys coriacea*) in the Atlantic Ocean (Dodge et al. 2014) to wood frogs (*Lithobates sylvaticus*) in northern Canada (Bishir et al. 2018).

A key feature of this study design is that the species data are obtained first and then environmental or habitat data obtained for the locations where the species was recorded. Note also that there are no survey locations that can be considered to represent species absence. Rather, locations representing availability are randomly selected either near to and paired with the presence locations (e.g., Winchell et al. 2017; Muller et al. 2018) or within some greater area encompassing the presence locations. This type of approach to analyzing habitat associations is sometimes referred to as a *use-availability* or *presence-availability* design (Section 7.4).

7.7 Effect of background density

By necessity a study of a species' habitat associations must occur within an area (local site, landscape, or region) where the species is known to occur and perhaps where it is relatively common, unless the focal species is rare and infrequent everywhere. Otherwise, surveying for the species would result in too many survey units in which the species is recorded as absent (Section 7.2). In a greater conceptual context, the background density of the species within the overall study area can influence the abundance or presence–absence data recorded for a set of survey units. This is true even if the survey units are randomly distributed and deployed in such a way that only areas of presumed and accessible habitat (broadly and inclusively defined) are sampled (Section 7.1). Moreover, within the greater overall study area, the density of the species may not be homogeneous. That is, the spatial dispersion of individuals of the species likely is neither uniform nor random, but rather it is aggregated to some extent. Importantly, intraspecific aggregation can be due to factors in addition to a non-random and spatially clumped dispersion of habitat. Section 9.2.1 discusses negative binomial regression as a method to account for intraspecific aggregation when conducting a habitat analysis. However, there is another effect of background density that is more difficult to control.

Consider that there is a continuum between the extreme scenario of a potential study area in which the species is too rare and infrequent to reliably conduct a habitat analysis (as discussed above). At the other end of the continuum the study area has a very high background density such that the population may be at or very near to carrying capacity. The relevance of this to an analysis of habitat is that some survey units may have an abundance of the species that is not reflective of the habitat conditions within the survey unit. That is, individuals might be "forced" to occupy areas of lower-quality habitat (and perhaps even non-habitat) if populations in other nearby areas of more suitable or appropriate habitat are at carrying capacity (van Beest et al. 2014; Matthiopoulos et al. 2015). This scenario invokes positive density-dependent dispersal (Matthysen 2005; Bowler and Benton 2005), source and sink dynamics, and the rescue effect; that is, some populations serving as the source of colonists to other populations that might not be capable of sustaining themselves if there were no dispersal from source to sink (Pulliam 1988; Dias 1996). Thus, some survey units might contain a substantial number of individuals even though the survey unit is sampling from a sink population that is in an area of marginally suitable habitat. The well-known consequence of this is that density can be a misleading indicator of habitat quality (Van Horne 1983). The explicit consequence for a habitat analysis is that the wrong environmental variables could be mistakenly identified as important habitat

characteristics. Regardless of the exact statistical analysis performed, the researcher might then subsequently define (infer) the habitat of the focal species incorrectly.

Another possibility is that background density is low enough that the typical correlative approach (Chapter 9) to conducting a habitat analysis and understanding species–habitat relationships is invalid. In such a scenario the abundance of the species in the survey units is determined by factors other than the environmental variables representing habitat conditions. That is, the abundance of the species is *limited* by factors such as predation or disease rather than the availability of suitable habitat and hence abundance is not correlated with any characteristic of the habitat (Mitchell 2005). For some authors, this suggests that a correlation of abundance with habitat can only exist if the population is near the maximal abundance (i.e., carrying capacity) for the particular type of habitat (Mitchell 2005). In my opinion, this is an overstated concern. One can legitimately perform a habitat analysis and identify meaningful habitat characteristics regardless of whether the sampled population(s) is at carrying capacity. If the abundance (or presence/absence) of the species in survey units is partly determined by other environmental variables (not considered as habitat per se), then identifying the important habitat characteristics is made more difficult but not impossible or illegitimate. These other factors could be abundance of a competitor or predator, food supply, and even recent weather events. Indeed, if data are available, then the effect of these other factors on species abundance can be incorporated into the habitat analysis as covariates (Chapter 9).

Concern over the effect of background density is warranted when a study has surveyed for the species at locations or sites spanning a substantial spatial extent (e.g., survey locations or units that are hundreds of kilometers apart). In this case, background density might vary among the study sites and, at this large scale, the heterogeneity in density is not due to local effects of intraspecific aggregation. Rather it is due to larger-scale macroecological processes that cause population size to vary in different parts of a species range and in particular decrease from the center to the periphery of the range (Hengeveld and Haeck 1982; Lawton 1993; Brown 1995; Gaston and Blackburn 2000). Further, within different parts of the range, a species might exist as different ecotypes and thus occupy different types of habitat (Murphy and Lovett-Doust 2007). Ecotypes can complicate the study of species–habitat relationships and yet ecologists need to be more aware of and take advantage of their existence in studying habitat. Morrison (2012) discusses this and related issues in depth. My main point is that researchers need to be cognizant of heterogeneity in background density when data are collected *over any spatial extent*. However, this heterogeneity does not preclude a habitat analysis. Further, such heterogeneity as caused by non-habitat factors can be accounted for in some types of statistical analysis.

7.8 Normality of the response and predictor variables

Many types of statistical analyses require that the data derive theoretically from some type of sampling distribution with known properties such as the mean and standard deviation that define or characterize the distribution. This is particularly true for the response variable. These properties are *parameters* of the distribution and hence this very broad class of statistical analysis has traditionally been referred to as "parametric statistics." Examples are ANOVA and least-squares regression. Alternative to this, we have "non-parametric" statistics, which can be truly non-parametric and non-distributional, such as all the statistical analyses based upon the analysis of ranks rather than the raw data (e.g., Mann–Whitney U-test, Kruskal–Wallis ANOVA, Spearman rank-correlation test). We also have a class loosely composed of randomization tests (sometimes referred to as Monte-Carlo tests) that are also non-parametric in that they do not assume or require any particular theoretical sampling distribution but still nonetheless utilize a hypothetical (randomized) distribution for statistical testing (Veech 2012). With regard to parametric statistics, many tests require that the response (dependent) variable and predictor (independent) variables are *normally* distributed, although parametric statistics can also be based on other theoretical sampling distributions. As many readers will already know, the

normal distribution is the classic bell-shaped curve (Fig. 7.1). Indeed, the perfect or ideal form of the normal distribution has the same numerical value for its mean, median, and mode. Further, 68.2 percent of observations in the distribution are within $\pm$ 1 SD unit of the mean, 95.4 percent are within $\pm$ 2 SD unit, and 99.8 percent are within $\pm$ 3 SD unit. The normal distribution is also sometimes called the Gaussian distribution (after Carl Friedrich Gauss, who discovered the distribution in 1809 and promoted its use in very early statistical analyses) particularly in the older statistical and scientific literature.

In a strict sense, parametric statistical tests do not so much require that the data (response and predictor variables) are normally distributed but rather that the error values or residuals have a normal distribution. In ANOVA and other parametric tests for differences in the mean response variable among two or more groups, the error (or unexplained variation) is the squared difference between an observation i and the mean of the group, $\varepsilon_i = (y_i - \bar{y})^2$. In least-squares regression, the residual is the difference between an observation i and the value predicted by the regression equation, $\varepsilon_i = y_i - \hat{y}$. If the response variable and independent variables are normally distributed, then it is more likely that the error terms and residuals are normally distributed and the test statistics (t, F, R^2) and corresponding P-values are valid (Neter et al. 1989; Sokal and Rohlf 1995).

Given the fundamental importance of the normal distribution to so many statistical analyses, an important first step in any habitat analysis (and good statistical practice in general) is to check whether response and predictor variables are normally distributed. There are many methods and metrics for assessing normality of a variable including construction of simple histograms followed by visual inspection for the classic bell-shaped distribution. A slightly more advanced visual technique is to create a q-q plot and visually examine whether the points of the plot fall on the line of unity (slope = 1). This technique is demonstrated on a hypothetical dataset in Section 8.1. There are many other ways of verifying the normality of variables, including metrics such as kurtosis and skewness that describe the "shape" of the distribution. Also, the

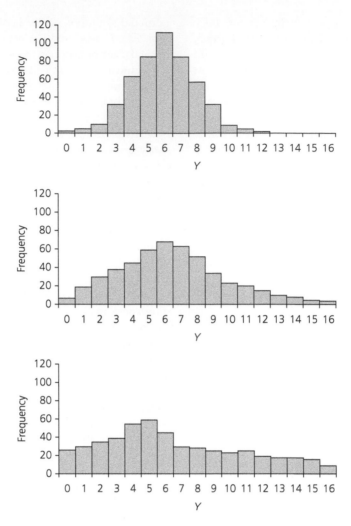

Figure 7.1 Classic bell-shaped normal distribution (top panel). The shape of a null distribution is partly determined by the mean and standard deviation. Distribution in top panel has mean = 6.0 and sᴅ = 2.0, middle panel mean = 6.5 and sᴅ = 3.3, bottom panel mean = 6.5 and sᴅ = 4.2. As the standard deviation increases, the distribution flattens out and begins to approach a uniform distribution. In the context of a habitat analysis, the response variable Y could represent species abundance recorded from surveyed plots.

Kolomogorov–Smirnov (K-S) test can be used to see if the observed data depart significantly from a normal distribution. The K-S test is a very flexible and useful technique for determining whether the observed distribution of a variable is different from any of various theoretical distributions. Normality of variables is such a common requirement for a wide variety of statistical tests that nearly every statistical textbook has a thorough and detailed discussion of techniques for assessing normality and transforming variables to achieve normality. For

further information the reader should consult their favorite statistics textbook or manual, I will not repeat any of that detail here. When normality cannot be satisfied for a given response variable, then the researcher may either choose to use nonparametric statistics in subsequent testing or the variable might be made to have a normal distribution through some sort of data transformation. For example, a log transformation (of any base) is often useful when the variable is non-normal due to some observations having a really large magnitude compared with the others. Consider that $\log_{10}(1) = 1$, $\log_{10}(100) = 2$, and $\log_{10}(1{,}000) = 3$, and so on; clearly a log transformation reduces the value of really large numbers and brings them in line with the majority of the other observations. In principle, a log transformation could also be used to bring normality to a variable in which there are a few observations of relatively small magnitude and a majority having larger magnitudes. There are other transformations intended for particular types of variables. For example, the arcsine transformation is useful for percentage and proportion data in that they are constrained to be between 0 and 100 or 0 and 1, respectively, and hence are often non-normal. Sokal and Rohlf (1995) have a good discussion of this and other transformations, as do other statistical textbooks. Again, the issue of normality is important enough that most statistical textbooks have a thorough discussion of its assessment as well as correction from non-normality to normality when needed. Also, statisticians differ in their opinions as to how critical normality is to statistical testing, particularly when sample sizes become really large (e.g., $N > 500$). Therefore, sometimes normality might best be thought of as a very useful property of data (independent and dependent variables) rather than a strict requirement.

7.9 Multicollinearity among predictor variables

Multicollinearity is the condition in which two or more predictor variables are correlated with one another. That is, the particular value of one is not completely statistically independent of the other. Many types of statistical analysis require that predictor variables are not excessively correlated in order for estimates of standard errors and significance testing to be valid. However, for many measured environmental variables, some amount of correlation (positive or negative) is to be expected and is unavoidable. For example, the amount of understory vegetation may be negatively correlated with tree canopy cover or density if a thick canopy hinders light penetration and prevents plant growth on the forest floor. The volume of cavities available for nesting may be positively correlated with tree height and trunk diameter (Joy 2000); even the

height of the cavity opening may be positively correlated with trunk diameter (Pakkala et al. 2018). In some types of ecological studies and analyses, multicollinearity might be the outcome of poor experimental design and protocol for measuring variables. However, with regard to a habitat analysis, multicollinearity among the measured environmental variables likely represents a useful reality. Although researchers will typically measure several or many potential habitat variables as though they operate independently of one another in affecting a response variable such as species presence/absence, abundance, or activity time, this empirical/statistical perspective on the data does not necessarily reflect an ecological reality. The habitat requirements of any species can be very complex, multi-faceted, temporally fluid, and dependent on spatial scale (Chapter 6). As such, a species may have ecologically meaningful associations with many different habitat variables, many of which are correlated with one another to some extent. The habitat of a species, as a real entity in nature, is a holistic integrated complex of environmental factors and conditions; it should not be defined solely as a list of environmental variables that a species associates with. More on this in Chapter 12.

As with many statistical issues, there is no *universal* rule on how much correlation is too much. The Pearson correlation coefficient (r) is perhaps the most straightforward way to assess the amount of correlation between two variables. Figure 7.2 presents my intuition on the amount of correlation between predictor variables that would be tolerable for a habitat analysis and the amount for which a researcher would need to take corrective action before proceeding with the statistical analyses. The cutoffs in Fig. 7.2 are in general accord with the categories of weak, moderate, and strong correlation recognized by other authors (Cohen 1988; Taylor 1990; Evans 1996; Dormann et al. 2013; Vatcheva et al. 2016).

The Pearson correlation coefficient measures the correlation between only two variables. The variance inflation factor (VIF) can be used to check for multicollinearity among the set of all P predictor variables. For a given variable, it is calculated as $\text{VIF} = 1/(1 - R^2)$ where R^2 is the multiple R^2 value from a regression of the focal variable (temporarily

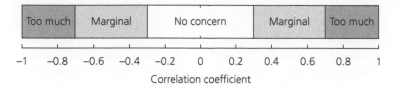

Figure 7.2 Range of Pearson's correlation coefficients (r) that indicate either too much correlation between two predictor variables, a marginal amount of correlation, and an amount that is minimal enough not to warrant any concern. When $-0.7 < r < -0.3$ or $0.3 < r < 0.7$, the two variables may be too highly correlated for some statistical analyses; the decision on whether to take corrective action or not is up to the researcher.

being treated as the response variable) against all the other $P - 1$ variables. VIF measures the extent to which the variance of the focal variable is inflated due to being correlated with the other variables (Neter et al. 1989). If there is no multicollinearity, then VIF = 1 for all variables, otherwise it is technically unlimited, although a practical limit is about 30–50 when $R^2 \approx 0.97$–0.98. An inflated variance for any predictor variable is problematic for estimating the regression coefficient and testing its statistical significance in a multiple regression model (see Box 8.1). Over the years, many authors have suggested VIF = 10 as a tolerable limit (to alleviate concern about multicollinearity) (Neter et al. 1989; Chatterjee et al. 2000; Hair et al. 2014), although there is no particular mathematical or statistical justification for this cutoff value or any other (Farrar and Glauber 1967; O'Brien 2007; Vatcheva et al. 2016).

If one decides that two or more variables are excessively correlated (for the particular analysis to be conducted) then correcting or adjusting for multicollinearity can be accomplished in several ways. Perhaps the easiest option is simply to exclude one or more of the variables that are correlated with the others, particularly if the correlation is due to the variables essentially measuring the same habitat characteristic or if one variable is a mathematical function of the other. For example, in their study of the habitat requirements of the violet click beetle (*Limoniscus violaceus*), Gouix et al. (2015) measured several variables including tree circumference at 30 and 130 cm above ground, and internal width, depth, and height of tree hollows. The beetle requires tree hollows for larval development. Obviously, circumference measurements of a tree are correlated at all heights and in particular the correlation is likely strong when the measurements are only 1 m apart. The two circumference values

for a given tree essentially measured the same habitat characteristic—tree size. The internal width, depth, and height of an internal tree hollow might be correlated to some extent and all are likely constrained by tree circumference. Note also that the volume of the hollow (as a variable) is a mathematical function of width, depth, and height. Gouix et al. (2015) checked for correlations among the variables and were careful not to include highly correlated variables (e.g., the two circumference measurements) in the same multiple regression models. Thus, excluding variables need not mean omitting one or more of the highly correlated variables from the *entire* study but rather not using such variables together in the same statistical model.

Another possibility is to combine highly correlated variables into a single composite variable through a variable reduction technique such as principal components analysis (PCA). Briefly, for each variable, PCA derives factor loadings ("weighting parameters") that are then multiplied by the observed values for the variables and then those products are summed to get the new variable, Z. For example, assume three environmental variables have been measured within some number of survey plots or locations. Further, Pearson correlation coefficients or the VIF reveals that those variables are highly correlated and so action must be taken to remedy this. Through PCA a new variable for each plot i can be calculated as $Z_i = x_1 y_{1i} + x_2 y_{2i} + x_3 y_{3i}$, where x_1, x_2, and x_3 are the factor loadings for variables 1, 2, and 3, and y_{1i}, y_{2i}, and y_{3i} are the observed (measured) values of variables 1, 2, and 3 for plot i. Section 9.5.1 explains PCA in more mathematical detail. PCA can also be used to combine variables into a single composite variable even when the variables are not highly correlated. This often makes sense when the variables represent the same basic habitat characteristic. For example, although Gouix et al. (2015) did not use PCA, their three measurements of hollow size along with height of the hollow entrance from the ground and stage of wood decay around the hollow are five variables that could be combined through PCA to create a new variable that is essentially a habitat characteristic representing the availability and accessibility of a resource important to larval development. Section 8.4.5 further illustrates the use of PCA in creating a new variable for subsequent analysis.

The correlation between two or more predictor variables may not be readily apparent or obvious prior to obtaining the data. As an example, again visualize a sampling plot of some size. The number of shrubs and nearest-neighbor distances (NND) of shrubs has been measured within each plot, thus the researcher has a variable representing shrub density (dens) and one representing shrub spacing. However, for a finite and constant plot size, density and spacing will almost always be correlated to some extent. This correlation could be as straightforward and simple as $NND = 1/(2 \times \sqrt{dens})$ (Clark and Evans 1954; Krebs 1999). Even if there is not a strong mathematical link, we would expect that NND would decrease with increasing density. Thus, in a statistical analysis of habitat, it is typically not appropriate to include both density (of some type of count object) and NND as predictor variables even though at a superficial glance they might seem to measure different characteristics of the habitat.

Another common form of multicollinearity among predictor variables occurs in some landscape-level analyses of habitat. This occurs when buffers of different sizes are centered on the survey unit or more typically the point location where a species was recorded as absent or present. These buffers might be circular areas of 100, 500, and 1,000 m radii extents in which GIS-based data processing of remote-sensing images (or data layers) allows one to measure an environmental variable (e.g., percentage forest cover) at each spatial extent. Clearly, measurements of the variable at different spatial extents are correlated given that smaller buffers are nested within larger ones, although the extent of the correlation can decline substantially as the smallest buffer becomes an increasingly diminished proportion of the largest buffer (e.g., the area of the smallest buffer composes only 1 percent of a buffer that has a radius that is 10× the radius of the smallest buffer). In most circumstances, it is best to avoid conducting statistical analyses (i.e., regression models) that simultaneously include variables derived from nested buffers. Graham (2003) and Dormann et al. (2013) discuss the general problem and consequences of multicollinearity and remedies for it, including ones that I have not presented.

7.10 Spatial and temporal autocorrelation

The effects of and remedies for spatial and temporal autocorrelation in data is a huge field of statistical research that has attracted the attention of ecologists in the past two decades (Legendre 1993; Koenig 1999; Legendre et al. 2002; Lichstein et al. 2002; Rangel et al. 2006; Dormann et al. 2007; Diniz-Filho et al. 2008; Kissling and Carl 2008; Gaspard et al. 2019). For spatial autocorrelation, this problem essentially is a manifestation of Tobler's First Law of Geography, literally stated as "everything is related to everything else, but near things are more related than distant things" (Tobler 1970, p. 236; Sui 2004, p. 269). The consequence of this is that the values of the environmental variables and species response variable measured at a location (or within a plot) are more likely to be similar to the values measured at other nearby plots than to values from plots farther away. The similarity is simply due to spatial proximity. It leads to spatial autocorrelation of a variable. This is a problem for most types of statistical analysis that assume observations are independent, particularly for response variables. An excess amount of spatial autocorrelation causes the residual (or error) variance to be distributed in a non-normal way. This compromises significance testing; that is, *P*-values may not be trustworthy if they are based on theoretical sampling distributions that assume normality (Section 7.8).

There are many different ways of avoiding or lessening the unwanted effects of spatial autocorrelation in a study of a species' habitat associations. First, during the planning stage and assuming that the researcher has control over the placement of the spatial units to be surveyed, the units could be spaced far enough apart to lessen spatial autocorrelation (of the predictor variables) to a sufficiently low level. This might require that the researcher has some prior knowledge about spatial gradients or patterns in the environmental variables that will be measured. This information could be used to set a minimum distance between survey units and still allow for a more or less random dispersion of the units. Of course this does not account for spatial autocorrelation in the species response variable. Such autocorrelation may be difficult to plan for prior to conducting the study because presumably the locations or plots have yet to be surveyed. However, knowledge about the species' behavior (particularly movement, Section 6.4) and basic ecology might help to decide a minimum separation distance with regard to the units being spatially independent. For example, if two or more units are so close together that the same individuals may be shared among the units (when individuals move back and forth), then those units are not as independent as they would be if they shared no individuals. Strict spatial independence of this sort may not always be required depending on how the statistical analysis is applied and the inferences that are made from the results.

Sometimes spatial autocorrelation is not discovered or suspected until after the environmental and species data are collected. When this is the case, one option is simply to randomly delete observations (survey locations or plots) that are too near to another observation. Such spatial thinning of the data can substantially lessen spatial autocorrelation but it might require somewhat subjective decisions about the minimum allowable separation distance and it obviously leads to a decrease in sample size. This type of data filtering is sometimes used when a habitat analysis is based upon previously collected species data such as that from citizen-science or monitoring programs (Section 7.6) (e.g., Muller et al. 2018) and data filtering in general is recommended for citizen-science data (Steen et al. 2019; Young et al. 2019; Robinson et al. 2020). In those studies, a researcher might decide a priori that locations of species occurrence must be x km apart from the next closest location in order to be retained for the habitat analysis. Similarly, the researcher can stipulate minimum separation distance for the randomly dispersed absence (or availability) locations.

Instead of filtering the data and omitting some observations prior to the analysis, the spatial autocorrelation in data could be allowed to remain and then addressed during the data analysis phase. There are many statistical methods for diagnosing, estimating, and controlling for spatial autocorrelation (Keitt et al. 2002; Diniz-Filho et al. 2003; Dormann et al. 2007; Bivand et al. 2008; Thayn and Simanis 2013). One of the most direct methods is autoregressive modeling. These are multiple regression models (see Sections 9.2 and 9.3) with a covariate added to estimate the effect of the response variable (e.g., species presence/absence or abundance) recorded at all locations $k = 1$ to N (except i) on the response variable at location i for each $i = k$ (Augustin et al. 1996; Osborne et al. 2001; Lichstein et al. 2002; Knapp et al. 2003; Kissling and Carl 2008; Crase et al. 2012; Bardos et al. 2015; Ver Hoef et al. 2018). Many methods for analyzing spatial autocorrelation take into account straight-line (Euclidean) distances among the locations where the data were collected; this is often accomplished in the form of a weighting term added to a statistical model (Dormann et al. 2007; Kissling and Carl 2008; Bardos et al. 2015). The effect of spatial autocorrelation can also be estimated and hence accounted for in occupancy models (e.g., Moore and Swihart 2005; Mohamed et al. 2013; Lee and Carroll 2014; Webb et al. 2014; Stanton et al. 2015; Gustafson and Newman 2016; Chen and Ficetola 2019; Wang et al. 2019). Occupancy modeling has become a widely used technique for species–habitat analysis (Section 9.6) and it is made even more useful by having the capability of modeling spatial autocorrelation.

Even apart from conducting a habitat analysis, there are various R code packages (e.g., Goslee and Urban 2007; Bivand et al. 2008; Broms et al. 2014; Vilela and Villalobos 2015; Carl et al. 2018) and stand-alone software (Fischer and Getis 2010; Rangel et al. 2010) for statistically analyzing data that are spatially autocorrelated. Likely, most of these techniques can accommodate analyses intended to quantify and reveal habitat associations. Detailed explanation of the statistical methods for handling spatial autocorrelation is beyond the scope of this book. Nonetheless the reader is encouraged to consult any of the references listed in this section to become familiar with the methods. In particular, Keitt et al. (2002) is an excellent discussion of spatial autocorrelation in an ecological and statistical context, and in reference to many different ecological patterns and processes, not just species–habitat associations.

It is also worth mentioning that modeling the effect of geographical placement (e.g., latitude and longitude of survey units as predictor variables) is not the same as examining spatial autocorrelation. The former examines how the spatially explicit geographic location of a survey unit affects the species response variable, but this does not necessarily control for spatial autocorrelation. If latitude or longitude has a significant effect on the response variable then the response variable (and perhaps other predictor variables) likely has some amount of spatial autocorrelation although such autocorrelation is not estimated or controlled for by simply having latitude and longitude as predictors in the statistical analysis (model). Thus, it is appropriate and often worthwhile to include latitude, longitude, or some other spatial reference in a statistical model to examine the effect of geography, but the model must also control for spatial autocorrelation (e.g., Cummings and Veech 2014).

Temporal autocorrelation is likely not as much of an issue for habitat analyses. With regard to the data used in an analysis of habitat this form of autocorrelation can primarily occur in two ways. Locations or plots are surveyed on more than one occasion and the surveys at a given location occur in relatively quick succession (i.e., days or weeks apart). The environmental variables (presumed habitat characteristics) likely would not have changed from one survey to the next and indeed constancy of habitat conditions actually might be necessary for some analyses unless the goal is to examine how the species responds to a changing habitat. The response variable (species presence/absence or abundance) is likely to be temporally autocorrelated at a location that is repeatedly surveyed. However, such autocorrelated structure in the data is actually used directly by some types of analyses designed to estimate a species detection probability (e.g., occupancy modeling, see Section 9.6). The real problem with temporal autocorrelation occurs when repeated observations at the same survey location are mistakenly taken to represent independent samples and hence are not distinguished from spatially

distinct samples. This is a conflation of temporal sampling with spatial sampling and essentially is a form of pseudo-replication (Hurlbert 1984). Pseudo-replication inflates the actual sample size and hence leads to faulty significance testing; that is, P-values that are not trustworthy.

Temporal autocorrelation can also occur when a habitat analysis is based on movement data obtained by GPS- or radio-telemetry tracking of individuals. As described previously (Section 7.6), habitat sampling plots are established a posteriori at locations where the individual has "stopped" or appeared to have selected a type of habitat. Temporal autocorrelation can occur in such data (for the same individual) when the stop locations are not separated by much time and hence are not temporally independent. Note that this type of study design can also involve spatial autocorrelation if the stop locations are not sufficiently far apart. Imagine the extreme scenario where two stop locations for a relatively vagile organism are only a day apart and 500 m distant from one another—are these two observations truly independent? The individual may not have even left the same basic habitat type during the 1-day intervening period, particularly if the habitat type is relatively homogeneous, contiguous, and unfragmented. As with spatial autocorrelation, there are statistical methods for controlling temporal autocorrelation.

More generally, there is an entire field of statistics that involves the analysis of time-series data. One classic and simple example is the repeated-measures ANOVA. In such analyses, it is assumed that response variables are temporally autocorrelated. In ecology, any analysis of change in population size over time requires, by necessity, the analysis of time-series data. Such analyses do not discount or remove temporal dependence in the data, rather they use it to uncover meaningful ecological pattern.

7.11 Standardization of predictor variables

Prior to conducting a habitat analysis, it is often a good idea to standardize the predictor variables; that is, the variables in the set [X] (Chapter 5). Given that these variables can represent a wide variety of different environmental factors and characteristics (Chapter 5), with different units of measurement and different variances, standardization is often essential to allow comparison of their effects. To standardize any variable, subtract each observation x_i from the mean and then divide by the standard deviation: $x_{i,std} = (x_i - \bar{x}) / \text{sd}(x)$. This produces a standardized distribution of x that has mean = 0 and SD = 1, with most values of $x_{i,std}$ between −3 and 3 if the initial distribution of x is normal. Standardized values of a variable are sometimes referred to as z-scores, particularly if they derive from a normal distribution. In that case, the distribution itself is referred to as the standard normal distribution. For a given set of variables [X], standardization puts each one on the same scale and thus allows for direct comparison. For example, consider a set of 1,000 m² survey plots among which mean grass height is 30 cm and the standard deviation is 12 cm. One particular plot has grass that is 60 cm high and thus the standardized grass height for the plot is $x_{i,std} = 2.5$. The same plot has four shrubs and among all plots mean number of shrubs is 4.1 and the standard deviation is 2.8. Thus for shrub density in this plot, $x_{i,std} = -0.036$. The plot has a grass height that is substantially above average whereas shrub density is almost exactly average. A different plot has 11 shrubs and hence $x_{i,std} = 2.46$. The standardized value for shrub density in this plot is approximately the same as the standardized value for grass height in the first plot. Imagine that we were interested in identifying the important habitat characteristics for a rare grasshopper species such that we also recorded grasshopper abundance in each plot. By standardizing grass height and shrub density among all the plots, we can compare their effects on grasshopper abundance in a subsequent statistical analysis such as multiple linear regression (Section 9.3).

The strict statistical definition of standardization is to convert a variable to have mean = 0 and SD = 1. However, this does not necessarily entail that a distribution of x that is initially non-normal will become normal. Standardization can help in making an initial distribution a bit more normal if the variable is already normally distributed. Standardization does not lead to a normal distribution if the initial distribution is multi-modal, highly skewed, and/or consists of many observations of the same numerical value. This is important because normality of the response variable is a requirement of many statistical analyses (Chapter 9), but standardization is often not the best data transformation to achieve that normality.

7.12 Number of predictor variables to include in the analysis

The number of environmental predictor variables to include in a habitat analysis depends on the goals of the study (see Chapter 4) and the purpose(s) for doing the analysis. As discussed in Chapter 5, there is a great variety of environmental variables over a wide range of spatial scales that can potentially be defining characteristics of a species habitat. The habitat analysis might be intended to obtain a complete and detailed description of habitat for the focal species in which case many variables should be examined. Alternatively, the analysis might be designed to identify a few most important variables that best define the habitat of the species. Also, the extent of our current knowledge of the habitat requirements of the focal species could help determine which environmental variables and how many are examined. If a given environmental factor is known to be an important characteristic of the habitat of the focal species then there may be no need to examine it, unless one wanted to compare other less studied variables with it. Thus a habitat analysis could include between three or four variables up to several dozen. Another important consideration is the number of observations (sample size) relative to the number of predictor variables. For many types of statistical analysis (e.g., multiple regression), the overall sample size sets a limit on the number of predictor variables that should (legitimately) be tested. There is no concrete statistical rule on this issue although general guidance established over the years by statisticians is that the ratio of sample size to number of predictors should be at least 5 and preferably >8. Although this may seem overly constraining, even a relatively small sample size of 50 survey plots or locations still allows the researcher to examine 6–10 predictor variables. Of course, multicollinearity among at least some of the predictor variables becomes more likely as more variables are included in the analysis, particularly if some of the variables are functionally redundant to one another (i.e., measure the same basic habitat characteristic). The number of predictor variables also is relevant to model fitting (and over-fitting) and the practical ease of comparing models (Section 10.1). Essentially there is no precise rule of thumb for the number of variables to include in a habitat analysis. However, the researcher should always practice intellectual honesty and scientific rigor by not testing variables that are known a priori definitely not to define the habitat of the species and variables that are an obvious component of the species' habitat. The interested reader should consult Anderson and Gutzwiller (2005) or pages 151–81 of Morrison et al. (2006) for a more detailed discussion of the environmental factors to consider when conducting a study of a species' habitat requirements.

7.13 Home range analysis

Ecologists, and particularly wildlife ecologists, have been interested in the home range as a concept and a measurable property of an individual(s) and species for a long time (Klugh 1927; Murie and Murie 1931; Hamilton 1937; Blair 1942; Haugen 1942; Burt 1943; Mohr 1947). Most readers are likely already familiar with it and perhaps it does not need to be defined. However, to be clear, I will define it as the area that an individual animal typically moves about in during its normal daily/nightly activities—this is basically the definition used by Morrison et al. (2006) and Fryxell et al. (2014) among others. The exact definition is not so critical; however, recognize that an individual need not move throughout its entire home range in a given day (or any other relatively short time period) and the size of its home range might change seasonally. Presumably, by definition, an individual uses all parts of its home range even if briefly or sporadically. Home range does not (should not) include movements that are dispersal events. Also, the home range is an expressed property of an individual that is partly due to the traits of the individual, but it is also by extrapolation a characteristic of the species. Thus, the size of a home range varies

somewhat among individuals, even those in the same population, and of course the sizes of typical home ranges vary among species (Mohr 1947; McNab 1963; Harestad and Bunnell 1979; Kelt and Van Vuren 2001; Ofstad et al. 2016). Home range and territory are not perfect synonyms as the latter entails active defense of the area against intrusion by other individuals, particularly conspecifics. However, this distinction is not important to the discussion that follows.

The home range of an individual can be spatially delineated and area subsequently estimated in a variety of ways (Mohr 1947; Hayne 1949; Schoener 1981; Anderson 1982; Worton 1989; Seaman and Powell 1996; Otis and White 1999; Getz and Wilmers 2004; Calenge 2006; Moorcroft and Lewis 2006). To do so requires that the researcher track the movement of the individual and obtain a sufficient number of spatially referenced locations (Fig. 7.3). This

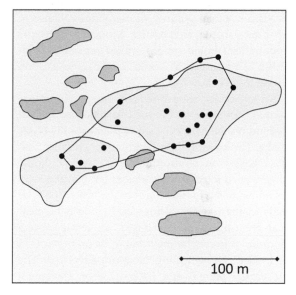

Figure 7.3 Home range of a hypothetical species inhabiting a landscape where there are three habitat types, grassland (yellow), forest type 1 (light green), and forest type 2 (dark green). The researcher used radio-telemetry or some other technique to record the locations (black dots) of an individual of the species tracked over some period of time. In this example, the home range is defined and delineated as the minimum convex polygon for the set of point locations. Note that the area encompassed by the home range is mostly composed of forest type 1, and 18 of the 21 recorded locations for the individual were in that forest type. Compositional Data Analysis (Section 11.3) could be used to determine if the proportions of forest types 1, 2, and grassland are different within the home range.

can be achieved through radio-telemetry, GPS tracking, and sometimes simple direct observation. For brevity I will not discuss all the various methods of acquiring movement data, its limitations and caveats (Swihart and Slade 1985; Börger et al. 2006; Laver and Kelly 2008; Kie et al. 2010; Fieberg and Börger 2012), and the statistical techniques for estimating home ranges (above references). For the purposes of this book, it is important to recognize that the area circumscribed by the home range provides a sampling frame for environmental data that can be used to conduct a habitat analysis (Fig. 7.3). In this case, each individual and its home range is an observation wherein the response variable is essentially presence of the individual and the predictor variables derive from the environmental characteristics of the home range. If one were to also have environmental data on hypothetical home ranges (i.e., areas not occupied or used by individuals of the focal species) taken as representing species absence then some of the methods explained in Chapter 9 could be applied. More precisely, the hypothetical home ranges would represent *availability* not necessarily *absence* of the species upon surveying (Section 7.4).

An important issue here is that the environmental data and the subsequent analysis of habitat depend on the method used to delineate the home range. Thus, researchers conducting a habitat analysis based on environmental data from home ranges need to carefully consider the method used to delineate the home ranges of the individual animals in their study. Moreover, it should be clear that this study design based on home ranges is quite different from what I have previously described. No sampling plots are surveyed. The individual animal and its pattern of movement define the "plots" or home ranges. This is roughly similar to the design where habitat sampling plots are established a posteriori at locations where a moving animal has stopped (Section 7.6). But the design based on home range delineation goes even further in that the entire area that an individual traverses through is used in obtaining the environmental data. There is a particular type of statistical analysis, Compositional Data Analysis, that is generally appropriate for this type of design (Section 11.3).

7.14 Difference between measuring habitat preference/selection versus habitat associations

Habitat selection occurs when an individual organism selects, settles into, and subsequently uses a particular type of habitat (defined based on one or a few environmental factors or more holistically as a comprehensive set of factors) at a rate above that of its general availability on the landscape or seascape. In Chapter 3, I discussed habitat selection extensively in the context of the Dispersal–Settlement–Establishment model and the habitat-cue hypothesis of species distribution. So, I will not repeat any of that here. The main point of this section is to emphasize that the experimental designs for measuring habitat preference and selection are different from the basic design for conducting a habitat analysis, and the two should not be confused. Indeed, in Section 4.4 I discussed how an analysis of a species' habitat associations can be a useful precursor to a study of habitat preference and selection. A habitat analysis (as portrayed in this book) is about discovering pattern, whereas studies of habitat preference and selection are about examining process.

A habitat analysis is essentially about identifying a correlation between the species response variable Y and a set of environmental variables [X] (Chapter 5). It is a search for habitat associations, not a cause-and-effect relationship. Studies of habitat preference and selection must go further than this. They typically must demonstrate experimentally that discrete alternative options (i.e., different habitat types or characteristics) are simultaneously available to an individual organism that then "decides" or chooses. Even better, the experiment should be designed with experimental trials repeated over and over again to determine if individuals exhibit a consistent habitat preference. In this way, the researcher can establish that the decision (or choice) is not random. A great example is the classic experiment by Wecker (1964) demonstrating habitat selection by prairie and woodland subspecies of deer mice (*Peromyscus maniculatus bairdi* and *Peromyscus maniculatus gracilis*, respectively).

When interpreting and communicating results of a habitat analysis, authors should use words carefully (Chapter 1) and not misrepresent their study as demonstrating habitat preference or selection. Habitat associations are often due to preference and selection, but they alone, no matter how strong, are not direct evidence of preference and selection. In my opinion, it is permissible to refer to *habitat use* and *habitat requirements* when discussing the results of a habitat analysis. The former term simply implies that an individual occurring in a given habitat is doing more than just residing there, and the latter term reasonably assumes that the habitat associations of a species (individual) are also requirements for its survival, growth, and reproduction. The actual measurement and statistical analysis of habitat preference and selection (given the correct study design) has some similarity to some of the methods used to analyze habitat associations. Hence, I discuss selectivity and preference indices in Section 11.2.

References

Abba, A.M., Zufiaurre, E., Codesido, M., and Bilenca, D.N. (2016). Habitat use by armadillos in agroecosystems of central Argentina: does plot identity matter? *Journal of Mammalogy*, 97, 1265–71.

Anderson, D.J. (1982). The home range: a new nonparametric estimation technique. *Ecology*, 63, 103–12.

Anderson, S.H. and Gutzwiller, K.J. (2005). Wildlife habitat evaluation. Pages 489–502 in *Techniques for Wildlife Investigations and Management*. Sixth edition. Braun, C.E. (editor). Wildlife Society, Bethesda, Md.

Augustin, N.H., Mugglestone, M.A., and Buckland, S.T. (1996). An autologistic model for the spatial distribution of wildlife. *Journal of Applied Ecology*, 33, 339–47.

Bardos, D.C., Guillera-Arroita, G., and Wintle, B.A. (2015). Valid auto-models for spatially autocorrelated occupancy and abundance data. *Methods in Ecology and Evolution*, 6, 1137–49.

Bishir, S.C., Hossack, B.R., Fishback, L., and Davenport, J.M. (2018). Post-breeding movement and habitat use by wood frogs along an Arctic-Subarctic ecotone. *Arctic, Antarctic, and Alpine Research*, 50, e1487657.

Bivand, R.S., Pebesma, E.J., and Gómez-Rubio, V. (2008). *Applied Spatial Data Analysis with R*. Springer, New York.

Blakesley, J.A., Noon, B.R., and Anderson, D.R. (2005). Site occupancy, apparent survival, and reproduction of California spotted owls in relation to forest stand characteristics. *Journal of Wildlife Management*, 69, 1554–64.

Blair, W.F. (1942). Size of home range and notes on the life history of the woodland deer-mouse and eastern chipmunk in northern Michigan. *Journal of Mammalogy*, 23, 27–36.

Börger, L., Franconi, N., De Michele, G., et al. (2006). Effects of sampling regime on the mean and variance of home range size estimates. *Journal of Animal Ecology*, 75, 1393–1405.

Bowler, D.E. and Benton, T.G. (2005). Causes and consequences of animal dispersal strategies: relating individual behaviour to spatial dynamics. *Biological Reviews*, 80, 205–25.

Broms, K.M., Johnson, D.S., Altwegg, R., and Conquest, L.L. (2014). Spatial occupancy models applied to atlas data show southern ground hornbills strongly depend on protected areas. *Ecological Applications*, 24, 363–74.

Brown, J.H. (1995). *Macroecology*. University of Chicago Press, Chicago, Ill.

Burt, W.H. (1943). Territoriality and home range concepts as applied to mammals. *Journal of Mammalogy*, 24, 346–52.

Calenge, C. (2006). The package "adehabitat" for the R software: a tool for the analysis of space and habitat use by animals. *Ecological Modelling*, 197, 516–19.

Carl, G., Levin, S.C., and Kühn, I. (2018). spind: an R package to account for spatial autocorrelation in the analysis of lattice data. *Biodiversity Data Journal*, 6, e20760.

Chatterjee, S., Hadi, A.S., and Price, B. (2000). *Regression Analysis by Example*. John Wiley and Sons, New York.

Chen, W. and Ficetola, G.F. (2019). Conditionally autoregressive models improve occupancy analyses of autocorrelated data: an example with environmental DNA. *Molecular Ecology Resources*, 19, 163–75.

Clark, P.J. and Evans, F.C. (1954). Distance to nearest neighbor as a measure of spatial relationships in populations. *Ecology*, 35, 445–53.

Cohen, L.H. (1988). Measurement of life events. Pages 11–30 in *Life Events and Psychological Functioning: Theoretical and Methodological Issues*. Cohen, L.H. (editor). Sage Publishing, Newbury Park, Calif.

Crase, B., Liedloff, A.C., and Wintle, B.A. (2012). A new method for dealing with residual spatial autocorrelation in species distribution models. *Ecography*, 35, 879–88.

Cummings, K. and Veech, J.A. (2014). Assessing the influence of geography, land cover and host species on the local abundance of a generalist brood parasite, the brown-headed cowbird. *Diversity and Distributions*, 20, 396–404.

Dias, P.C. (1996). Sources and sinks in population biology. *Trends in Ecology and Evolution*, 11, 326–30.

Diniz-Filho, J.A.F., Bini, L.M., and Hawkins, B.A. (2003). Spatial autocorrelation and red herrings in geographical ecology. *Global Ecology and Biogeography*, 12, 53–64.

Diniz-Filho, J.A.F., Rangel, T.F., and Bini, L.M. (2008). Model selection and information theory in geographical ecology. *Global Ecology and Biogeography*, 17, 479–88.

Dodge, K.L., Galuardi, B., Miller, T.J., and Lutcavage, M.E. (2014). Leatherback turtle movements, dive behavior, and habitat characteristics in ecoregions of the northwest Atlantic Ocean. *PLoS ONE*, 9, e91726.

Dormann, C.F., McPherson, J.M., Araújo, M.B., et al. (2007). Methods to account for spatial autocorrelation in the analysis of species distributional data: a review. *Ecography*, 30, 609–28.

Dormann, C.F., Elith, J., Bacher, S., et al. (2013). Collinearity: a review of methods to deal with it and a simulation study evaluating their performance. *Ecography*, 36, 27–46.

Evans, J.D. (1996). *Straightforward Statistics for the Behavioral Sciences*. Brooks and Cole Publishing, Pacific Grove, Calif.

Evans, B.E.I., Ashley, J., and Marsden, S.J. (2005). Abundance, habitat use, and movements of blue-winged macaws (*Primolius maracana*) and other parrots in and around an Atlantic Forest reserve. *Wilson Bulletin*, 117, 154–64.

Farrar, D.E. and Glauber, R.R. (1967). Multicollinearity in regression analysis: the problem revisited. *Review of Economics and Statistics*, 49, 92–107.

Fieberg, J. and Börger, L. (2012). Could you please phrase "home range" as a question? *Journal of Mammalogy*, 93, 890–02.

Fischer, M.M. and Getis, A. (2010). *Handbook of Applied Spatial Analysis*. Springer, New York.

Fryxell, J.M., Sinclair, A.R.E., and Caughley, G. (2014). *Wildlife Ecology, Conservation, Management*. Third edition. Wiley Blackwell, Oxford.

Gaspard, G., Kim, D., and Chun, Y. (2019). Residual spatial autocorrelation in macroecological and biogeographical modeling: a review. *Journal of Ecology and Environment*, 43, e19.

Gaston, K.J. and Blackburn, T.M. (2000). *Pattern and Process in Macroecology*. Wiley Blackwell, Oxford.

Getz, W.M. and Wilmers, C.C. (2004). A local nearest-neighbor convex-hull construction of home ranges and utilization distributions. *Ecography*, 27, 489–05.

Gjerdrum, C., Elphick, C.S., and Rubega, M.A. (2008). How well can we model numbers and productivity of saltmarsh sharp-tailed sparrows (*Ammodramus caudacutus*) using habitat features. *The Auk*, 125, 608–17.

Goslee, S.C. and Urban, D.L. (2007). The ecodist package for dissimilarity-based analysis of ecological data. *Journal of Statistical Software*, 22, e7.

Gouix, N., Sebek, P., Valladares, L., Brustel, H., and Brin, A. (2015). Habitat requirements of the violet click beetle (*Limoniscus violaceus*), an endangered umbrella species of basal hollow trees. *Insect Conservation and Diversity*, 8, 418–27.

Graham, M.H. (2003). Confronting multicollinearity in ecological multiple regression. *Ecology*, 84, 2809–15.

Gustafson, K.D. and Newman, R.A. (2016). Multiscale occupancy patterns of anurans in prairie wetlands. *Herpetologica*, 72, 293–302.

Hair, J.F., Black, W.C., Babin, B.J., and Anderson, R.E. (2014). *Multivariate Data Analysis*. Seventh edition. Pearson, New York.

Hamilton, W.J. (1937). Activity and home range of the field mouse, *Microtus pennsylvanicus pennsylvanicus*. *Ecology*, 18, 255–63.

Harestad, A.S. and Bunnell, F.L. (1979). Home range and body weight—a reevaluation. *Ecology*, 60, 389–402.

Haugen, A.O. (1942). Life history studies of the cottontail rabbit in southwestern Michigan. *American Midland Naturalist*, 28, 204–44.

Hayne, D.W. (1949). Calculation of size of home range. *Journal of Mammalogy*, 30, 1–18.

Heemeyer, J.L., Williams, P.J., and Lannoo, M.J. (2012). Obligate crayfish burrow use and core habitat requirements of crawfish frogs. *Journal of Wildlife Management*, 76, 1081–91.

Hengeveld, R. and Haeck, J. (1982). The distribution of abundance. I. measurements. *Journal of Biogeography*, 9, 303–16.

Hilje, B. and Aide, T.M. (2012). Calling activity of the common tink frog (*Diasporus diastema*) (Eleutherodactylidae) in secondary forests of the Caribbean of Costa Rica. *Tropical Conservation Science*, 5, 25–37.

Hurlbert, S.H. (1984). Pseudoreplication and the design of ecological field experiments. *Ecological Monographs*, 54, 187–11.

Humes, M.L., Hayes, J.P., and M.W. Collopy. (1999). Bat activity in thinned, unthinned, and old-growth forests in western Oregon. *Journal of Wildlife Management*, 63, 553–61.

Iwai, N., Yasumiba, K., and Akasaka, M. (2018). Calling-site preferences of three co-occurring endangered frog species on Amami-Oshima Island. *Herpetologica*, 74, 199–206.

Johnson, D.H. (2008). In defense of indices: the case of bird surveys. *Journal of Wildlife Management*, 72, 857–68.

Johnson, C.J., Nielsen, S.E., Merrill, E.H., McDonald, T.L., and Boyce, M.S. (2006). Resource selection functions based on use-availability data: theoretical motivation and evaluation methods. *Journal of Wildlife Management*, 70, 347–57.

Joy, J.B. (2000). Characteristics of nest cavities and nest trees of the red-breasted sapsucker in coastal montane forests. *Journal of Field Ornithology*, 71, 525–30.

Keating, K.A. and Cherry, S. (2004). Use and interpretation of logistic regression in habitat-selection studies. *Journal of Wildlife Management*, 68, 774–89.

Keitt, T.H., Bjørnstad, O.N., Dixon, P.M., and Citron-Pousty, S. (2002). Accounting for spatial pattern when modeling organism-environment interactions. *Ecography*, 25, 616–25.

Kelt, D.A. and Van Vuren, D.H. (2001). The ecology and macroecology of mammalian home range area. *American Naturalist*, 157, 637–45.

Kie, J.G., Matthiopoulos, J., Fieberg, J., et al. (2010). The home-range concept: are traditional estimators still relevant with modern telemetry technology? *Philosophical Transactions of the Royal Society B*, 365, 2221–31.

Kirkpatrick, C. and Conway, C.J. (2010). Importance of montane riparian forest and influence of wildfire on nest-site selection of ground-nesting birds. *Journal of Wildlife Management*, 74, 729–38.

Kissling, W.D. and Carl, G. (2008). Spatial autocorrelation and the selection of simultaneous autoregressive models. *Global Ecology and Biogeography*, 17, 59–71.

Klugh, A.B. (1927). Ecology of the red squirrel. *Journal of Mammalogy*, 8, 1–32.

Knapp, R.A., Matthews, K.R., Preisler, H.K., and Jellison, R. (2003). Developing probabilistic models to predict amphibian site occupancy in a patchy landscape. *Ecological Applications*, 13, 1069–82.

Koenig, W.D. (1999). Spatial autocorrelation of ecological phenomena. *Trends in Ecology and Evolution*, 14, 22–26.

Koshkina, A., Grigoryeva, I., Tokarsky, V., et al. (2020). Marmots from space: assessing population size and habitat use of a burrowing mammal using publicly available satellite images. *Remote Sensing in Ecology and Conservation*, 6, 153–67.

Krebs, C.J. (1999). *Ecological Methodology*. Second edition. Benjamin Cummings Publishing, Menlo Park, Calif.

Laver, P.N. and Kelly, M.J. (2008). A critical review of home range studies. *Journal of Wildlife Management*, 72, 290–98.

Law, B.S., Brassil, T., Gonsalves, L., Roe, P., Truskinger, A., and McConville, A. (2018). Passive acoustics and sound recognition provide new insights on status and resilience of an iconic endangered marsupial (koala, *Phascolarctos cinereus*) to timber harvesting. *PLoS ONE*, 13, e0205075.

Lawton, J.H. (1993). Range, population abundance and conservation. *Trends in Ecology and Evolution*, 8, 409–13.

Lee, M.-B. and Carroll, J.P. (2014). Relative importance of local and landscape variables on site occupancy by avian species in a pine forest, urban, and agriculture matrix. *Forest Ecology and Management*, 320, 161–70.

Legendre, P. (1993). Spatial autocorrelation: trouble or new paradigm? *Ecology*, 74, 1659–73.

Legendre, P., Dale, M.R.T., Fortin, M.-J., Gurevitch, J., Hohn, M., and Myers, D. (2002). The consequences of spatial structure for the design and analysis of ecological field surveys. *Ecography*, 25, 601–15.

Li, P. and Martin, T.E. (1991). Nest-site selection and nesting success of cavity-nesting birds in high elevation forest drainages. *The Auk*, 108, 405–18.

Lichstein, J.W., Simons, T.R., Shriner, S.A., and Franzreb, K.E. (2002). Spatial autocorrelation and autoregressive models in ecology. *Ecological Monographs*, 72, 445–63.

Liner, A.E., Smith, L.L., Golladay, S.W., Castleberry, S.B., and Gibbons, J.W. (2008). Amphibian distributions within three types of isolated wetlands in southwest Georgia. *American Midland Naturalist*, 160, 69–81.

Manly, B.F.J., McDonald, L.L., Thomas, D.L., McDonald, T.L., and Erickson, W.P. (2002). *Resource Selection by Animals, Statistical Design and Analysis for Field Studies*. Second edition. Kluwer Academic Publishers, Boston, Mass.

Matthiopoulos, J., Fieberg, J., Aarts, G., Beyer, H.L., Morales, J.M., and Haydon, D.T. (2015). Establishing the link between habitat selection and animal population dynamics. *Ecological Monographs*, 85, 413–36.

Matthysen, E. (2005). Density-dependent dispersal in birds and mammals. *Ecography*, 28, 403–16.

McNab, B.K. (1963). Bioenergetics and the determination of home range size. *American Naturalist*, 97, 133–40.

Mitchell, S.C. (2005). How useful is the concept of habitat?—a critique. *Oikos*, 110, 634–38.

Mohamed, A., Sollmann, R., Bernard, H., et al. (2013). Density and habitat use of the leopard cat (*Prionailurus bengalensis*) in three commercial forest reserves in Sabah, Malaysian Borneo. *Journal of Mammalogy*, 94, 82–89.

Mohr, C.O. (1947). Table of equivalent populations of North American small mammals. *American Midland Naturalist*, 37, 223–49.

Moorcroft, P.R. and Lewis, M.A. (2006). *Mechanistic Home Range Analysis*. Princeton University Press, Princeton, NJ.

Moore, J.E. and Swihart, R.K. (2005). Modeling patch occupancy by forest rodents: incorporating detectability and spatial autocorrelation with hierarchically structured data. *Journal of Wildlife Management*, 69, 933–49.

Morrison, M.L. (2012). The habitat sampling and analysis paradigm has limited value in animal conservation: a prequel. *Journal of Wildlife Management*, 76, 438–50.

Morrison, M.L., Marcot, B.G., and Mannan, R.W. (2006). *Wildlife-Habitat Relationships: Concepts and Applications*. Third edition. Island Press, Washington, DC.

Muller, J.A., Veech, J.A., and Kostecke, R.M. (2018). Landscape-scale habitat associations of Sprague's pipits wintering in the southern United States. *Journal of Field Ornithology*, 89, 326–36.

Murphy, H.T. and Lovett-Doust, J. (2007). Accounting for regional niche variation in habitat suitability models. *Oikos*, 116, 99–110.

Murie, O.J. and Murie, A. (1931). Travels of *Peromyscus*. *Journal of Mammalogy*, 12, 200–209.

Neter, J., Wasserman, W., and Kutner, M.H. (1989). *Applied Linear Regression Models*. Irwin Publishing, Homewood, Ill.

Nichols, J.D. (1992). Capture-recapture models. *Bioscience*, 42, 94–102.

O'Brien, R.M. (2007). A caution regarding rules of thumb for variance inflation factors. *Quality and Quantity*, 41, 673–90.

Ofstad, E.G., Herfindal, I., Solberg, E.J., and Sæther, B.-E. (2016). Home ranges, habitat and body mass: simple correlates of home range size in ungulates. *Proceedings of the Royal Society B*, 283, e20161234.

Osborne, P.E., Alonso, J.C., and Bryant, R.G. (2001). Modelling landscape-scale habitat use using GIS and remote sensing: a case study with great bustards. *Journal of Applied Ecology*, 38, 458–71.

Otis, D.L. and White, G.C. (1999). Autocorrelation of location estimates and the analysis of radiotracking data. *Journal of Wildlife Management*, 63, 1039–44.

Pakkala, T., Tiainen, J., Piha, M., and Kouki, J. (2018). Nest tree characteristics of the old-growth specialist three-toed woodpecker *Picoides tridactylus*. *Ornis Fennica*, 95, 89–102.

Plumb, R.E., Anderson, S.H., and Knopf, F.L. (2005). Habitat and nesting biology of mountain plovers in Wyoming. *Western North American Naturalist*, 65, 223–28.

Pollock, K.H. (1991). Modeling capture, recapture, and removal statistics for estimation of demographic parameters for fish and wildlife populations: past, present, and future. *Journal of the American Statistical Association*, 86, 225–38.

Pulliam, H.R. (1988). Sources, sinks and population regulation. *American Naturalist*, 132, 652–61.

Punjabi, G.A., Chellam, R., and Vanak, A.T. (2013). Importance of native grassland habitat for den-site selection of Indian foxes in a fragmented landscape. *PLoS ONE*, 8, e76410.

Rangel, T.F., Diniz-Filho, J.A., and Bini, L.M. (2006). Towards an integrated computational tool for spatial analysis in macroecology and biogeography. *Global Ecology and Biogeography*, 15, 321–27.

Rangel, T.F., Diniz-Filho, J.A., and Bini, L.M. (2010). SAM: a comprehensive applications for spatial analysis in macroecology. *Ecography*, 33, 46–50.

Robinson, O.J., Ruiz-Gutierrez, V., Reynolds, M.D., Golet, G.H., Strimas-Mackey, M., and Fink, D. (2020). Integrating citizen science data with expert surveys increases accuracy and spatial extent of species distribution models. *Diversity and Distributions*, 26, 976–86.

Sanz-Pérez, A., Ordiz, A., Sand, H., et al (2018). No place like home? A test of the natal habitat-biased dispersal hypothesis in Scandinavian wolves. *Royal Society Open Science*, 5, e181379.

Schoener, T.W. (1981). An empirically based estimate of home range. *Theoretical Population Biology*, 20, 281–25.

Seaman, D.E. and Powell, R.A. (1996). An evaluation of the accuracy of kernel density estimators for home range analysis. *Ecology*, 77, 2075–85.

Seber, G.A.F. (1982). *The Estimation of Animal Abundance and Related Parameters*. Second edition. Macmillan Publishing, New York.

Sokal, R.R. and Rohlf, F.J. (1995). *Biometry*. Third edition. W.H. Freeman Publishing, New York.

Stanton, R.A., Thompson, F.R., and Kesler, D.C. (2015). Site occupancy of brown-headed nuthatches varies with habitat restoration and range-limit context. *Journal of Wildlife Management*, 79, 917–26.

Steen, V.A., Elphick, C.S., and Tingley, M.W. (2019). An evaluation of stringent filtering to improve species distribution models from citizen science data. *Diversity and Distributions*, 25, 1857–69.

Suarez-Rubio, M., Ille, C., and Bruckner, A. (2018). Insectivorous bats respond to vegetation complexity in urban green spaces. *Ecology and Evolution*, 8, 3240–53.

Sui, D.Z. (2004). Tobler's first law of geography: a big idea for a small world? *Annals of the Association of American Geographers*, 94, 269–77.

Swihart, R.K. and Slade, N.A. (1985). Influence of sampling interval on estimates of home-range size. *Journal of Wildlife Management*, 49, 1019–25.

Taylor, R. (1990). Interpretation of the correlation coefficient: a basic review. *Journal of Diagnostic Medical Sonography*, 6, 35–39.

Thayn, J.B. and Simanis, J.M. (2013). Accounting for spatial autocorrelation in linear regression models using spatial filtering with eigenvectors. *Annals of the Association of American Geographers*, 103, 47–66.

Thorn, M., Green, M., Bateman, P.W., Waite, S. and Scott, D.M. (2011). Brown hyaenas on roads: estimating carnivore occupancy and abundance using spatially autocorrelated sign survey replicates. *Biological Conservation*, 144, 1799–07.

Tobler, W.R. (1970). A computer movie simulating urban growth in the Detroit region. *Economic Geography*, 46, 234–40.

Troy, J.R., Holmes, N.D., Joyce, T., Behnke, J.H., and Green, M.C. (2016). Characteristics associated with Newell's shearwater (*Puffinus newelli*) and Hawaiian petrel (*Pterodroma sandwichensis*) burrows on Kauai, Hawaii, USA. *Waterbirds*, 39, 199–204.

van Beest, F.M., Uzal, A., Vander Wal, E., et al. (2014). Increasing density leads to generalization in both coarse-grained habitat selection and fine-grained resource selection in a large mammal. *Journal of Animal Ecology*, 83, 147–56.

Van Horne, B. (1983). Density as a misleading indicator of habitat quality. *Journal of Wildlife Management*, 47, 893–901.

Vatcheva, K.P., Lee, M., McCormick, J.B. and Rahbar, M.H. (2016). Multicollinearity in regression analyses conducted in epidemiologic studies. *Epidemiology*, 6, e2.

Veech, J.A. (2012). Significance testing in ecological null models. Theoretical Ecology, 5, 611–16.

Ver Hoef, J.M., Peterson, E.E., Hooten, M.B., Hanks, E.M., and Fortin, M.-J. (2018). Spatial autoregressive models for statistical inference from ecological data. *Ecological Monographs*, 88, 36–59.

Vilela, B. and Villalobos, F. (2015) letsR: a new R package for data handling and analysis in macroecology. *Methods in Ecology and Evolution*, 6, 1229–34.

Wang, B., Rocha, D.G., Abrahams, M.I., et al. (2019). Habitat use of the ocelot (*Leopardus pardalis*) in Brazilian Amazon. *Ecology and Evolution*, 9, 5049–62.

Webb, M.H., Wotherspoon, S., Stojanovic, D., et al. (2014). Location matters: using spatially explicit occupancy models to predict the distribution of the highly mobile, endangered swift parrot. *Biological Conservation*, 176, 99–108.

Wecker, S.C. (1964). Habitat selection. *Scientific American*, 211, 109–17.

White, G.C. (2005). Correcting wildlife counts using detection probabilities. *Wildlife Research*, 32, 211–16.

Winchell, K.M., Carlen, E.J., Puente-Rolón, A.R., and Revell, L.J. (2017). Divergent habitat use of two urban lizard species. *Ecology and Evolution*, 8, 25–35.

Worton, B.J. (1989). Kernel methods for estimating the utilization distribution in home-range studies. *Ecology*, 70, 164–68.

Young, B.E., Dodge, N., Hunt, P.D., Ormes, M., Schlesinger, M.D., and Shaw, H.Y. (2019). Using citizen science data to support conservation in environmental regulatory contexts. *Biological Conservation*, 237, 57–62.

Zollner, P.A., Smith, W.P., and Brennan, L.A. (2000). Microhabitat characteristics of sites used by swamp rabbits. *Wildlife Society Bulletin*, 28, 1003–11.

CHAPTER 8

Analysis of the Habitat Associations of a Hypothetical Beetle Species

In this chapter I apply five methods of habitat analysis to an example dataset. The goal here is to provide the reader with a gradual introduction and comparison of the methods without getting bogged down in the mathematical details. I walk the reader through the methods of analysis pointing out particular decisions and assessments to undertake. The hypothetical example is then followed by Chapter 9, which describes each method in more technical detail and in general terms. Most of the methods are standard tools of data analysis and hypothesis testing used in a wide variety of research fields and for various purposes, although of course I emphasize the use of each method for identifying habitat associations. The methods presented in Chapters 8 and 9 are not meant to comprise an exhaustive list. There are other techniques not explicitly covered in these chapters, although many of those techniques are not strictly directed at uncovering the strength of the relationship between environmental variables [Y] and species response [X]; that is, the general framework of habitat analysis described in Chapter 5. I briefly discuss these additional extended techniques in Chapter 11. The methods of analysis discussed below are relatively straightforward and easily applicable to determining whether species presence, abundance, or any measure of demographic output or habitat use is significantly associated with any number of measured environmental variables,

and further each method allows for a quantitative estimate of the strength of the association.

8.1 Survey design and data collected

The example is a dataset for a hypothetical ground-dwelling beetle species sampled in 100 10 × 10 m plots randomly located throughout a forest. Imagine that the researcher deployed some set number of pitfall traps for some set length of time in each plot. Number of beetles captured ranged from 0 (60 of the plots) to 15, with only 9 plots having 10 or more beetles captured (Appendix 8.1). This is fairly typical for abundance or count data of a surveyed species; many sampling or survey units fail to reveal the species. For most plots where the species was present, only a few individuals are captured (recorded) and only a few plots have a relatively high abundance of the species. In this hypothetical example, the response variable for the habitat analysis could be either presence/absence (coded as 1/0) or the observed abundance (counts).

Assume that the researcher had some knowledge of the basic ecology of the beetle species such that the measured environmental variables were an informed selection. The following variables were measured for each plot: percentage canopy cover, depth of leaf litter, volume of woody debris, ratio of oak to non-oak trees, and soil type (Appendix 8.1). The exact details on how these variables were measured are not important, except to say that they were

Habitat Ecology and Analysis. Joseph A. Veech, Oxford University Press (2021). © Joseph A. Veech. DOI: 10.1093/oso/9780198829287.003.0008

measured with a minimal amount of error. Perhaps volume of woody debris was determined by gathering up all the downed wood and placing it in a large container for measurement. The oak:non-oak ratio was obtained by counting trees in some larger plot (perhaps 40 × 40 m) centered on the smaller 10 × 10 m plot. The soil type of each plot could have been obtained from a soil map or direct sampling. Note that soil type was recorded as 1, 2, 3, or 4. As such, it could be treated as a categorical (and qualitative) variable, a categorical but ordinal (quantitative) variable, or perhaps a numerically indexed value. Regardless, it is certainly not a *continuous* numerical variable. For the sake of the example, assume that soil type is a quantitative indicator of a physical property of the soil such as its texture (particle size). Thus, for statistical analysis, soil type can be treated as a discrete numerical variable. All the measured environmental variables are quantitative yet different in form.

One of the variables (leaf litter depth) is a one-dimensional measurement that scales from 0.55 to 17.45 cm. Canopy cover is a percentage (derived from a measurement of area) and hence is constrained between 0 and 100, with actual values (for the hypothetical example) ranging from 20 to 99 percent. One of the variables is a ratio derived from counts that are of course discrete. Woody debris is measured as a volume ranging from 0.01 to 2.47 m³. Soil type is a numerical indicator variable limited to values of 1, 2, 3, or 4 as explained above. Figure 8.1 shows scatterplots for each environmental variable and beetle abundance. Plots such as these are a useful way of obtaining a first glimpse of possible habitat associations. For each variable there is substantial scatter in the data; however, note that for canopy cover and leaf litter depth there appear to be minimum threshold values below which the species is absent (Fig. 8.1). The goal of the habitat analysis is to determine if any of the environmental variables (alone or in combination) define the habitat of the hypothetical beetle species. That is, does the statistical method of habitat analysis reveal a significant association between beetle presence/absence or abundance and any of the environmental variables? And, if so, what is the strength of the association?

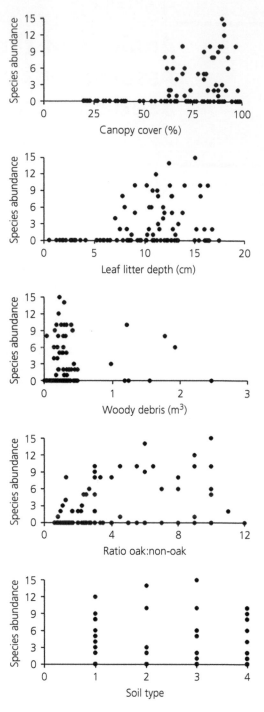

Figure 8.1 Scatterplots of beetle species abundance versus each of five environmental variables measured in 10 × 10 m sampling plots (*N* = 100).

8.2 Checking for normality of the response and predictor variables

As discussed in Section 7.8, prior to conducting a habitat analysis (i.e., statistical testing) one should check whether the response variable and predictor variables are normally distributed. This is because many types of statistical analyses assume and require normality of the variables (particularly the response variable), except for the so-called non-parametric (non-distributional) tests, which by definition do not require any particular distribution of the variables. A reliable and often-used technique for assessing normality is to create a q-q plot and visually examine whether the points of the plot fall on the line of unity (slope = 1). For a given variable, a q-q plot is constructed by calculating the quantiles

of the data wherein the maximum number of quantiles ($N - 1$, with N = number of observations or sample size) is used. Thus for the beetle dataset, there are 99 quantiles and each quantile is defined to have one observation. Essentially, this means that the observations for the given variable are put in ascending order and then plotted against random (computer-generated) observations that derive from a hypothetical normal distribution with the same mean and standard deviation as the distribution of observed values.

Figure 8.2 shows the q-q plots for four of the environmental variables in the beetle dataset. The plots reveal that each of the four variables has an approximate normal distribution. The least-squares (i.e., best fit) regression lines indicate that there is almost a one-to-one match between the observed

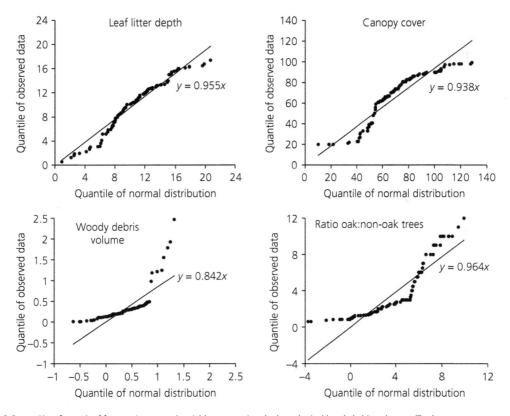

Figure 8.2 q-q Plots for each of four environmental variables composing the hypothetical beetle habitat dataset. The least-squares regression line and equation are shown in each plot. All of the variables (except woody debris volume) appear to be normally distributed as indicated by the regression line being very similar to the line of unity ($y = x$). For woody debris volume and the ratio oak:non-oak trees, the low end of the normalized quantile range is <0 because the mean of the observed values was low and the standard deviation was relatively high. A q-q plot for soil type is not shown as it is not very informative. Recall that soil type is constrained to values of 1, 2, 3, and 4; hence comparison of observed quantiles to quantiles from a normal distribution (which is continuous) is meaningless.

quantiles and the quantiles from a "perfectly" normal distribution having the same mean and standard deviation (Fig. 8.2). Another way of saying this is that the regression line is very close to being the line of unity. One possible exception is woody debris volume ($y = 0.842x$). For this variable the observed quantiles depart from the normal expectation at really low and high values. Also note that woody debris volume and ratio oak:non-oak trees have some normalized quantiles with negative values (the x-axes in Fig. 8.2). This is no cause for concern; this apparent oddity comes about because these two variables have relatively small means close to zero ($\bar{x} = 0.36$ and 3.61, respectively) and comparatively large standard deviations (sp = 0.39 and 2.97), and thus a normal distribution generated with these parameters includes some values <0. Of course, the observed values (quantiles) for woody debris and ratio oak:non-oak cannot be <0. Hence the q-q plots for woody debris and oak:non-oak trees indicate that there is some departure from normality in the distributions of these two variables due to a small mean combined with a relatively large sp. Also note in the q-q plot for canopy cover that the observed quantiles cannot exceed 100 (Fig. 8.2) because canopy cover is measured as a percentage between 0 and 100. This physical constraint on canopy cover might make it a little bit less normal than it would be otherwise. Lastly, I did not perform any assessment of normality for soil type. It is clearly not a continuous variable and really is not even a discrete variable. The normal distribution only directly applies to continuous variables although it can sometimes be approximated for discrete variables that have a wide numerical range. Nonetheless, soil type as a measured variable does not appear to have characteristics that would be problematic for subsequent statistical analyses. There are roughly equal numbers of observations with values of 1 ($n = 24$), 2 ($n = 20$), 3 ($n = 23$), and 4 ($n = 33$) such that the mean ($\bar{x} = 2.65$) is approximately mid-way between 1 and 4.

8.3 Checking for multicollinearity among the predictor variables

A second important step, prior to a statistical analysis of habitat, is to check for multicollinearity among the predictor variables (Section 7.9). The lack of excessive correlation among predictor variables is an assumption and requirement for some statistical analyses, such as multiple regression (Box 8.1). For the hypothetical beetle data, examination of the correlation coefficients and variance inflation factors (Section 7.9) clearly indicate substantial correlation between canopy cover and leaf

Box 8.1 Multicollinearity and multiple regression

Whenever you have a statistical software package or R code perform a multiple regression (and many other statistical tests), behind the scenes the computer is running algorithms to solve equations. These equations often involve matrix algebra (for an excellent explanation of some common mathematical operations on matrices, see pp. 447–58 in Gotelli and Ellison, 2004). For multiple regression, there is a vector (a matrix with one column and N rows) holding the data for the response variable and an $N \times P$ matrix (where N = number of observations and P = number of predictor variables plus an extra column of 1s for estimating the y-intercept) containing the data for the measured predictor variables. These matrices are often symbolized as $\mathbf{Y}$ and $\mathbf{X}$ (bolding indicates a matrix). These two matrices are used to obtain a $P \times 1$ vector of regression coefficients (often symbolized as $\mathbf{b}$) by the following equation: $\mathbf{b} = (\mathbf{X'X})^{-1}\mathbf{X'Y}$. Note that $\mathbf{X'X}$ is itself an $N \times P$ matrix obtained by multiplying $\mathbf{X'}$ (the transposition of $\mathbf{X}$, switching of rows and columns) by $\mathbf{X}$, and then inverting the $\mathbf{X'X}$ matrix to get $(\mathbf{X'X})^{-1}$. (The inverse of a matrix is defined as that matrix that when multiplied by the original matrix yields the identity matrix which is a matrix filled with 1s.) Lastly, matrix $(\mathbf{X'X})^{-1}$ is multiplied by vector $\mathbf{X'Y}$. Given that $\mathbf{Y}$ is a $P \times 1$ vector, so is $\mathbf{X'Y}$ and hence $\mathbf{b}$ is a $P \times 1$ vector as well. By now, you can probably plainly see how multiple regression involves a lot of matrix algebra. Unfortunately, some operations (e.g., inversion) in matrix algebra can be very sensitive to correlations among the predictor variables (i.e., among the elements in different columns of matrix $\mathbf{X}$) and thus lead to incorrect estimates of the regression coefficients (i.e., the elements in $\mathbf{b}$) (Neter et al. 1989). In effect, this means that really small changes in the data within $\mathbf{X}$ can lead to very different estimates of regression coefficients (i.e., the elements of $\mathbf{b}$). This is not good; the outcome of any statistical estimation procedure should not greatly change with small changes in the data, or else the procedure is unreliable.

Table 8.1 Correlation coefficients and variance inflation factors (VIF) for the environmental variables measured in 10 × 10 m plots for a study of the habitat associations of a hypothetical beetle species.

	Canopy cover	Leaf litter depth	Woody debris volume	Ratio oak:non-oak	Soil type	VIF
Canopy cover	–					10.03
Leaf litter depth	0.899	–				9.92
Woody debris volume	0.264	0.253	–			1.40
Ratio oak:non-oak	0.055	0.033	−0.085	–		1.23
Soil type	−0.046	−0.029	−0.049	0.138	–	1.18

litter depth (Table 8.1). This is not surprising in that one might expect greater rates of leaf fall and hence greater depth of leaf litter in those 10 × 10 m plots having the greatest amount of canopy foliage. Once a correlation like this is uncovered, the analyst must decide whether to correct for it prior to the subsequent habitat analysis or allow the correlation to remain while accepting that caveats or disclaimers will need to be provided upon interpreting the results of the habitat analysis. In this hypothetical example, either approach is legitimate, particularly given that the correlation between canopy cover and leaf litter depth is real (not due to peculiarities of measuring the variables), ecologically sensible, and not mathematically predetermined (as would occur if one variable were a mathematical function of the other or somehow constrained mathematically). The greatest challenge for the analyst is to figure out which variable has the greater importance in defining habitat for the species, assuming that is a question of interest or one of the goals of the habitat analysis. For this hypothetical ground-dwelling beetle species, canopy cover and leaf litter depth might both be important components of its habitat, thus I retained both in all subsequent analyses. Further, the correlation between these two variables is not due to some aspect of their measurement, physical constraint, or any type of mathematical commonality (i.e., neither is a mathematical function of the other). Rather canopy cover and leaf litter might synergistically define the habitat of the beetle and this argues for including both variables in statistical analyses of habitat associations. In Chapter 12, I further discuss this perspective of habitat as an integrated holistic entity in the context of statistical analysis.

Another way to explain the adverse effects of multicollinearity is to consider how different the estimates of **b** are in the presence and absence of collinearity between two predictor variables, X_1 and X_2. The regression coefficients in **b** are typically denoted as β_0, $\beta_1, \beta_2, \ldots \beta_P$ (with β_0 being the estimated y-intercept) when presented in the equation for a multiple linear regression model (Section 9.2). Prior to all the matrix algebra, if the data in matrix **X** are first standardized (i.e., rescaled to have a mean = 0 and standard deviation = 1), then the elements in **b** are standardized partial regression coefficients, often symbolized with a prime as β' (although in this case the prime does not indicate a transposition). Given that we have only two predictor variables, X_1 and X_2, β'_1 can also be estimated as $\beta'_1 = (r_{y1} - r_{y2}r_{12}) / (1 - r_{12}^2)$ where r_{y1} is the Pearson correlation coefficient between the response variable and X_1, r_{y2} is the correlation coefficient between the response variable and X_2, and r_{12} is the correlation between X_1 and X_2 (i.e., collinearity). Hence, note that if there is no collinearity then by definition, $r_{12} = 0$, and thus $\beta'_1 = r_{y1}$ (Neter et al. 1989; Graham 2003). Thus, any small amount of correlation between X_1 and X_2 does alter β'_1 and β_1 (as it would also alter β'_2 and β_2). Moreover, the statistical significance of a regression coefficient and hence the effect of predictor variable X is typically assessed with a t-test on the null hypothesis, H_0: $\beta = 0$ (or $\beta' = 0$). The test statistic, t, can be determined as $t = (r_{y1} - r_{y2}r_{12}) / \sqrt{MSE}$, where MSE is the mean squared error or $\Sigma(y_i - \hat{y}_i)^2 / (n-1)$ (Neter et al. 1989; Graham 2003). Therefore, as r_{12} increases, t decreases and hence becomes less likely to be significant. Essentially, multicollinearity inflates the standard errors of the β estimates leading to t-tests with reduced power for finding significant effects.

Another problem due to multicollinearity can be seen when we consider the variance–covariance matrix of the estimated regression coefficients. This is always a $P \times P$ (square) matrix that has the variances of estimated $\beta_0, \beta_1, \ldots \beta_P$ along the diagonal and covariances as the off-diagonal elements. The standard error and hence statistical significance of each regression coefficient is influenced by its variance and its covariance with all the other predictor variables in the regression model. Indeed, a β estimate itself depends on the variances of X and Y and their covariance. When there is substantial multicollinearity among predictor variables, covariances (numerical values) within the variance–covariance matrix can vary substantially depending on which predictor variables are in the model (Neter et al. 1989; Kraha et al. 2012). Adding a particular predictor variable to a model can change the structure of the variance–covariance matrix, even to the point of changing the sign (or direction) of regression coefficients for variables already in the model. This may also result in a change in the statistical significance of regression coefficients. Thus, in some situations, the effect of a particular predictor variable on a response variable will depend more on the other variables that are included in the model than on a pure direct effect (Neter et al. 1989).

As explained above, excessive correlation among predictor variables is problematic for multiple regression for various reasons: incorrect estimation of regression coefficients within any single model, flawed statistical testing of regression coefficients, and context- or model-dependency of the regression coefficients and their significance. Quinn and Keough (2002, pp. 127–30) present an excellent discussion of multicollinearity and procedures to diagnose and remedy it, and an even more thorough and mathematical explanation is provided by Neter et al. (1989, pp. 295–305). For a thorough explanation, the reader should also see Graham (2003), Dormann et al. (2013), Vatcheva et al. (2016), or Winship and Western (2016); they all used simulated data to demonstrate the problems associated with multicollinearity. Also, be aware that some authors refer to multicollinearity simply as "collinearity."

8.4 Methods of habitat analysis as applied to the hypothetical beetle species

In this section, I describe each of five different methods for conducting a habitat analysis. As a way of illustration, each method is applied to the beetle dataset. I provide a more detailed technical explanation of each method in the next chapter.

8.4.1 Comparison among group means—beetle example

In this example, the researcher recorded abundance of the beetle species within 100 10 × 10 m plots. However, presence/absence is the simplest response variable for an analysis of habitat associations and of course abundance data are always easily converted to presence/absence. The beetle species was present (at least 1 individual captured) in 40 plots and apparently absent from 60 plots. Hence, presence and absence plots naturally compose two groups that can be compared with regard to the five environmental variables that were measured. Therefore, in the method of comparison among group means, presence/absence is not used as a response variable (strictly speaking) but rather as a grouping variable. I performed five separate *t*-tests (one for each environmental variable) comparing the mean of the presence plots to that of the absence plots. These tests revealed the presence plots to have significantly greater ($P < 0.05$) canopy cover, deeper leaf litter, and a higher ratio of oak to non-oak tree species than the absence plots (Table 8.2).

In addition to or as an alternative to comparing presence and absence plots, we can also define multiple groups based upon abundance. Recorded abundance ranged from 0 to 15 individuals. Thus, I split the data (plots) into four groups: high (≥ 10 individuals), medium (6–9), low (1–5), and zero individuals. Note that the groups need not have the same range of abundance nor do they need to have the exact same sample size: high ($n = 9$ plots), medium ($n = 11$), low ($n = 20$), and zero ($n = 60$). Of most importance, each group should have sufficient size so as to get fairly accurate and precise estimates

Table 8.2 Comparison of group means as a method of analyzing habitat associations of the hypothetical beetle species. Table gives mean values for each of the five environmental variables for groups composed of sampling plots where beetles were present and those where beetles were absent; those means are compared with a t-test. In a separate analysis, ANOVA was applied to each variable to compare the means among groups representing plots with high beetle abundance ($\geq$ 10 individuals), medium (6–9), low (1–5), and zero (same as "absence").

Group	N	Canopy cover (%)	Leaf litter depth (cm)	Woody debris (m³)	Oak:non-oak trees	Soil type
Presence	40	79.9	11.8	0.40	4.61	2.63
Absence	60	61.4	9.0	0.33	2.94	2.67
t		4.30	3.57	0.84	2.85	0.17
P		0.00004	0.0006	0.40	0.005	0.86
High	9	85.8	12.9	0.38	7.06	3.00
Medium	11	77.3	11.2	0.53	5.12	2.45
Low	20	78.7	11.6	0.34	3.22	2.55
Zero	60	61.4	9.0	0.33	2.94	2.67
$F_{3,96}$		6.40	4.56	0.78	7.29	0.41
P		0.0005	0.005	0.51	0.0002	0.74

for the means of the environmental variables to be analyzed. Also, it is worth emphasizing that the absence plots should generally form their own exclusive group rather than being placed in the low-abundance group given that absence plots might be uniquely different from the other plots (as was revealed by the t-tests). Five separate single-factor ANOVAs revealed significant differences ($P < 0.05$) in mean canopy cover, leaf litter depth, and oak:non-oak tree ratio among the four abundance groups, but no significant differences in woody debris volume or soil type (Table 8.2). Further, the significant differences were generally as expected (based on results of the t-test comparing presence and absence groups): high-abundance plots had the greatest canopy cover, deepest leaf litter, and greatest oak:non-oak tree ratio. Of course, a significant P-value from an ANOVA only indicates that there are differences among the groups. It does not indicate which groups are significantly different from one another. To obtain that information, one would need to do follow-up tests based on pairwise comparisons of groups; this can be accomplished using Tukey's test or any number of similar post-hoc tests. I did not conduct any pairwise comparisons as it is fairly clear that the main differences are between the high-abundance and zero-abundance groups (Table 8.2).

8.4.2 Multiple linear regression—beetle example

The beetle data can also be analyzed using multiple linear regression with abundance as the response variable and each of the five environmental variables as predictor variables. An immediate decision has to be made with regard to whether to include the zero-abundance data (i.e., the 60 plots where beetles were apparently absent) in the analysis. In any type of regression, problems can sometimes arise when identical values of the response variable comprise a large proportion of all observations. Specifically, the main problem is that the residuals from the regression may not be normally distributed because the response variable itself is not normally distributed (Section 7.8), violating an assumption of most types of regression (Section 9.2). In order to illustrate this problem, I conducted a full regression model (all five environmental variables as predictors) on the abundance data, including the 60 plots representing zero abundance. I then plotted the residuals against the predicted values of the response variable, a common diagnostic visualization to assess unusual pattern in the residuals.

As shown in Fig. 8.3 (top panel), the pattern reveals that the residuals are not normally distributed; they correlate somewhat with the predicted

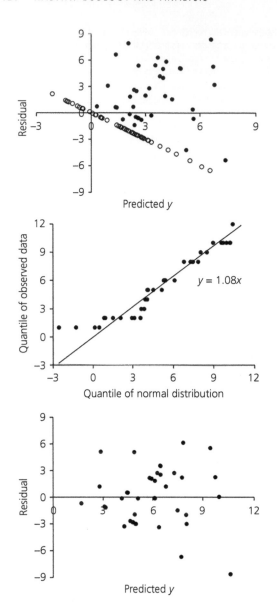

values. If the residuals were normally distributed and uncorrelated with the predicted values, then the scatter of points would be relatively even (uniform) along the entire range of the predicted values of the response variable (x-axis in the figure). Inclusion of the zero-abundance data (absence plots) is the main reason that the residuals have non-random structure, and this is because the response variable (abundance) is far from a normal distribution when $y = 0$ comprises 60 percent of the observations. The residual plot also reveals heteroscedasticity in the residuals; that is, the error $(\varepsilon_i = y_i - \hat{y}$; Section 7.8) in the multiple regression equation or model is not random with regard to the response variable. Another assumption of multiple regression is homoscedasticity: error that is uncorrelated with the response variable. In essence, heteroscedasticity (correlated error) means that the regression model has a better fit for some y-values than for others—that is not a good outcome. Again, for these hypothetical beetle data, absence plots should not be included in a multiple regression model.

If we exclude the absence plots then we are left with less data overall (smaller sample size) but a potentially better fitting and statistically legitimate multiple regression model. Prior to applying the full regression model to the $N = 40$ plots having a non-zero recorded beetle abundance, I constructed a q-q plot to check for normality. Beetle abundance as a response variable is normally distributed (Fig. 8.3, middle panel). As such, the residuals appear to be normally distributed and homoscedastic (Fig. 8.3, bottom panel). The multiple linear regression model revealed the ratio of oak:non-oak trees as the only variable with a statistically significant coefficient ($P < 0.05$) (Table 8.3). The table also gives the standardized regression coefficients. These coefficients allow for direct comparison of predictor variables (Section 9.2), which might have been of interest if more than one had been statistically significant. Note how in Table 8.3, the ratio of oak:non-oak trees has the largest standardized coefficient ($\beta_{std.} = 0.477$) but not the largest raw coefficient. Of most importance, the coefficient is positive indicating that the abundance of the beetle species increases with an increasing ratio of oak to non-oak trees in the area immediately surrounding a 10 × 10 m

Figure 8.3 Assessment of the full multiple regression model predicting beetle abundance in survey plots as a function of the five environmental predictor variables. In the top panel, the residuals are shown to clearly not have a normal distribution and are heteroscedastic. They are correlated with the predicted abundance values mostly due to the survey plots with zero recorded abundance (open circles). In the middle panel, a q-q plot indicates that beetle abundance in the presence plots ($N = 40$) is normally distributed; the line through the plot is very close to unity. In the bottom panel, the full multiple regression model has been applied only to the survey plots with beetles present, and the residuals are normally distributed and homoscedastic.

Table 8.3 Results of using multiple linear regression to analyze the habitat associations of the hypothetical beetle species. Response variable is beetle abundance excluding the 60 sampling plots in which no beetles were found (zero-abundance data). Table gives the regression coefficients (β) and their standardized values. The overall regression is statistically significant ($F_{5,34} = 2.73$, $P = 0.035$, $R^2_{adj.} = 0.18$).

Variable	β	$\beta_{std.}$	t	P
Intercept	−4.14	–	−0.84	0.41
Canopy cover	0.111	0.325	1.42	0.16
Leaf litter depth	−0.134	−0.090	−0.40	0.69
Woody debris	1.160	0.119	0.78	0.44
Ratio oak:non-oak	0.567	0.477	3.18	0.003
Soil type	−0.123	−0.040	−0.25	0.80

Table 8.4 Results of using multiple logistic regression to analyze the habitat associations of the hypothetical beetle species. Response variable is beetle presence/absence (coded as 1/0). Table gives the regression coefficients (β) and their standardized values. The overall regression is statistically significant ($F_{5,94} = 7.44$, $P < 0.00001$, $R^2_{adj.} = 0.25$); however, Hosmer and Lemeshow (1989) suggest not reporting adjusted R^2 values from logistic regressions because they are not based on a sum-of-squares calculation—see Section 10.1 for more details.

Variable	β	$\beta_{std.}$	z	P
Intercept	−4.79	——	−3.57	0.0004
Canopy cover	0.058	2.71	2.26	0.02
Leaf litter depth	−0.052	−0.43	−0.39	0.69
Woody debris	0.076	0.06	0.14	0.89
Ratio oak:non-oak	0.231	1.39	2.71	0.007
Soil type	−0.066	−0.158	−0.33	0.74

sampling plot. This relationship between beetle abundance and the oak:non-oak ratio is not visually evident in the scatterplot of the raw data (Fig. 8.1); however, it is revealed by multiple regression and other statistical analyses (*t*-tests and ANOVA). It is important to emphasize that this habitat analysis based upon multiple regression does not include the absence plots although we know (from results of the *t*-tests) that these plots are important in revealing habitat associations of the beetle species. Beetles are absent from plots with <50 percent canopy cover and leaf litter depth 7 cm or less (Fig. 8.1). The absence plots are ecologically and statistically important and thus should not be ignored or excluded from an analysis of habitat.

8.4.3 Multiple logistic regression—beetle example

There is one particular type of statistical analysis that is specifically designed for making use of absence data (i.e., survey sites or sampling plots where the species was recorded as absent). This is multiple logistic regression (Section 9.3). Again using all the environmental variables, I applied a full multiple logistic regression model to the beetle presence–absence data. The response variable for the model was presence–absence coded as 1 and 0, respectively. For any type of logistic regression, the response variable is constrained to be between 0 and 1 (although it need not take only values of 0

and 1; Section 9.3), hence the multiple logistic regression model could not be applied to the beetle abundance data. The results clearly show that the ratio of oak:non-oak trees affects whether beetles are present in a sampling plot (Table 8.4). The importance of this variable was also revealed in the previously discussed analyses. But now, using multiple logistic regression, we see that canopy cover also has a statistically significant ($P < 0.05$) regression coefficient (Table 8.4) that was not revealed by the multiple *linear* regression. Recall that the *t*-test was also able to identify this effect of canopy cover (Table 8.2). Note that the multiple logistic regression model did not reveal a significant effect of leaf litter depth (Table 8.4), although this effect was uncovered by the *t*-test (Table 8.2). A "perfect" logistic regression curve has the classic S-shape when fitted to the response variable (Section 9.3), although this form of the curve need not be manifested even when the regression coefficients are significant. Figure 8.4 shows the relationship between beetle presence–absence and canopy cover and ratio oak:non-oak trees. The fitted "lines" are certainly not S-shaped although they show hints of that form; also, they are definitely not linear. The lines and the positive regression coefficients indicate that the probability of the beetle species being present in a plot increases with percentage canopy cover and the ratio of oak:non-oak trees (Table 8.4;

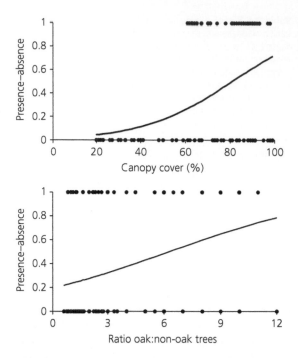

Figure 8.4 The relationship between beetle presence–absence and canopy cover (top panel) and ratio oak:non-oak trees (bottom panel) as revealed by multiple logistic regression. The lines shown in each plot derive from a multiple logistic regression model having only those two variables, not all five environmental variables as in the full regression model (Table 8.4). The coefficients are slightly different ($\beta_0 = -4.90$, $\beta_{canopy} = 0.051$, and $\beta_{oak} = 0.226$) than in the full model (compare with Table 8.4). Also, for each line, the variable not plotted was held constant at its mean value (ratio oak:non-oak = 3.61, top panel; percentage canopy cover = 68.77, bottom panel).

Fig. 8.4). This is another nice property of logistic regression, the output or predicted y-value can be interpreted as a probability of a species being able to use a certain location, although it is not strictly the probability of species presence (see Sections 9.3 and 11.1).

8.4.4 Classification and regression trees (CART)—beetle example

Classification and regression trees (CART) are another method of habitat analysis that can be applied to presence–absence as well as abundance data. When the response variable is categorical (e.g., presence–absence), then a classification tree is constructed, whereas if the response variable is quantitative (continuous or discrete) as in abun-

dance data, then a regression tree is constructed (Section 9.4). Otherwise, CARTs are mathematically similar. In both techniques, algorithms progressively split the observations (all the data in [X]) into groups that are increasingly homogeneous. As such, CART identifies threshold values of the predictor variables and those threshold values define the groups. In the tree analogy, the threshold values represent branching points. The output of an analysis based on CART is a tree with a particular number of final groups or terminal nodes on the tips of the branches. In the context of a habitat analysis, those environmental or predictor variables that have the greatest influence in determining the branching nodes and terminal nodes are also (by inference) the variables that are most important in defining the habitat associations of the species.

For the hypothetical beetle species, the classification tree revealed percentage canopy cover and ratio of oak:non-oak trees as important variables. Canopy cover defined five of the eight branching points (splitting criteria) including the first branching point and the two most terminal points (Fig. 8.5). Not surprisingly, the first split was based on a canopy cover value of 60 percent; all 25 absence plots having a value <60 percent were assigned to a group and the 75 plots with canopy cover ≥60 percent were assigned to a group that was further subdivided. This group of 75 plots consisted of 40 presence plots and 35 absence plots hence $\bar{y} = 0.53$ (Fig. 8.5). Further splitting of this group resulted in seven final groups that were mixes of presence and absence plots. For example, Group 8 defined by canopy cover ≥80.5 percent and oak:non-oak tree ratio ≥2.88, consisted of 13 presence and 1 absence plot (Fig. 8.5). Each of the final eight groups was characterized by a unique range of canopy cover percentages and oak:non-oak tree ratios (Table 8.5). When the response variable is presence–absence (coded as 1/0), the mean value of the response variable for a group also can be interpreted as the probability that a plot with the particular environmental conditions (i.e., habitat characteristics) has the beetle species present. However, strictly speaking, this probability assumes an unlimited supply of beetles and that all plots are equally accessible. That is, beetles being present (or absent) in a given plot does not affect the probability that another plot contains

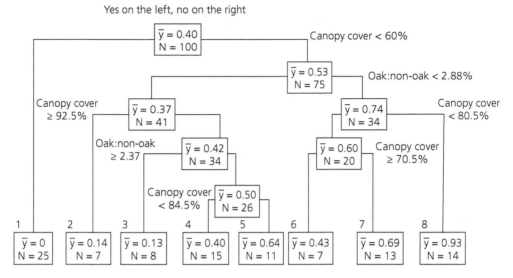

Figure 8.5 Regression tree diagram showing habitat associations for the hypothetical beetle species. The tree is based on presence/absence (coded as 1/0) in each of 100 sampling plots. The splitting criterion is indicated at each node. Sample plots satisfying the particular criterion ("yes") are always in the group on the left side of the split and those not satisfying the criterion ("no") are to the right. The group at each node is indicate by a box that contains the mean value of the response variable for those observations (sample plots) in the group and the number of observations. Given that the response variable is presence/absence, the mean represents the proportion of the samples plots where the species was recorded as present. In this tree, there are seven splits and hence eight terminal nodes. Trees are typically shown in an inverted orientation as in this figure.

the species. It is also not the probability that any given individual beetle will be present in the plot or that any given individual plot will have a beetle. Nonetheless, the probabilities can be informative. When canopy cover was 60–70.4 percent and ratio oak:non-oak trees ≥2.88, then *P(species present)* = 0.69, and this increased to 0.93 when canopy cover increased to 80.5 percent or more (Groups 7 and 8 in Table 8.5). One odd result of the classification tree was that *P(species present)* was relatively low (0.43) when canopy cover was in the range 70.5–80.4 percent even with a large ratio of oak:non-oak trees at ≥2.88 (Group 6 in Table 8.5).

The regression tree also indicated an effect of canopy cover and oak:non-oak tree ratio on beetle abundance. Oak:non-oak tree ratio defined the first split and canopy cover was the splitting criterion at four of seven branching points (Fig. 8.6). However, soil type and depth of leaf litter were also represented in the regression tree unlike in the classification tree. This indicates that beetle abundance may not depend as strictly on canopy cover and tree ratio alone, compared with beetle presence–

absence. In fact, the regression tree revealed that beetle abundance is zero when leaf litter <7.48 cm and very low at $\bar{y} = 0.14$ beetles per plot when soil type = 4 (Groups 5 and 2 in Fig. 8.6; Table 8.5). The greatest abundance occurred when canopy cover was ≥80.5 percent and oak:non-oak tree ratio was ≥2.88 (Fig. 8.6), which also corresponded to the cut-off values for defining the group with the greatest probability of species presence (Fig. 8.5). Overall there was substantial agreement between the CARTs particularly with regard to how canopy cover and oak:non-oak tree ratio were used to define groups.

The analysis based on CART revealed knowledge of habitat associations of the hypothetical beetle species that other methods of analysis (e.g., comparison of group means, multiple linear regression, multiple logistic regression) could not. In particular, CART identified critical levels (values) of some habitat variables that presumably influence the probability of a species' presence and its abundance in the 10 × 10 m plots. The other methods are not intended to or capable of uncovering such

Table 8.5 Results of a classification and regression trees (CART) analysis applied to the presence–absence (PA) and abundance (abd.) data of hypothetical beetle species surveyed in 100 plots in which the following environmental variables were also measured: percentage canopy cover, ratio of oak:non-oak trees, leaf litter depth, amount of woody debris, and soil type.

	Terminal node (group)[1]							
	1	**2**	**3**	**4**	**5**	**6**	**7**	**8**
Splitting criteria (P/A)								
Canopy cover (%)	<60	≥92.5	60–92.4	60–84.4	84.5–92.4	70.5–80.4	60–70.4	≥80.5
Oak:non-oak	–	<2.88	2.37–2.88	<2.37	<2.37	≥2.88	≥2.88	≥2.88
P(species present)	0	0.14	0.13	0.40	0.64	0.43	0.69	0.93
Expected grouping[2]	Absent	Absent	Absent	Absent	Present	Absent	Present	Present
MSE (P/A)	0	0.12	0.11	0.24	0.23	0.25	0.21	0.07
Correctly classified as absent	25/25	6/7	7/8	9/15	–	4/7	–	–
Correctly classified as present	–	–	–	–	7/11	–	9/13	13/14
Incorrectly classified as absent	–	–	–	–	4/11	–	4/13	1/14
Incorrectly classified as present	0/25	1/7	1/8	6/15	–	3/7	–	–
Splitting criteria (abd.)								
Canopy cover (%)	<77	≥77	≥85.5	77–85.4	<80.5	70.5–80.4	<70.4	≥80.5
Oak:non-oak	<2.88	<2.88	<2.88	<2.88	≥2.88	≥2.88	≥2.88	≥2.88
Soil type	–	=4	=1–3	=1–3	–	–	–	–
Leaf litter (cm)	–	–	–	–	<7.5	≥7.5	≥7.5	–
Mean abundance	0.24	0.14	1.29	3.43	0	1.29	4.31	8.71
MSE (abundance)	0.67	0.12	3.06	7.39	0	4.20	14.5	15.8

[1] Terminal nodes refer to the tree diagrams in Figs. 8.5 and 8.6.

[2] Expected grouping refers to whether the majority of observations in the group represented species presence or absence.

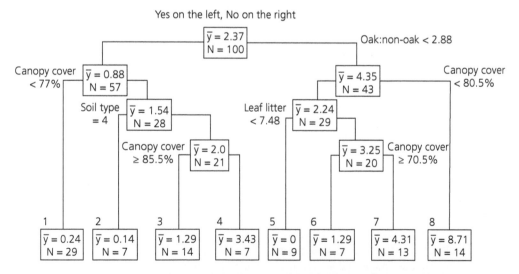

Figure 8.6 Regression tree diagram showing habitat associations for the hypothetical beetle species. The tree is based on abundance in each of 100 sampling plots. The splitting criterion is indicated at each node. Sample plots satisfying the particular criterion ("yes") are always in the group on the left side of the split and those not satisfying the criterion ("no") are to the right. The group at each node is indicated by a box that contains the mean value of the response variable for those observations (sample plots) in the group and the number of observations. In this tree, there are seven splits and hence eight terminal nodes. Trees are typically shown in an inverted orientation as in this figure.

intricate relationships. This unique capacity of CART can be seen in a scatterplot of the important variables along with the cutoff values for defining the groups (Fig. 8.7). The horizontal and vertical lines of the figure nicely show a structured pattern with regard to how canopy cover and oak:non-oak tree ratio influence the presence–absence of the beetle species. For example, Group 8 is clearly very different from Group 1 (Fig. 8.7). CARTs are also predictive models. Table 8.5 contains information (e.g., mean squared error, MSE) on the predictive accuracy of the trees; this can be thought of as an assessment of model fit, so I discuss it in more detail in Section 10.1.

8.4.5 Multivariate analysis—beetle example

I used principal components analysis (PCA) to examine whether groups composed of beetle presence plots and beetle absence plots differed from one another with regard to the five environmental variables. As is ordinary practice for PCA, the variables were first standardized; that is, converted to have mean = 0 and SD = 1. PCA, like other multi-

variate ordination techniques, combines multiple response variables into composite variables, each of which "explains" a proportion (or percentage) of the total variance within and among the variables (Section 9.5.1). A composite variable is referred to as an axis or principal component (PC). In any PCA, the number of axes equals the number of response variables. Hence, for the beetle dataset, there were five axes. Each axis has a contribution from each variable, known as factor loadings (Section 9.5.1) and sometimes broadly called coefficients. For the beetle dataset, PC 1 accounted for 40.7 percent of the overall variance. Its factor loadings were canopy cover (−0.666), leaf litter depth (−0.664), woody debris volume (−0.334), oak:non-oak tree ratio (−0.021), and soil type (0.061). Canopy cover and leaf litter depth had the strongest loadings therefore PC 1 is primarily characterized by these two variables. Recall that percentage canopy cover and leaf litter depth were correlated, which is why their loadings were very similar. PC 2 accounted for 23.3 percent of the overall variance with factor loadings of canopy cover (0.089), leaf litter depth (0.087), woody debris volume (−0.278), oak:non-oak tree ratio (0.712), and soil type (0.632). Thus this PC axis was primarily based on oak:non-oak tree ratio and to a lesser extent soil type.

Note that up to this point, the PCA applied to the beetle dataset has not actually used either the presence–absence or the abundance data. The PC loadings and their use in calculating PC scores do not take into account beetle abundance within a survey plot. Again, PCA is primarily a way to derive *composite* environmental variables that can then be used in further analyses (Sections 7.9 and 9.5.1). PCA and related multivariate ordination techniques are often used to find pattern or order in a dataset. I used the factor loadings to calculate the PC 1 and PC 2 scores for each of the 100 hypothetical beetle survey plots (Section 9.5.1). I then plotted the scores and separately applied Welch's *t*-test (assumes unequal variances of the groups being compared) to PC 1 and PC 2 scores to test for a difference between presence and absence plots. Figure 8.8 shows a clear separation between presence and absence plots along PC 1, although there are some absence plots that had negative PC 1 scores as did most of the presence plots. For PC 2, presence and absence plots are not

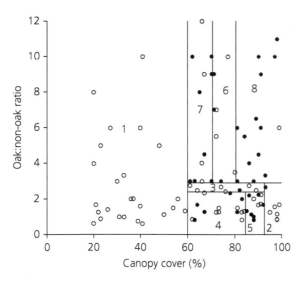

Figure 8.7 Scatterplot of percentage canopy cover and ratio of oak:non-oak trees for the sampling plots where the hypothetical beetle species was recorded as present (closed circles) and absent (open circles). Vertical and horizontal lines indicate the cutoff values (branching points) identified in the classification and regression trees (CART) analysis. The analysis revealed eight final groups or terminal nodes on the tree. See Figure 8.5 for details.

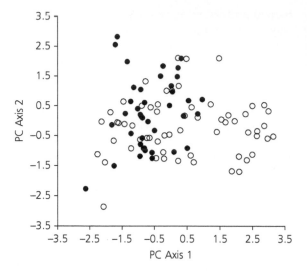

Figure 8.8 Plot of principal components (PC) scores for the beetle presence plots (filled circles, $N = 40$) and beetle absence plots (open circles, $N = 60$).

as clearly differentiated. The t-tests revealed significant differences between the presence and absence plots for both PCs (PC 1: mean score presence plots = −0.64, absence plots = 0.43, $t = -4.44$, $P_{two-tail}$ = 0.00002; PC 2: mean score presence plots = 0.28, absence plots = −0.18, $t = 2.04$, $P_{two-tail}$ = 0.04). By the way, I also applied t-tests to PCs 3, 4, and 5; all were non-significant ($P > 0.05$).

This habitat analysis based on a multivariate technique (PCA) indicated a joint effect of canopy cover and leaf litter depth on beetle presence–absence. More precisely, the overall variance–covariance structure of the five environmental variables was largely determined by the variances and covariance of canopy cover and leaf litter as well as variances and covariance of oak:non-oak tree ratio and soil type (PCs 1 and 2 together explained 64 percent of the overall variance–covariance of the five variables). In addition, as shown in Figure 8.8 and the t-tests, presence and absence plots differed with regard to the composite variables (PC 1 and 2), which were themselves primarily influenced by canopy cover and leaf litter depth (PC 1), and by oak:non-oak tree ratio and soil type (PC 2). That is, the factor loadings were greatest for these two variables on either PC 1 or 2.

Discriminant function analysis (DFA or sometimes just DA) is another multivariate technique that is potentially useful in identifying the environmental (i.e., habitat) variables that a species associates with. Unlike PCA, DA classifies observations (e.g., sampling plots) based upon the environmental variables (Section 9.5.2). In this regard, DA is superficially similar to CART. I applied a linear discriminant analysis (LDA) to the beetle dataset. I used presence–absence as the grouping variable. [Note that this is a difference between DA and either PCA or CART. The latter two methods do not require any type of grouping variable.] As with PCA, the five environmental variables were standardized prior to the LDA. Given that there were only two predefined groups (plots with beetles present and plots with beetles absent) there was only one discriminant function: LDF 1 = 1.061 × canopy cover − 0.173 × leaf litter depth + 0.014 × woody debris volume + 0.618 × oak:non-oak tree ratio − 0.082 × soil type + 0.218. Based on the coefficients, canopy cover and oak:non-oak tree ratio appeared to be the two most important variables discriminating presence plots from absence plots. As with CART, LDA can be thought of as a predictive model. As such, the classification accuracy of an LDF can be assessed (Sections 9.5.2 and 10.1).

8.5 Inference from the beetle example

The example of a hypothetical ground-dwelling beetle species surveyed in small 10 × 10 m plots was intended to show you five different methods of analysis that can be applied to assess whether a species associates with one or more measured environmental variables. If any of the analyses indicates a positive association with one or more variables then by inference we can conclude that those variables define (at least in part) the habitat of the species. They are habitat characteristics. In this example, canopy cover and the presence of oak trees clearly define habitat for the hypothetical beetle species. The inference is that the beetles like oak trees, or at least they like to be under them. Of course, for most studies, a general statement such as this is typically followed by more interpretation in an ecological context. For instance, our beetle species might be thermally constrained and unable to cope with high ground surface temperatures hence its positive association with canopy cover that provides shade. Further, it may depend on oak trees for some reason, perhaps related to its diet or reproduction. It also would appear to not be saproxylic as the volume of woody debris in a plot had no apparent effect on presence–absence or abundance.

The methods of habitat analysis differ with regard to how the multiple environmental variables are treated. In the simplest method (comparison among group means) each variable is examined in isolation of the others. Opposite of this, multivariate techniques (PCA and DFA) do not treat any variable as an isolated stand-alone effect but rather examine the environmental variables as a combined effect. Techniques such as multiple regression (linear or logistic) and CART allow for examination of individual environmental variables while also having the flexibility to assess an overall additive effect of the multiple variables (such as when comparing the overall fit of a full regression model to a model including only a subset or one predictor variable—see Section 10.2). Of course, even the multivariate techniques permit some assessment of the effects of individual variables. This occurs whenever an analyst compares the factor loadings of the different variables on a given PC axis—essentially these loadings are analogous to standardized partial regression coefficients of a multiple regression model. Operationally, multivariate techniques and multiple regression differ in that in the former the multiple variables are examined simultaneously, whereas in the latter the variables are examined sequentially. My main point is that none of the methods of habitat analysis is *inherently* better or more appropriate than the others. They are different in operation and have different strengths and weaknesses—and this entails that for a given dataset one might be more appropriate than another. If a given set of environmental variables truly characterizes the habitat of a given species and a sufficient dataset has been obtained, then to some extent *all* of the methods of analysis should identify those variables as affecting

presence–absence, abundance, or some other measured species response.

From the example of the hypothetical beetle species, it should be clear to you that there was general agreement among the different methods with regard to identifying the most important variables defining the habitat of the species. All five methods revealed an effect of percentage canopy cover and oak:non-oak tree ratio on either beetle presence–absence or abundance. To a slightly lesser extent, there may even be an effect of leaf litter depth, although it is difficult to tease apart its effect separate from canopy cover given that these two variables were highly correlated. Correlation among environmental variables (i.e., putative habitat characteristics) is not necessarily to be avoided when designing a study and collecting the environmental data. Some amount of correlation is natural and to be expected (e.g., percentage canopy cover and depth of leaf litter) and represents the conceptual possibility and empirical reality that the habitat of a species is defined by multiple integrated environmental factors (Chapter 12). In the beetle example, it is possible that soil type and oak:non-oak tree ratio could have been correlated if the species of oak tree only grew on a particular type of soil (although I intentionally did not structure the data in that way) and in such a case soil type would have been found by at least a few of the analyses to have an effect on presence–absence or abundance. This could be true even if the beetle species itself did not directly require or rely on a particular soil type. Nonetheless, soil type would be an important factor in characterizing its habitat. It is also important to recognize that no study can measure every possible factor that could be important in defining the habitat of the species. In the beetle example, there were only five environmental variables. Perhaps there are others that were not included and yet greatly influence where the beetle occurs (then again it is just a fictional species). In the next chapter, I present the mechanics of each method of habitat analysis in more detail, point out strengths and weaknesses, and discuss some select studies of real species in which each method was used.

Appendix 8.1 Dataset for a study of the habitat associations of a hypothetical beetle species whose abundance was recorded in 100 sampling plots. Five environmental variables were also measured for each plot. "P/A" is presence/absence coded as 0/1. "Forest" indicates the location of the plot in one of four forests (labeled A–D).

Plot	P/A	Abundance	Leaf litter depth (cm)	Canopy cover (%)	Woody debris (m³)	Ratio oak:non-oak trees	Soil type	Forest
1	1	15	15.1	90	0.23	10.00	3	A
2	1	14	12.45	91	0.30	6.00	2	A
3	1	12	11.15	91	0.22	9.00	1	A
4	1	10	16.35	97	0.26	10.00	4	A
5	1	10	14.0	90	0.33	4.50	4	A
6	1	10	12.55	89	0.30	6.50	3	A
7	1	10	15.6	84	0.40	5.50	2	A
8	1	10	9.0	70	0.19	9.00	4	B
9	1	10	10.2	70	1.21	3.00	4	D
10	1	9	10.8	88	0.42	3.00	1	A
11	1	9	11.25	81	0.16	6.00	4	B
12	1	8	11.35	93	0.29	3.33	1	A
13	1	8	12.7	86	0.21	4.00	1	A
14	1	8	15.65	85	1.78	1.33	1	B
15	1	8	12.75	65	0.32	8.00	4	B
16	1	8	7.75	61	0.04	3.00	4	C
17	1	6	11.85	93	0.29	2.67	3	A
18	1	6	10.65	71	0.15	7.00	3	A
19	1	6	10.45	65	1.93	8.00	4	B
20	1	6	8.0	62	0.20	10.00	1	C
21	1	5	13.8	82	0.27	2.50	3	A
22	1	5	12.75	81	0.33	3.00	1	A
23	1	5	12.0	78	0.28	2.33	1	B
24	1	5	9.0	70	0.23	10.00	3	C
25	1	4	11.85	81	0.20	1.29	1	A
26	1	4	7.05	67	0.15	1.29	4	D
27	1	3	13.4	90	0.98	2.25	1	C
28	1	3	11.35	87	0.44	1.13	2	C
29	1	2	16.0	98	0.21	11.00	4	A
30	1	2	16.8	92	0.49	1.67	3	B
31	1	2	10.4	88	0.29	1.00	3	B
32	1	2	15.2	86	0.22	2.20	2	B
33	1	2	12.85	83	0.37	2.00	1	B
34	1	2	8.2	74	0.34	3.00	1	C
35	1	2	7.5	64	0.40	1.67	2	C
36	1	2	11.45	63	0.45	3.00	4	C
37	1	1	13.1	88	0.34	0.83	4	A

Plot	P/A	Abundance	Leaf litter depth (cm)	Canopy cover (%)	Woody debris (m³)	Ratio oak:non-oak trees	Soil type	Forest
38	1	1	8.65	71	0.30	9.00	3	A
39	1	1	10.85	67	0.30	4.50	4	B
40	1	1	10.25	63	0.18	0.83	4	D
41	0	0	15.05	99	0.46	1.67	2	A
42	0	0	17.45	99	0.33	6.00	1	A
43	0	0	16.5	98	0.46	0.83	1	A
44	0	0	16.3	98	0.48	1.14	4	A
45	0	0	12.0	98	0.43	0.86	2	A
46	0	0	15.45	97	0.33	1.57	3	A
47	0	0	13.35	95	0.28	1.25	3	A
48	0	0	16.05	91	0.29	2.25	4	B
49	0	0	15.05	91	0.39	1.71	3	B
50	0	0	13.2	90	0.20	1.67	4	B
51	0	0	12.65	89	1.24	2.50	4	B
52	0	0	16.2	88	1.18	2.00	2	B
53	0	0	10.8	86	0.48	2.75	2	B
54	0	0	10.9	84	0.28	1.33	4	B
55	0	0	13.25	83	0.11	1.29	4	B
56	0	0	10.15	83	0.03	0.83	1	B
57	0	0	9.25	81	2.47	1.50	1	B
58	0	0	11.2	80	0.01	3.50	4	B
59	0	0	13.35	77	0.29	10.00	2	B
60	0	0	12.7	76	0.37	2.40	2	B
61	0	0	10.1	74	0.21	2.67	2	B
62	0	0	12.2	74	0.36	2.67	4	C
63	0	0	8.35	73	0.25	1.25	3	C
64	0	0	9.45	73	1.55	1.50	4	C
65	0	0	13.0	72	0.16	1.25	1	C
66	0	0	9.6	72	0.31	7.00	3	C
67	0	0	8.6	72	0.32	5.50	4	C
68	0	0	12.25	67	0.46	2.33	2	C
69	0	0	10.15	67	0.19	9.00	4	C
70	0	0	11.6	66	0.37	3.00	2	C
71	0	0	8.5	66	0.04	4.00	3	C
72	0	0	7.2	66	0.39	12.00	3	C
73	0	0	7.2	64	0.26	2.50	1	C
74	0	0	12.3	62	0.44	0.86	2	C
75	0	0	8.25	61	0.05	2.75	3	C
76	0	0	10.95	59	0.19	1.33	1	C

(Continued)

Appendix 8.1 (Continued)

Plot	P/A	Abundance	Leaf litter depth (cm)	Canopy cover (%)	Woody debris (m³)	Ratio oak:non-oak trees	Soil type	Forest
77	0	0	9.6	56	0.49	2.00	4	C
78	0	0	5.2	54	0.19	1.50	3	C
79	0	0	10.1	50	0.11	1.13	1	D
80	0	0	7.8	48	0.10	5.00	3	D
81	0	0	6.55	41	0.28	10.00	4	D
82	0	0	3.15	41	0.17	0.63	2	D
83	0	0	6.3	40	0.35	1.60	4	D
84	0	0	3.1	40	0.31	6.00	2	D
85	0	0	6.55	39	0.19	0.75	4	D
86	0	0	5.95	37	0.26	2.00	4	D
87	0	0	3.6	36	0.12	2.00	4	D
88	0	0	1.95	33	0.37	1.00	2	D
89	0	0	2.65	33	0.15	3.33	4	D
90	0	0	5.55	31	0.24	1.00	1	D
91	0	0	2.8	30	0.05	3.00	1	D
92	0	0	1.85	27	0.13	6.00	1	D
93	0	0	5.2	26	0.01	1.40	3	D
94	0	0	5.85	23	0.19	0.88	1	D
95	0	0	1.25	23	0.10	5.00	3	D
96	0	0	1.6	22	0.13	1.25	3	D
97	0	0	0.55	21	0.14	1.67	3	D
98	0	0	3.0	20	0.11	0.63	4	D
99	0	0	2.2	20	0.08	8.00	2	D
100	0	0	4.1	20	0.01	4.00	2	D

References

Dormann, C.F., Elith, J., Bacher, S., et al. (2013). Collinearity: a review of methods to deal with it and a simulation study evaluating their performance. *Ecography*, 36, 27–46.

Gotelli, N.J. and Ellison, A.M. (2004). *A Primer of Ecological Statistics*. Sinauer Associates, Sunderland, Mass.

Graham, M.L. (2003). Confronting multicollinearity in ecological multiple regression. *Ecology*, 84, 2809–15.

Hosmer, D.W. and Lemeshow, S. (1989). *Applied Logistic Regression*. Wiley, New York.

Kraha, A., Turner, H., Nimon, K., Reichwein-Zientek, L., and Henson, R.K. (2012). Tools to support interpreting multiple regression in the face of multicollinearity. *Frontiers in Psychology*, 3, e44.

Neter, J., Wasserman, W., and Kutner, M.H. (1989). *Applied Linear Regression Models*. Irwin Publishing, Homewood, Ill.

Quinn, G.P. and Keough, M.J. (2002). *Experimental Design and Data Analysis for Biologists*. Cambridge University Press, Cambridge.

Vatcheva, K.P., Lee, M., McCormick, J.B., and Rahbar, M.H. (2016). Multicollinearity in regression analyses conducted in epidemiologic studies. *Epidemiology*, 6, 227.

Winship, C. and Western, B. (2016). Multicollinearity and model misspecification. *Sociological Science*, 3, 627–49.

CHAPTER 9

Statistical Methods for Analyzing Species–Habitat Associations

In this chapter I discuss six methods for analyzing species–habitat associations, starting with the simplest (comparison of group means) and ending with the most complex (occupancy modeling). I occasionally refer back to the example of the hypothetical beetle species presented in Chapter 8. The technical and mathematical aspects of each method are presented in enough detail to give the reader a basic introductory understanding of the method. Indeed, some of the methods (e.g., ANOVA and multiple linear regression) are so common and widely used that most readers will already be generally familiar with them, and thus some very elementary details are left out. My goals in this chapter are to (1) explain the general mechanics of each method, (2) discuss the application of each method in the context of habitat analysis, (3) point out strengths and weaknesses of each method, and (4) explain and simplify the technical jargon associated with many of the methods. Overall, these goals are intended to encourage the reader to think carefully about their choice of method when conducting a habitat analysis and even more importantly provide the reader with some initial confidence to delve into the statistical literature and learn more about these methods. Lastly, to facilitate the actual use of these statistical methods, at the end of the chapter, I present an appendix that provides information on R packages and code to implement the various methods (Appendix 9.1).

9.1 Comparison of group means

This is the most straightforward and basic method of habitat analysis. It involves the use of t-tests, ANOVA, or their non-parametric equivalents to compare the mean values of environmental variables among two or more groups (of spatial survey units such as sampling plots or transects, study sites) defined by presence–absence of the species or abundance. ANOVA is such a fundamental and common method of statistical analysis that I will forgo a detailed mathematical description—most introductory statistical textbooks have a lucid explanation. Rather than abundance, the groups to compare could also be based on activity level, visitation rate, or any variable indicating the species' association with or use of the spatial survey unit. The variable could even be a parameter of demographic "performance" such as reproductive rate or survival rate (Chapter 5), although such data are much more difficult to collect than simply presence–absence, abundance, or activity. Rather than being a response (dependent) variable, the species data are used as a grouping variable. In this way, comparison among group means is different from most of the other methods that I am discussing. The actual response variables are the environmental data that have been collected in the survey units. Therefore, it is particularly important to confirm normality of the environmental variables (Section 7.8) and transform the variables if they are not normally distributed.

Checking for multicollinearity among the environmental variables is not as important (as confirming normality) in that comparing among group means involves applying a statistical test (e.g., t-test, ANOVA) *separately* to each variable. However, prior to conducting a statistical test for each environmental variable, a MANOVA can be applied

Habitat Ecology and Analysis. Joseph A. Veech, Oxford University Press (2021). © Joseph A. Veech. DOI: 10.1093/oso/9780198829287.003.0009

simultaneously to all the environmental variables (Box 9.1). Although the MANOVA does not directly indicate or control for the possibility that two or more of the response variables are correlated, a significant *P*-value (from the MANOVA) indicates that two or more of the variables might be correlated and also differ among two or more of the groups. Thus, interpretation of the results of a set of *t*-tests or ANOVAs (one for each environmental variable) would need to proceed with caution because the tests are not necessarily independent of one another. This means that a significant result for one of the variables makes finding statistical significance for the other variable(s) more likely because the two or more variables are correlated. This is essentially what statisticians refer to as study-wide Type I error—finding false statistical significance because a large number of non-independent statistical tests are being performed. The practical problem is that it is difficult for the analyst to know which *P*-values are trustworthy and which represent Type I error.

To be clear, performing a MANOVA is not a required step that must be performed prior to conducting a set of mean comparisons involving multiple environmental variables. In the hypothetical beetle example, I did not do a MANOVA prior to either the *t*-tests or the ANOVA. At least two of the variables (canopy cover and leak litter depth) were

known to be correlated (Table 8.1) prior to my conducting the *t*-tests or ANOVAs. Hence a strict statistical perspective would say that the two *t*-tests for these two variables are not statistically independent, nor are the two ANOVAs (Table 8.2). In reality, one of the variables might be significantly different among the presence and absence groups (and among the abundance groups) simply because it is correlated with the other variable that is significantly different. In other words, it may be difficult to know which habitat characteristic the beetles are really responding to. This hints at one of the shortcomings of using comparison among group means as the method of habitat analysis—it treats each environmental variable in isolation from the others. Throughout this book I've stressed that the habitat of a species (or individual) is a multi-faceted whole composed of potentially many factors (characteristics) that are integrated with one another (Chapter 12).

Although comparison among group means may be somewhat simplistic and overly reductionist, it can be a very powerful way to identify which environmental variables are more important than others in defining the habitat of a species. The holistic integrated perspective of habitat does not entail that all environmental variables are *equally* important aspects of a species' habitat. A species might respond to or require some habitat characteristics more so than others, especially when dispersing and attempting to locate suitable habitat (Chapter 3).

9.1.1 Different types of ANOVA

Typically, ANOVA is not as simple as the analysis applied to the beetle data. There is a plethora of different "types" of ANOVA that are distinguished based upon properties of the grouping variables (treatment factors), number of grouping variables, whether grouping variables are nested, and whether interaction effects between grouping variables are to be tested. The computation of an ANOVA also differs depending on the study design, such as whether blocking is used, the extent to which the replicates are truly randomized, and whether the replicates represent repeated measures. In addition, the researcher might be confronted with unbalanced data (unequal sample sizes for the groups)

Box 9.1 MANOVA

As some readers will know, a MANOVA is a multivariate test that is often used as an a priori assessment of whether the means of two or more response variables (which might be correlated) differ among groups. Importantly, MANOVA examines two (or more) response variables simultaneously in that it is actually comparing centroids (or *n*-dimensional "centroids" for more than two response variables where *n* = number of response variables) rather than means, which are one-dimensional. In MANOVA, the multiple response variables can be thought of as being formed into a combination represented by a single value. Thus, a one-factor MANOVA (one grouping variable) returns just a single *P*-value indicating whether the groups differ in their "average" combination. Quinn and Keough (2002) provide an excellent explanation of the mechanics of MANOVA and its various uses.

and this might determine the type of ANOVA that is used. For the most part, the different types of ANOVA are necessary because the issues just listed affect the structure of the variance within and among groups (regardless of the actual response variables being examined), and every type of ANOVA is essentially a comparison of the variance within groups to that among the groups. Again, most statistical textbooks or manuals have excellent and thorough descriptions of the different types of ANOVA (e.g., Snedecor and Cochran 1989; Sokal and Rohlf 1995; Glover and Mitchell 2002; Quinn and Keough 2002; Vik 2014; Crawley 2015).

One main distinction among the different types of ANOVA concerns whether the grouping variable (or factor) represents a fixed effect or a random effect. If the variable is under the control of or somehow determined by the researcher, then it is a fixed-effect variable, otherwise it is a random-effect variable because presumably its values represent random (more or less) draws from some larger sampling population. More importantly, the actual values (or levels) of a fixed-effect variable are intentionally set or chosen by the researcher and do not necessarily represent a random sample but rather a treatment difference that the researcher is interested in. An ANOVA conducted on a fixed factor is often referred to as a Class I or Model I ANOVA, whereas an ANOVA conducted on a random factor is Class or Model II. Model III or mixed effects describes ANOVAs having both fixed factors and random factors (Sokal and Rohlf 1995; Quinn and Keough 2002). When a single-factor (one-way) ANOVA is conducted, as in each of the ANOVAs applied to the beetle data in which there was only one grouping variable (although there may be more than two groups), the distinction between Model I and II has no bearing on the calculations. That is, the sum of squares, mean squares, and F-ratio are calculated in the same way. In this scenario, the only real difference between Models I and II pertains to the inference from the results. In Model I, inference is limited to discussing treatment differences; that is, the extent to which the mean of the response variable Y differs (or does not differ) based upon how different the groups or treatments are in the grouping variable (or levels of X). In Model II, inference can be put in terms of variation in X (the grouping variable) leading to differences or additional variation in the response variable. That is, inferences about the relationship of X to Y can be cast over the entire range of X including levels or values of X that were not involved in defining the groups. This is why Model II ANOVA has strong conceptual and mathematical similarities to linear regression.

In most cases, a simple one-way ANOVA used in a habitat analysis (as in the beetle example) will involve a grouping variable that is a random factor. At first glance, it may seem like species abundance represents a fixed factor because I decided the cut-off points for grouping the plots. However, beetle abundance in a plot was in no way set or determined by me (even if I had really conducted this hypothetical study) but was instead a variable measured for each plot based upon how many beetles fell into the pitfall traps (Section 8.1). Thus, abundance and presence–absence are both random factors in the beetle example and likely in a vast majority of species–habitat datasets. Further, as explained above, for a one-way ANOVA the distinction between Model I and II is unimportant to the math of the ANOVA.

If a two-way or higher ANOVA is being conducted then the distinction between Models I, II, and III does matter to the math. This is because the F-ratios are calculated in a different way for each model; Gotelli and Ellison (2004, pp. 317–22) provide an excellent explanation of these differences. Recall that P-values and hence statistical significance (or the lack of it) directly derive from the F-statistic. Again, for most habitat analyses, a grouping variable based on species abundance in sampling or survey plots is a random factor. Therefore, a random- or mixed-effects ANOVA (Model II or III) is most appropriate when a continuous variable (e.g., abundance) has been shoehorned into a categorical variable for the purpose of conducting an ANOVA (Gotelli and Ellison 2004).

9.1.2 Multi-way ANOVA applied to beetle data

In the beetle example I used only one grouping variable in each ANOVA (Chapter 8). The grouping variable represented the species response (either presence–absence or abundance level). However, it is possible to use multiple grouping variables and conduct a multi-way ANOVA. As an example, assume that the 100 beetle survey plots were distributed among four different forests or study sites, wherein each one had 25 plots (Appendix 8.1). The forests might actually differ in one or more of the environmental variables and this could potentially affect the outcome of the habitat analysis. To illustrate this, I conducted a nested ANOVA in which presence–absence was nested within forest and percentage canopy cover was the response variable. This ANOVA examined the difference in mean canopy cover between presence and absence groups *within a given* forest as well as the differences among the four forests. There was a significant effect of forest; that is, the mean percentage canopy cover of the 10 × 10 m beetle survey plots differed among the forests (Table 9.1). Given that the "forest effect" was statistically significant; the comparison of presence to absence plots is most appropriately conducted with this factor nested within forest. Nesting the presence–absence grouping factor within forest effectively removes or controls for the effect of forest in "producing" differences in percentage canopy cover. The nested ANOVA revealed a significant difference in mean canopy cover between presence and absence plots in each forest (Table 9.1). Note that this result is more valid than the non-nested one-way ANOVA (Table 9.1), although the latter also revealed a significant difference in mean

canopy cover between presence and absence plots without taking into account the forests where the plots were located. Of course, nesting the species response (e.g., presence–absence) within another variable (e.g., forest) could also reveal a lack of statistical significance. As an example, I conducted a nested ANOVA on oak:non-oak tree ratio. Recall that the one-way ANOVA found a significant difference between presence and absence plots for this variable (Table 9.1). However, this significance disappeared when presence–absence was nested within forest (Table 9.2). Further note that this nesting was necessary because there was a significant direct effect of forest on oak:non-oak tree ratio. On average, plots in Forest A had a greater ratio of oak to non-oak trees than did plots in the other three forests (Table 9.2). Thus the proper inference is that presence and absence plots *do not differ* in oak:non-oak tree ratio when forest location is taken into account. This result partly comes about because Forest A tended to have the greatest ratio of oak to non-oak trees (6.68) and a disproportionate number of presence plots (18 out of 25), whereas Forest D had the lowest oak:non-oak tree ratio (2.11) and a very small number of presence plots (3 out of 25). So the test for an association between beetle presence–absence and oak:non-oak tree ratio is compromised because the forests differ in the extent that oak trees dominate. This does not necessarily mean that oak trees are unimportant in defining the habitat of the hypothetical beetle

Table 9.1 Results of a nested ANOVA applied to canopy cover percentage of plots in four forests (A–D) surveyed for a hypothetical beetle species. Presence–absence as a grouping variable is nested within forest.

Factor	df	SS	MS	F	P
Forest	3	5,715	1,905.0	4.27	0.007
Forest:pres-abs	4	5,029	1,257.2	2.82	0.03
Error	92	41,074	446.5		
	Mean canopy cover percentage of plots in Forest				
	A	B	C	D	
Presence	80.9	79.3	83.3	66.7	
Absence	72.1	58.1	66.1	56.4	
Combined	78.5	67.4	71.6	57.6	

Table 9.2 Results of a nested ANOVA applied to oak:non-oak tree ratio of plots in four forests (A–D) surveyed for a hypothetical beetle species. Presence–absence as a grouping variable is nested within forest.

Factor	df	SS	MS	F	P
Forest	3	351.1	117.0	21.05	<0.00001
Forest:pres-abs	4	10.1	2.52	0.45	0.77
Error	92	511.5	5.56		
	Mean oak:non-oak tree ratio of plots in Forest				
	A	B	C	D	
Presence	6.42	3.88	2.61	1.71	
Absence	7.33	3.33	1.81	2.17	
Combined	6.68	3.57	2.07	2.11	

species, but it does suggest that collection of new data in a different study design might be worthwhile, such as doubling or tripling the number of sampling plots in Forests A and D followed by a separate one-way ANOVA on each forest.

In the hypothetical beetle example, forest could be considered a design variable, particularly if the researcher had control over which forests were to be used in the study. In species–habitat studies, there may be other such design variables that could affect the habitat analysis because of some effect on or relationship to either the species-response variable or the environmental variables (Chapters 6 and 7). In addition to spatial location, such design variables might include time, date, or season during which the data are collected (e.g., consider that leaf litter depth might vary seasonally) and method of data collection (e.g., perhaps pitfall traps were used on plots with minimal leaf litter, whereas Berlese funneling of litter samples was used on plots with substantial leaf litter). To some extent, the unwanted effects of some design variables can be taken care of with a better, more standardized survey or sampling protocol. But if not, then using a nested ANOVA with the species-response grouping variable nested with the design variable could be useful for obtaining valid results from a habitat analysis.

9.1.3 ANOVA on groups defined by an environmental variable

As I have explained it so far, the method of comparing group means uses an ANOVA (or similar test) in which a species variable such as presence–absence or abundance is used to create groups and an environmental property is the response variable. However, an ANOVA could be applied in the reverse direction, wherein an environmental variable is used to create groups and the response variable is abundance. In concept, this approach is more in line with ANOVA as a method of testing whether certain factors have effects on the response variable. Logically, it makes more sense to think of canopy cover as affecting beetle abundance than vice versa. However, the problem here is that most environmental variables cannot easily be divided into groups, particularly if the variable

is continuous (e.g., percentage canopy cover, leaf litter depth) and has a sampling distribution with no natural breaks. Abundance is essentially a count and hence a discrete variable, which perhaps means that it is a bit easier to divide it into groups.

Environmental variables that are categorical could serve as grouping variables in an ANOVA, even if the categories are ordinal and numerical. Returning to the beetle example, soil type is such a variable. I applied an ANOVA to test whether beetle abundance differed among the four soil types. To be clear, soil type was the grouping variable and abundance was the response variable. Given that this response variable included many zeros it had a non-normal distribution and hence I used a Kruskal–Wallis ANOVA, the non-parametric equivalent of a regular ANOVA. Considering the results of the previous analyses it is not surprising that mean beetle abundance did not vary significantly among soil types (K-W test statistic = 2.81, df = 3, P = 0.42). Mean *abundance* appears to vary among the four soil types (1–3.0 beetles, 2–1.3, 3–2.2, 4–0.3) although the K-W test is applied to mean *ranks* and these show less variation (1–29.6, 2–36.3, 3–33.0, 4–39.5) because each soil type was represented by many plots having zero beetle abundance and these plots all had a rank of 41.

The test of whether beetle abundance varies among the four soil types essentially recognizes that each soil type could be considered a discrete type of habitat. This type of test represents an approach to habitat analysis that has often been used, although more so as a way of testing for habitat selection (or preference) rather than an analysis to identify a set of environmental variables (i.e., habitat characteristics) that the species might associate with. That is, the researcher knows and specifies a priori the habitat types that are to be examined. Indeed, this is part of the study design; intentionally establishing a desired number of survey plots (or other spatial sampling units) within each type of habitat, followed by an ANOVA to determine if abundance or some other species-response variable differs among the different types of habitat. For example, Beck et al. (2004) tested whether abundance and survival of the large-headed rice rat (*Hylaeamys megacephalus*, formerly *Oryzomys*) differed between treefall gaps and closed-canopy forests. In their ANOVA, they also included rainy vs. dry season as a factor nested within habitat type; essentially "season" was a design variable as I discussed in the previous section. They found that during the rainy season rice rats were more abundant in treefall gaps than in closed-canopy forests but during the dry season rat abundance did not differ between the two habitat types. Opposite of this, survival of rice rats was greater in the closed-canopy forests during the dry season whereas there was no between-habitat difference in survival during the rainy season. Beck et al. (2004) concluded that closed-canopy forest habitat was likely more important to rice rats in that their survival was greater in that habitat type, at least during the dry season. It is important to note that the study did not directly examine the specific characteristics or properties of each habitat type and thus it is not an example of a habitat analysis as I am presenting in this book. Nonetheless, such studies can be very useful in understanding how individuals of a species select and use habitat.

9.1.4 Creating groups for an ANOVA

In using the method of comparison among group means (e.g., ANOVA) it is important for the analyst to exercise restraint in creating abundance groups, or any other groups based on a non-categorical variable. There is some amount of subjectivity in deciding how many groups to recognize and specifying their boundaries (cutoff points), although this need not mean that the groups and their boundaries are arbitrary and illogical. Certainly, it is poor statistical practice for the analyst to "experiment" (via trial and error) with group number and size so as to try to obtain some preconceived outcome for differences among the groups in the mean values of a particular environmental variable(s). Finding an ecologically and statistically meaningful association of a species with a particular habitat characteristic (environmental variable) should not depend on how the abundance data are divided into groups.

9.2 Multiple linear regression

When many abundance groups are used, the comparison among group means method begins to superficially resemble linear regression in that the goal is to determine the relationship between abundance and the given environmental variable with greater and greater precision. Simple linear regression examines how a unit increase in a single predictor (i.e., independent, x) variable leads to a unit increase or decrease in the response (i.e., dependent, y) variable. The regression coefficient, β, or slope of the regression line expresses this change in the y-variable: $\hat{y} = \beta_0 + \beta X$, where β_0 is the y-intercept or constant "amount" of y that gets added (or subtracted) to arrive at the predicted value of the response variable (symbolized as $\hat{y}$) for a given value of the predictor variable. When there is more than one predictor variable then simple linear regression becomes multiple regression. There is still a y-intercept and each of 1 to N predictor variables has a regression coefficient: $\hat{y} = \beta_0 + \beta_1 X_1 + \beta_2 X_2 + \ldots \beta_N X_N$. These coefficients ($\beta_1$ to β_N) are sometimes referred to as partial regression coefficients, beta-weights, or beta-primes. Each one describes the unit increase (or decrease) in y that occurs with a change in the corresponding x variable *while the other x variables are held constant*. For example, assume we have three predictor variables with these coefficients: $\beta_1 = 0.5$, $\beta_2 = 12.0$, and $\beta_3 = 0.02$. The predicted value of the response variable, $\hat{y}$, increases by 0.5 units for each unit increase of variable X_1 when there is no change in X_2 and X_3. In the multiple regression equation, the terms $\beta_2 X_2$ and $\beta_3 X_3$ would both be constants, so the only way that $\hat{y}$ would change is through a change in X_1. This is what makes multiple regression such a useful statistical tool. The partial regression coefficients quantify the effect of X_i on $\hat{y}$ when the effects of all the other predictor variables are statistically (mathematically) controlled. However, note that this form of statistical control does not take care of any multicollinearity that might exist among the x variables. Indeed, if the x variables are correlated with one another, then it is impossible conceptually and in practice to have all the variables remain constant while only one changes. Multicollinearity can be a serious problem for multiple regression (Box 8.1).

Often the analyst would like to compare the effects of the predictor variables in a multiple regression model. The coefficients, $\beta_1 = 0.5$, $\beta_2 = 12.0$, and $\beta_3 = 0.02$ are in a raw unstandardized form and hence should not be directly compared with one another. For example, β_3 might be substantially smaller than β_1 and β_2 simply due to the units of measurement for X_1, X_2, and X_3. Perhaps X_3 is measured in m² (values from 1,000–10,000), whereas X_1 is a percentage (values 0–100) and X_2 is a count (values 1–10). The units of measurement influence the raw regression coefficients. If X_3 were measured in hectares, then $\beta_3 = 200$ because 1 ha = 10,000 m². In order to remove this influence, raw regression coefficients should always be standardized. This can be accomplished by multiplying the raw coefficient by the standard deviation of the x variable and then dividing by the standard deviation of y: $\beta_i' = \beta \times SD(x_i) / SD(y)$. Alternatively, the response variable and all of the predictor variables could be standardized prior to conducting the multiple regression (Section 7.11). Thus, depending on the values of $SD(x_i)$ and $SD(y)$, the unstandardized coefficients $\beta_1 = 0.5$, $\beta_2 = 12.0$, and $\beta_3 = 0.02$ might actually convert to $\beta_1' = 0.255$, $\beta_2' = 0.634$ and $\beta_3' = 0.881$, indicating that variable X_3 has the greatest effect on y even though it had the smallest raw unstandardized regression coefficient.

Standardized partial regression coefficients also have the useful property of accounting for differences in the variances of the x variables, which is often necessary even when the variables are measured in the same units. For example, again assume that we have three predictor variables with the following raw regression coefficients and range of values: $\beta_1 = 0.5$ $(0 \le x \le 100)$, $\beta_2 = 0.5$ $(20 \le x \le 80)$, and $\beta_3 = 0.5$ $(45 \le x \le 55)$. Although the raw coefficients are identical, the standardized coefficients are not: $\beta_1' = 0.834$, $\beta_2' = 0.543$, and $\beta_3' = 0.104$. This is because the variances of X_1 (26.1), X_2 (16.9), and X_3 (3.2) differ. In this example, the greatest standardized coefficient corresponds to the variable with the greatest variance, X_1, which also has the greatest spread from 0–100. This is another general feature

of regression and many other statistical techniques. The greater the spread of values in the predictor variable or treatment levels, the more likely it is an effect on the response variable will be revealed. In the context of multiple regression this makes sense—there needs to be sufficient variation in the predictor variables in order to determine how well they explain the variance in the response variable.

Mechanically, multiple regression proceeds by attempting to find a set of regression coefficients that minimizes the residual sum of squares, $SS_{res} = \Sigma(y_i - \hat{y}_i)^2$ for $i = 1$ to n observations (or data points). The total sum of squares is $SS_{tot} = \Sigma(y_i - \bar{y})^2$, which if divided by $n - 1$, we would obtain the variance of the response variable; note that SS_{tot} does not include any effect of the predictor variables ($\hat{y}$ is not a part of the summation). The sum of squares accounted for by the estimated regression coefficients (i.e., the fitted regression equation) is $SS_{reg} = \Sigma(\hat{y}_i - \bar{y})^2$. Given that $SS_{tot} = SS_{reg} + SS_{res}$, and SS_{tot} is a fixed value dependent only on the y data, then if SS_{res} gets minimized then SS_{reg} must be maximized without exceeding SS_{tot}. Indeed this is what is happening when a multiple regression analysis produces an F-statistic. It is calculated as $F = MS_{reg}/MS_{res}$, where MS_{reg} is the mean square from the effect of the regression coefficients, $MS_{reg} = SS_{reg}/N$, MS_{res} is the mean square of the residuals, $MS_{res} = SS_{res}/(n - N - 1)$ with n = total number of observations and N = total number of predictor variables. By the way, the denominators are also the degrees of freedom for each MS value. As F gets larger, P-values get smaller, and hence it is easy to see that statistical significance of a regression model becomes more likely as SS_{reg} gets larger compared with SS_{res}. That is, in conducting a multiple regression analysis, we want the regression coefficients to explain as much of the variance in y as possible. A few comments on terminology: MS_{res} is sometimes labeled as MSE or mean squared error (it is the "error" in the regression equation). Also this approach to regression (as described above) is often referred to as OLS (ordinary least squares) or LS regression given that it is based on the least or minimal value of SS_{res}. That is, least-squares estimation finds the values of the regression coefficients (β_X for all predictor variables) that together minimize SS_{res}.

Estimating the statistical significance of individual regression coefficients is a separate task from estimating the overall fit or significance of the entire equation (model). Also, in conducting a habitat analysis, we might be more interested in identifying those environmental variables that seem to be most important in defining habitat for the species rather than assessing the overall fit of a regression model without distinguishing the effects of individual variables (but see Section 10.1). This is essentially a test of the null hypothesis: $\beta_X = 0$ conducted separately as a t-test for each predictor variable. That is, we use $t = \beta_X / SE(\beta_X)$ where $SE(\beta_X)$ is the standard error of the coefficient and obtained from $MS_{res} \times (X'X)^{-1}$ (Box 8.1). If the t-statistic has a corresponding P-value that is significant then we can conclude that the predictor variable has a significantly positive effect on abundance or whatever species-response variable (except presence–absence, see Section 9.3) is being used. That is, the species associates with the environmental factor being measured by X and hence X (or particular values of X) is also a characteristic of the species habitat. The sign of the regression coefficient is important. If positive then the species abundance increases with increasing values of X (at least within the range of values examined), otherwise if β_X is negative then the species abundance decreases with increasing values of X. As an example, consider percentage canopy cover, unlike our hypothetical beetle species, a different species might actually require a very open type of habitat with minimal or no canopy cover in which case we would expect β_{canopy} to be significantly < 0.

Multiple regression models can also include predictor variables such as ancillary or design variables as discussed for ANOVA in the previous section (e.g., forest as a factor possibly influencing abundance of the hypothetical beetle species). Such variables might best be considered covariates given that the researcher might not necessarily be interested in their effects but rather just want to account for their contribution to the overall variance of the response variable. It is also possible to include non-linear predictors such as squared and cubed variables. Consider that the preferred or most suitable habitat condition for a given species might actually consist of intermediate values of a habitat characteristic. That is, intermediate values rather than extremely large or small values define the habitat. As an example, a species might require intermediate canopy cover in the range of 30–70 percent. In that case, a regression model that includes canopy + canopy^2 (and maybe even canopy^3) would fit the data better than a model lacking the squared variable (Fig. 9.1) (Smith et al. 2009; Holoubek and Jensen 2015; Reidy et al. 2014; Roach et al. 2019). As such, the regression coefficients for both variables would be significant. However, in this example, note that among the sampled plots we would need a range of canopy cover that included high and low values in order to fit the squared variable. In other words, X would need to span a large enough range of values so as to make the intermediate "hump" apparent (Fig. 9.1). It is also possible that above a certain threshold level of a habitat variable, a species abundance does not increase. In such a case we would expect an asymptotic effect.

In principle, we could also include interaction effects in a multiple regression model (i.e., a predictor variable that is $X_1 \times X_2$). This would represent testing whether the effect of X_1 (on the response variable) depends on the particular value or level of X_2, and vice versa. An example could be the possibility that the hypothetical beetle species can tolerate less canopy cover as the ratio of oak to non-oak trees increases. In such a case, beetle abundance might be relatively high in a plot surrounded mostly by oak trees even though the actual canopy cover in that plot is relatively low. For some species that maintain a balance or "trade-off" in particular habitat requirements, the modeling of interactions among the environmental variables might be worthwhile and insightful. However, in the initial stages of studying the habitat of a species, it is probably best if the habitat analysis avoids including interactions among predictor variables simply due to parsimony and

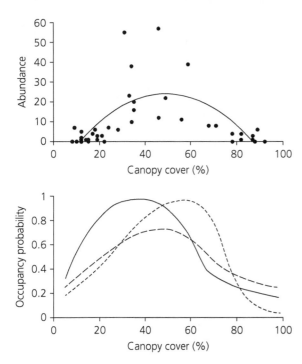

Figure 9.1 Non-linear relationship between species abundance and percentage canopy cover. In the top panel there are $N = 40$ data points (e.g., hypothetical sampling plots where the number of individuals of the focal species were counted and canopy cover was measured) in which the relationship between canopy cover and abundance is best described by a quadratic function ($y = -0.017x^2 + 1.66x - 16.6$, $R^2 = 0.43$). The linear function has a very poor fit ($y = 0.007x + 9.3$, $R^2 = 0.0002$); it is essentially a horizontal line (not shown). The bottom panel shows the relationship between occupancy probability (see Section 9.6) and percentage canopy cover for three hypothetical species. Each of the three has its greatest probability of occupancy (i.e., occurring at a given location) for locations that have an intermediate amount of canopy cover. See Holoubek and Jensen (2015) for similar line plots constructed on real data.

because the interpretation of such interactions can sometimes be complicated.

9.2.1 Negative binomial regression

As discussed previously (Chapter 3), conspecific density can be an important factor influencing settlement decisions by a dispersing organism. Hence, in nature, species often have aggregated or clumped spatial dispersions (Lloyd 1967;

Brown et al. 1995; Shorrocks and Sevenster 1995; He et al. 1997; Condit et al. 2000; Veech et al. 2003; Myers et al. 2013). Moreover, completely random dispersions are very rare (Taylor et al. 1978). Individuals of a species obviously have similar or nearly identical habitat requirements and so they will tend to be found in the same places. Opposite of this, if intraspecific competition is exceptionally strong (and an important resource is limited), then individuals of a species might have an overdispersed or repulsed spatial dispersion. All of this means that the counts of individuals in survey or sampling plots (or other spatial units) are often non-random even apart from the direct effects due to differences in the environmental conditions (i.e., habitat characteristics) among the plots. This effect of non-random spatial dispersion on count or abundance data can be accounted for by conducting negative binomial regression. In concept, negative binomial regression can be thought of as an extension of multiple linear regression. There is a set of predictor variables ($X_1 \ldots X_N$) along with a parameter (θ, or sometimes denoted as k) that estimates the amount of clumping of individuals among the sampling plots: $\hat{y} = \beta_0 + \beta_1 X_1 + \beta_2 X_2 + \ldots \beta_N X_N + f(\theta)$. In reality, estimation of the regression coefficients and the effect of $f(\theta)$ is more complicated than is depicted in this equation. Rather than least-squares estimation (as described in previous paragraphs), estimating the parameters of a negative binomial regression model typically involves maximum likelihood algorithms (Lawless 1987; Zeileis et al. 2008; Hilbe 2011; Cameron and Trivedi 2013) (Box 9.2). In order to conduct negative binomial regression as a habitat analysis, one need not know the mathematical and computational details of maximum likelihood. However, some knowledge of negative binomial distributions is helpful in understanding θ (Box 9.3).

I applied a negative binomial regression model to the hypothetical beetle data (non-zero abundance) and obtained the following standardized partial regression coefficients and P-values: $\beta_{canopy} = 0.059$ ($P = 0.13$), $\beta_{litter} = -0.020$ ($P = 0.60$), $\beta_{woody} = 0.027$ ($P = 0.27$), $\beta_{oak} = 0.085$ ($P = 0.0004$), and $\beta_{soil} = -0.007$ ($P = 0.78$). Compare these values to the coefficients

Box 9.2 What is a maximum likelihood estimate?

For a given observed outcome of a probabilistic event (e.g., rolling a six-sided die five times and obtaining a "2" three out of the five times), a likelihood is a number that indicates the most likely probability for the event. In this example, the event is the roll of the die and if the die is fair (balanced and not lopsided) then we expect $P(rolling\ a\ 2) = 1/6 = 0.167$, but typically the probability of the event leading to an observed outcome is not known. We can identify the most likely probability as the one corresponding to the maximum or highest likelihood. The likelihood of getting a "2" three out of five times (and in any order) is $L = P(rolling\ a\ 2)^3 \times P(not\ rolling\ a\ 2)^2$. The likelihood is only 0.0032 when $P(rolling\ a\ 2) = 0.167$. The likelihood is maximum at $L = 0.0346$ when $P(rolling\ a\ 2) = 0.6$. This probability is the one that would most likely lead to the outcome that we observed. Because 0.6 is quite different from 0.167, we could conclude that something is wrong with the die. In roughly the same way, when conducting a statistical model, we observe the data (more specifically the response and predictor variables) and wish to find the values of the model parameters (e.g., regression coefficients) associated with the highest likelihood. Many of the algorithms for finding maximum likelihood estimates are iterative processes that converge onto the solution. Also, maximization of the *log-likelihood* is often being sought rather than the likelihood itself. For example, the multiple logistic regression model of Table 8.4 has a maximum log-likelihood = −53.984 that corresponds to the estimated regression coefficients given in the table. In a systematic procedure, the algorithm arrived at this solution by repeatedly trying slightly different values for the regression coefficients, calculating the log-likelihood, and then finally stopping when the log-likelihood could not be made any larger (more negative). At that point, the algorithm had arrived at the maximum likelihood estimates for the model's regression coefficients. Note that there are direct mathematical ways (that do not require convergence algorithms) of acquiring maximum likelihood estimates although these can be very difficult to program into computer code unless the model includes only a few predictor variables.

Box 9.3 Negative binomial distributions

A negative binomial distribution of numbers (specifically counts) is based on a binomial process (or Bernoulli trial) in which there are only two possible outcomes, "success," which occurs with probability p, and "failure," which occurs with probability $q = 1 - p$. Success and failure are not meant to be taken literally, they simply represent whether a particular predefined outcome (e.g., rolling a "2" on a die) was obtained during the Bernoulli trial (e.g., rolling the die) or not. Further, defining the event representing a success (or failure) typically allows us to know or at least have a good guess for the value of p (for example, we know that $p = 1/6$ or 0.167 for rolling a "2" on a die, if it is a fair die). There are several ways to formulate (represent) the negative binomial distribution. One way is by the number of r successes obtained in n trials. Given that these numbers are counts, the negative binomial distribution is discrete, not continuous. The probability mass function is

$$P(X=r) = \binom{n-1}{r-1} \times (1-p)^{n-r} \times p^r.$$ This gives the probability

of obtaining exactly r successes in n trials without regard for the order of the successes. The factorial term represents the number of ways of taking $r - 1$ items out of a total set of $n - 1$. So far it may not seem very obvious how or why data on the abundance of a species distributed among sampling plots can be described by a negative binomial distribution. Well consider that there is an underlying binomial process at work. A given individual either is or is not in a given plot (conversely, a given plot either does or does not have a given individual); "in" or "out" are the only two possible outcomes of the trial that is essentially the process of the individual "deciding" to be in the plot or not. This process occurs for every individual and plot and hence the count data accumulate as a negative binomial distribution. But what about the clumping?

The probability equation above does not include the dispersion parameter, k. However, a more general formulation of the negative binomial distribution does include

k, $P(X=x) = \binom{k+x-1}{k-1} \times \left(\frac{k}{m+k}\right)^k \times \left(\frac{m}{m+k}\right)^x$ where m is the

mean of the counts and x is a particular count. Thus, $P(X = x)$ is the probability of a count $= x$ occurring (e.g., number of individuals in a sampling plot) for a given mean (e.g., average number of individuals among sampling plots) and spatial dispersion of the individuals, k. The parameter k must be ≥ 1; further, as k increases, the counts transition from a negative binomial to a Poisson distribution (which depicts a random spatial dispersion). This formulation of the negative binomial has been known for a while, particularly as applied to count data of animal species dispersed among sampling units (Anscombe 1949; Bliss and Fisher 1953; Ross and Preece 1985; Young et al. 1999). Recognize that the mean count when combined with the total number of sampling units (including those with zero individuals) gives the total number of individuals "available" to be dispersed among the units and k is essentially specifying how they are dispersed. The mean and k are related through the variance of the counts, $\mathrm{var}(X) = m + (m^2/k)$. Indeed, as k increases for a constant m, $\mathrm{var}(X)$ decreases. (Note that algebraic rearrangement gives $k = \dfrac{m^2}{\mathrm{var}(X) - m}$.) An ideal Poisson distribution (large k relative to m) has a mean equal to the variance, whereas in a negative binomial distribution the mean is always less than the variance.

As an aside, the name "negative binomial" comes about in the following way. A "positive" binomial, $(p + q)^n$, describes how n number of items are assigned to either of two "groups" represented by the probabilities p and q. These are the only two groups, so $p + q = 1$. For certain sets of numbers (e.g., a series of counts recorded for a set of sampling units) presumed to follow a binomial distribution, p can be estimated as the variance divided by the mean. However, when the variance > mean then p will be >1 and hence q will be negative (to satisfy $p + q = 1$). Further the product, npq, must equal the variance (always a positive number) and as such n must be a negative number, leading to a negative binomial of the form $(p - q)^{-n}$ (Student 1907; Bliss and Fisher 1953; Young et al. 1999). Of course, n as a number of items cannot really be <0, p cannot be >1 and q cannot be <0 since both are probabilities. However, these odd constraints do lead to the quantity, $(p - q)^{-n}$, always being a real number between 0 and 1. Despite this odd derivation, the negative binomial distribution is a useful description of count data that reflects some type of clumping or aggregating process. The negative binomial distribution can describe many different types of count data in a wide variety of scenarios (biological and non-biological), not just counts of individuals among sampling plots. It may have first been used to describe the spatial dispersion of an organism when the statistician W. S. Gosset (writing under the pseudonym "Student") applied it to yeast cells counted in a hemocytometer. By the way, Gosset worked for the Guinness Brewing Company, hence his interest in yeast.

from the regular multiple regression model (Table 8.3). You'll notice that the negative binomial model gave lower estimates for the coefficients. This is because the negative binomial model is also estimating an effect of clumping on the count data; that is, $\theta = 6.72$. By including the clumping effect in the model (and θ in the equation) there is less overall variance in the count data to be explained by the five predictor variables. Nonetheless, there is still a significant effect of the oak:non-oak tree ratio.

We can also consider whether the θ estimate seems to be indicative of a negative binomial distribution (recall that as θ or k increases, the distribution of count data becomes Poisson or random). The θ estimate had a standard error of 3.40, which is only about half of 6.72 indicating that the estimate is fairly precise. Further evidence that the non-zero count data represent a negative binomial distribution (rather than Poisson) is provided by the mean (5.93 beetles per plot) and variance (14.69); the mean is substantially less than the variance (Box 9.3).

From Box 9.3, we can also estimate θ using the mean and variance of the counts, $\theta = k = \dfrac{m^2}{\mathrm{var}(X) - m} = 5.93^2 / (14.69 - 5.93) = 4.01$. This estimate is somewhat less than the value of 6.72 estimated by maximum likelihood. The difference, 4.01 vs. 6.72, is probably not very meaningful. It could be due simply to the much more complicated mathematical algorithms needed to get the maximum likelihood estimate. However, another interpretation could be that the mean and variance of the observed counts suggest a distribution that is a bit more clumped (lower k) than it actually is, assuming the maximum likelihood estimate is more accurate.

To some extent, the "clumpiness" of the counts is evident in the raw data. Of the 40 survey plots with beetles present, 9 plots had ≥10 individuals and 12 plots had ≤2 individuals (Appendix 8.1) (Fig. 9.2). If these counts with the same mean had a Poisson distribution, then 4 plots would have ≥10 individuals and 3 plots ≤2 individuals on average (data simulated in EXCEL®). The negative binomial distribution arises because some plots have an excess of individuals and some plots are deficient in individuals (Fig. 9.2). If the beetle count data had a Poisson distribution (Fig. 9.2), then I could apply Poisson regression. This is a form of regression conceptually

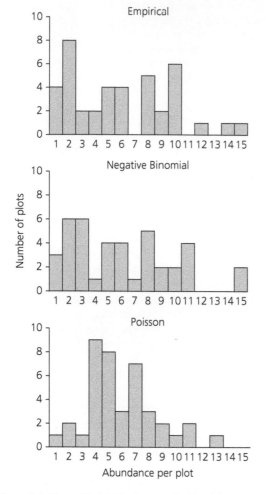

Figure 9.2 The empirical distribution, negative binomial distribution, and Poisson distribution of abundance for the hypothetical beetle species distributed among 40 survey plots (Chapter 8). For the negative binomial and Poisson distributions, mean abundance per plot was specified as 5.93 individuals. For the negative binomial distribution, θ = 6.72 was used although θ = 4.01 would give a similar distribution. Both theoretical distributions were constrained to not have values exceeding 15, which is the maximum abundance value in the empirical distribution. Poisson distributions can be generated in EXCEL® and negative binomial distributions can be generated with either the *rnbinom* or *rnegbin* functions in R.

similar to negative binomial regression. That is, the regression model (or equation) takes into account that some of the variation in the count data may be due to the inherent way in which the species is dispersed among sampling plots irrespective of the environmental or habitat conditions within the

plots. Again, the Poisson distribution describes a spatially random dispersion of individuals (or whatever items are being counted). In Poisson regression, the count data are log-transformed; hence Poisson regression models are often called log-linear models because the counts are also assumed to be a linear function (outcome) of the predictor variables, as in all forms of linear regression.

The benefit of using negative binomial regression as a method of habitat analysis is made clear when one recognizes that there can be a non-random dispersion of individuals among plots (i.e., clumping) beyond the non-randomness that would be induced by the species having certain habitat preferences (requirements) *and* the plots varying in these requirements. Similarly with Poisson regression, one is trying to take into account the amount of the variance in the count data due to an inherently random spatial dispersion relative to the amount of variance owing to the plots having different environmental conditions. Think about this in the context of conspecific attraction (or intraspecific aggregation). Both of these terms describe a process or pattern in which individuals of the same species are attracted to one another or somehow aggregate together possibly owing to factors inherent to the organism itself (e.g., release of and response to pheromones). In applying negative binomial regression, we are assuming that some form of conspecific attraction has influenced the count data and we must take that into account in estimating the "pure and direct" effects of the environmental predictor variables. Poisson regression basically assumes the opposite: individuals of the same species are dispersed independently of one another; that is, the beetles do not attract one another or in any way influence one another with regard to which plot gets occupied.

So, how was the negative binomial regression actually conducted on the beetle count data? I have not yet mentioned this except to say that estimation of the regression coefficients is more mathematically complicated than in typical least-squares estimation. Negative binomial regression can be carried out through generalized linear modeling (GLM). This is a very flexible and broad technique for applying a wide variety of regression models (Box 9.4).

Box 9.4 Generalized linear modeling

Generalized linear modeling (GLM) has its greatest utility in that it relaxes the assumption that residuals from a regression must be normally distributed and homoscedastic (McCullagh and Nelder 1983). GLM proceeds like many other regression techniques in that equations are solved to obtain a set of "ideal" regression coefficients that best account for the variance in the response variable. However, in GLM, regression coefficients are obtained through maximum likelihood algorithms not by least-squares estimation. This is in part why GLM does not require normality of response and predictor variables—it does not involve the matrix algebra described in Box 8.1 and earlier in this section. To be sure, in GLM the algorithms to find the maximum likelihood values for the regression coefficients are not computationally simple. However, in concept, GLM is relatively straightforward. A GLM has three components: response variable Y, a set of predictor variables [X], and a link function g. The response variable has a particular distribution (e.g., normal, binomial, negative binomial, Poisson, gamma) specified by the analyst when building the model, and this in turn determines the "type" of regression that is conducted (see below). The link function connects $\mathbf{X\beta}$ (recall that $\mathbf{X}$ is an $n \times p$ matrix of the measured values for the n observations of the p predictor variables and $\mathbf{\beta}$ is an $p \times 1$ vector holding the regression coefficients) to the *expected* values of the response variable, represented as μ, which is an $n \times 1$ vector. Thus, $\mu = \mathbf{X\beta}$; McCullagh and Neter (1983) referred to $\mathbf{X\beta}$ as the systematic component of a GLM given that is completely specified by measured variables and estimated coefficients.

The random component consists of the measured values of the response variable each of which has an expected value, so $E(Y) = \mu$. However, the link function now comes into play so as to estimate the values in μ without having to worry about normality, $g(\mu) = \mathbf{X\beta}$.

One way to think about GLM is that it uses the link function to "adjust" the predicted values of the response variable. Importantly, this adjustment is done as part of the maximum likelihood process of estimating $\mathbf{\beta}$. This then gives the regression equation used to derive $E(Y)$. The link function can take several different forms. It can be a log function such that $g(\mu_i) = \log(y_i) = \beta_0 + X_1\beta_1 + \ldots X_p\beta_p$ for $i = 1$ to n observations and p predictor variables. The logit link is $g(\mu_i) = \log[y_i \div (1 - y_i)] = \beta_0 + X_1\beta_1 + \ldots X_p\beta_p$. The identity link is $g(\mu_i) = y_i = \beta_0 + X_1\beta_1 + \ldots X_p\beta_p$. The choice of link function and probability distribution for Y depends on the type of regression that one is intending to do. The identity link combined with Y being modeled as a normal or Gaussian variable is equivalent to the typical least-squares regression model. Logistic regression is equivalent to using a logit link and modeling Y as a binomial distribution (Section 9.3). Quinn and Keough (2002) present a thorough discussion of different types of regression (e.g., logistic, negative binomial, Poisson) in the context of GLM. Myers and Montgomery (1997) is a good source for learning more about the computational details of applying a GLM. Lastly, a generalized linear model should not be confused with a *general linear* model. The latter simply describes the use of ordinary least squares (OLS) to conduct a regression.

9.2.2 Zero-inflated regression models

GLM can even be applied to count data in which there are a lot of zeros. Recall that having 60 percent of the beetle survey plots classified as absence plots (or zero abundance) kept me from being able to apply (legitimately) a multiple regression model to the entire dataset. However, there is a class of models called *zero-inflated* that can accommodate data that has excess zeros; two such models are the zero-inflated negative binomial model (ZINB) and the zero-inflated Poisson model (ZIP) (Martin et al. 2005; Sileshi 2006; Zeileis et al. 2008; Zuur et al. 2009; Blasco-Moreno et al. 2019). Each of these models actually involves two separate regression models, one for estimating counts and the other for

estimating zeros (Zeileis et al. 2008). The *count-estimating* model examines the effect of the predictor variables on the count data, while the *zero-estimating* model examines the effect of the predictor variables on the probability of a zero count occurring. The count-estimating model is the negative binomial or Poisson model as described in the previous paragraphs. The zero-estimating model is simply a binary function (i.e., value of the response variable is 0 or 1 indicating a zero count or a non-zero count) of the predictor variables. Importantly, the actual (observed) zero counts can be attributed to two sources. A zero may occur due to the spatial dispersion (clumped or random) of the individuals among sampling plots—and this is estimated by the count-estimating model. A zero may also occur as a result of the binary process (an individual either being in or out of a plot) that is estimated by the zero-estimating model. Essentially, the zero-estimating model is estimating the probability of a zero recorded abundance in a sampling plot beyond the probability that zeros occur due to the spatial dispersion of individuals. The zero-estimating model estimates the *excess zeros* as a function of the predictor variables. Thus, in conducting a ZINB or ZIP regression model, one obtains two sets of partial regression coefficients, one from the count-estimating model and one from the zero-estimating model.

I applied ZINB regression to the hypothetical beetle data, in this case using all 100 plots, which includes the 60 plots in which no beetles were recorded. Again, this is the benefit of zero-inflated models—they can be applied to datasets in which there is a substantial proportion of zeros. Table 9.3 shows regression coefficients for the count-estimating and zero-estimating models. Again we see a significant effect of oak:non-oak ratio on the counts (i.e., beetle abundance) as the other analyses have revealed. There are also significant effects of oak:non-oak ratio and canopy cover on the probability of a zero count occurring. In other words, these two variables are partly responsible for the excess (inflated number) of zeros in the dataset. These could be real ecological effects or it could be that there's something peculiar about low canopy cover plots with a low ratio of oak trees that leads to the pitfall traps not being very effective in capturing beetles. In applying zero-inflated models to count data, the researcher is implicitly acknowledging that there may be more zeros in the count data than

was initially expected and this excess of zeros could be due to some aspect of the data collection protocol. Thus, in using zero-inflated regression models, one is primarily interested in the coefficients from the count-estimating model.

The so-called *hurdle* models are conceptually similar to zero-inflated models (Zeileis et al 2008; Lewin et al. 2010; McMahon et al. 2017). The main difference is that a hurdle model literally splits the data into two portions: $y = 0$ and $y > 0$. Then two regression models (similar to those described for zero-inflated models above) are fit to the two data portions. Because of this split in the data, the count-estimating model is applied only to $y > 0$. That is, zero counts are modeled as coming from only one source, the binary process simulated by the zero-estimating model. For comparison with the ZINB, I also applied a negative binomial hurdle model to the beetle data. The results were very similar for the count-estimating models. Partial regression coefficients for a given predictor variable were nearly identical between the ZINB and NB hurdle regressions. However, coefficients for the zero-estimating models were substantially different. The coefficients from the hurdle regression were different in sign and somewhat different in magnitude than those presented for the ZINB regression in Table 9.3. This is not surprising given that zero-inflated models and hurdle models handle zero-count data in different ways. Moreover, in a habitat analysis conducted as either a zero-inflated model or a hurdle model, the coefficients from the zero-estimating model are not nearly as informative as those from the count-estimating model. The reason for

Table 9.3 Results of using zero-inflated negative binomial (ZINB) regression to analyze the habitat associations of the hypothetical beetle species. Response variable is beetle abundance in 100 sampling plots (60 plots had zero abundance or a count of 0). Table gives the regression coefficients (β) and their standardized values for the count-estimating model and the zero-estimating model.

Variable	Count-estimating model				Zero-estimating model			
	β	$\beta_{std.}$	z	P	β	$\beta_{std.}$	z	P
Canopy cover	0.023	0.142	1.53	0.14	−0.055	−2.548	−2.01	0.04
Leaf litter depth	−0.037	−0.040	−0.59	0.55	0.044	0.371	0.33	0.75
Woody debris	0.308	0.032	1.15	0.25	−0.026	−0.021	−0.046	0.96
Ratio oak:non-oak	0.111	0.087	3.15	0.002	−0.217	−1.306	−2.46	0.01
Soil type	−0.020	−0.006	−0.21	0.83	0.064	0.152	0.30	0.76

conducting either a zero-inflated model regression or a hurdle model regression is simply to use the entire dataset (zero and non-zero data) in a multiple regression model rather than a dataset that includes just non-zero data. Given that both zero-inflated models and hurdle models can be performed as GLMs (Box 9.4) the lack of normality in the distribution of the residuals is not a problem in getting valid estimates of regression coefficients as it would be if a regular multiple regression were applied to a dataset with a substantial proportion (e.g., >10 percent) of zeros (Section 7.8).

9.2.3 Interpreting regression coefficients

From a strict statistical perspective, if a particular environmental variable does not have a significant regression coefficient then it should not be interpreted as a meaningful characteristic of the species habitat, even if the effect size of the variable seems to be substantial (Box 10.1). However, this is not always a black-and-white decision. For any given predictor variable, the numerical value of its regression coefficient may depend on the other variables that are included in the regression model. Even the statistical significance of the coefficient may depend on one or more other variables either being present or absent from the model. This potential inconsistency in the regression coefficient and its statistical significance may seem like a weakness of multiple regression but it actually highlights the flexibility of the method. That is, for any given set of predictor variables, many different models can be conducted and compared (Section 10.2). Further, it should also be fairly obvious that a particular regression coefficient might change as more data are added to the dataset being analyzed. In general, an increase in sample size leads to a more powerful analysis (ability to detect a real effect of a predictor variable). Thus, it is possible that as sample size increases, a non-significant regression coefficient becomes statistically significant even though the magnitude of the coefficient may decrease (i.e., get closer to zero). As with many types of statistical tests, a sufficiently large sample size can often provide a statistically significant result even though the effect size is small enough to bring into question whether the predictor variable really is ecologically meaningful

(Box 10.1). None of these issues with interpreting regression coefficients argues against the use of multiple linear regression as a method of habitat analysis. Indeed, these issues also apply to coefficients that are estimated via the other methods of analysis.

9.3 Multiple logistic regression

As a method of habitat analysis, multiple logistic regression is ideal when the response variable is species presence–absence (coded as 1,0). The regression equation is:

$$\hat{y} = e^{\beta_0 + \beta_1 x_1 + \ldots \beta_N x_N} / (1 + e^{\beta_0 + \beta_1 x_1 + \ldots \beta_N x_N})$$

Importantly, this equation always yields a number between 0 and 1 (note that the numerator and denominator are the same except for the addition of 1 to the denominator). As a consequence, the equation can only be used to model a response variable (i.e., can only be fit to data) that is confined to the 0,1 interval, or $0 \leq y \leq 1$. In addition to presence–absence data, such variables could include nest success defined as at least one fledgling being produced from the nest. This response variable is often used in studies of the nesting habitat of various bird species (Dalley et al. 2008; Peak and Thompson 2014; Inselman et al. 2015). The response variable can also be continuous between 0 and 1, rather than strictly binary. Examples include survival probability of individuals at given locations, that is, in different habitats (Paterson et al. 2014; Johnson et al. 2015; Bonar et al. 2016), amount of time spent by individuals within a survey unit (habitat) as a proportion of total time available, and perhaps even probability that an individual reproduces at a given location. Another benefit of using multiple logistic regression is that one can examine the effects of environmental (habitat) variables in addition to variables such as body size and age of the individual as these might also affect the probabilities of survival and reproduction (Trexler et al. 1992; Paterson et al. 2014).

Survival, time allocation, and activity data are often collected by monitoring (tracking) of known individuals or continual surveillance of survey locations. This approach is sometimes used in studies of habitat selection; this makes sense because marked individuals are being tracked or monitored over time and hence would appear to make active habitat choices (i.e., selection of habitat). However, the data generated by tracking and recording the fate of individuals are also appropriately examined in the more general framework of a habitat analysis as described in this book. Any response variable that can be expressed as a probability or a proportion and that also somehow reflects the species occurrence, use, or demographic output at a particular location or within a given survey area is amenable to analysis with logistic regression. Most importantly, though, the data need to be independent. For example, it would be completely inappropriate to analyze a response variable that is the proportion of a given individual's time spent at different survey locations. This is because the proportions at different locations are not independent and hence the locations are not independent.

Multiple logistic regression can be conducted as a GLM by using the logit function as the link and specifying a binomial distribution for Y. More specifically, the logit is expressed as the natural log (i.e., base e) so that the regression equation is $\ln[\hat{y}/(1-\hat{y})] = \beta_0 + \beta_1 X_1 + \ldots \beta_N X_N$. Given that one of the laws of logs is that $e^{\ln(x)} = x$, we get $\hat{y}/(1-\hat{y}) = e^{\beta_0 + \beta_1 X_1 + \ldots \beta_N X_N}$. Because $\hat{y}$ can be considered a probability, the term on the left side of the equation is an odds ratio. It gives the *likelihood* of event A occurring relative to event B when $\hat{y}$ is the probability of A occurring (e.g., if $\hat{y} = 0.5$, then the odds ratio is 1:1, for every occurrence of B there should be one occurrence of A). We convert an odds ratio to a probability by dividing the odds ratio by itself + 1, therefore we get $\hat{y} = e^{\beta_0 + \beta_1 X_1 + \ldots \beta_N X_N}/(1 + e^{\beta_0 + \beta_1 X_1 + \ldots \beta_N X_N})$, the multiple logistic regression equation.

In applying multiple logistic regression to presence–absence data as a method of habitat analysis, it is important to recognize that the estimated value of the response variable ($\hat{y}$) is a probability, but it is *not* the probability of species presence even when a particular value of $\hat{y}$ is tied to a particular survey (sampling) plot or location. Consider that the species either was or was not at the location ($y = 1$ or 0) and this is known because after all a set of observed or recorded y values composed the data

for the response variable used in fitting the multiple logistic regression equation. For the moment, set aside the possibility that in some survey protocols, some 1s could represent false positives, and for a lot of surveys, 0s may often actually be false negatives. The value $\hat{y}$ as obtained from a multiple logistic regression applied to presence–absence data is the probability that the species (at least one individual) *could* use or occur at a location having the particular combination of environmental variables (and their values) represented by X_1 to X_N and as determined by the regression coefficients β_1 to β_N. It is not the probability that the species actually does occur at the location. Also, the location is unspecified, it could be anywhere (within obvious limits). The probability does not take into account spatial location, either implicitly or explicitly. Another way to see this limitation on the probability is to consider that over some large spatial extent (relative to the dispersal or movement ability of the organism) that encompasses the survey locations, there may be a limited number of individuals, hence dispersal limitation might keep the species from saturating all the available habitat. Further without also knowing the physical accessibility of survey locations, it is impossible for a logistic regression equation alone to give the probability that a particular (spatially explicit) location could have the species. In Chapter 11, in the context of resource selection functions and species distribution models, I further discuss some of these constraints on how logistic regression is applied to data and limitations on inference.

Regardless of the actual response variable (whether it is strictly a binary 1,0 variable such as presence–absence or a continuous variable constrained between 0 and 1), a "perfect" logistic regression model or equation displays the classic sigmoidal form when plotted (Fig. 9.3) if there is a wide enough range in the predictor variable (plotted along the x-axis). Further the exact shape of the curve depends on the regression coefficient (β_x) and y-intercept (β_0) (Fig. 9.3). The sigmoidal shape is due to the logistic regression equation involving an exponentiated number (in this case e^x); any function with an exponent will have curvature except when the exponent is 1. The constraint between 0 and 1 is due to the +1 term in the denominator. For any

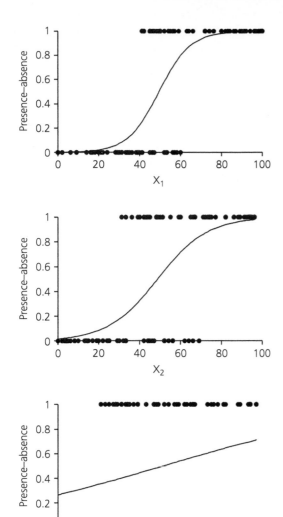

Figure 9.3 Logistic regression curves for species presence–absence data or any binary response variable. Observed values of the response variable are shown as black dots. For predictor variable X_1, the observations of a species absence had values ranging from $X_1 = 0$ to 60 and observations of species presence had values ranging from $X_1 = 41$ to 100. For variable X_2, the range was 0 to 70 for absence observations and 31 to 100 for presence observations. For variable X_3, the range was 0 to 80 for absence observations and 21 to 100 for presence observations. In all three cases, the number of absence and presence observations were equal ($n = 100$ overall). The sigmoidal shape of the logistic regression curve depends on the estimated values of β_0 and β_X (variable X_1: $\beta_0 = -6.44$ and $\beta_1 = 0.131$; variable X_2: $\beta_0 = -4.01$ and $\beta_2 = 0.082$; variable X_3: $\beta_0 = -1.01$, $\beta_3 = 0.019$). Note that the curves derive from single-factor (only one predictor variable) logistic regression models. Also, if the presence and absence data were swapped then the curves would have a reverse S-shape; the signs of β_0 and β_X would switch.

logistic regression model, as β_X gets closer to 0 there is less of an effect of predictor variable X on response variable Y (just as in linear regression). If $\beta_X = 0$ then $\hat{y} = e^{\beta_0} / (1 + e^{\beta_0})$, and β_0 is estimated such that $\hat{y} =$ mean y. That is, as the effect of X_1 becomes weaker, the S-shaped logistic regression curve begins to flatten out and becomes a horizontal line at the mean of the response variable over the entire range of X. This makes intuitive sense. If there is no effect of X on Y, then a predicted $\hat{y}_i$ is simply the proportion of observations recorded as species presence (if the response variable is presence–absence). One way to understand this is to recognize that if the environmental variables in a multiple logistic regression model provide no useful information for prediction (all $\beta_X = 0$), then the best that the model can do is to predict $\hat{y}_i$ as the mean y.

In the logistic regression curves of Fig. 9.3, X_1 has the strongest effect on Y and X_3 has the weakest effect. This is evident in that the values of X_3 are more equally divided between presence and absence observations than are the values of X_1 (i.e., in the plots there is more horizontal overlap of the black dots in the plot of Y vs. X_3 than in the plot of Y vs. X_1). Another inference is that X_3 is not as good at distinguishing presence and absence observations as is X_1. However, logistic regression models revealed statistical significance ($P < 0.05$) of β_0 and β_X for each variable (not shown), including X_3 even though the "curve" does not have the classic sigmoid shape (Fig. 9.3). A multiple regression model that included all three predictor variables revealed a significant effect of X_1 and X_2 but not X_3 (not shown). Importantly, each plot of Fig. 9.3 is for a single-factor logistic regression. The effect of a given variable in the context of the multiple logistic regression model could also be shown as a plotted curve (using β_0 and β_X from the multi-factor model instead of the single-factor), except with the two non-focal variables held at a constant value (typically their means), as exemplified in Fig. 8.4.

The sigmoidal shape of a logistic regression curve is intriguing in the context of a habitat analysis. When it is strong and well defined then there is a narrow range of values of the environmental (predictor) variable in which the predicted response variable ($\hat{y}$) changes rapidly. Assuming the response variable is species presence–absence, this suggests that the species probability of occurring at a given location (subject to the caveats explained above) can quickly transition from being much less than 0.5 to being much greater than 0.5 (or vice versa) even though the environmental variable has changed only slightly. This raises the possibility that some environmental factors (variables) have strong threshold effects that determine or define the habitat of the species, and these can be discovered by applying logistic regression. The most dramatic instantaneous change in $\hat{y}$ occurs at the inflection point. For logistic regression curves, the inflection point is always defined by $\hat{y} = 0.5$, which corresponds to $x' = -1 \times \beta_0/\beta_X$. The inflection point for a given predictor variable X_j in the context of a multiple logistic regression model is obtained as $x'_j = -1 \times (\beta_0 + M)/\beta_j$ where $M = \Sigma(\beta_m \times x_m)$ for all other m predictor variables in the model except predictor j and for each corresponding x_m as a user-defined constant unique for each of the other predictor variables (as mentioned previously x_m is usually set equal to the mean of the predictor variable for plotting purposes [as in Fig. 8.4] but could be set equal to any value). Again, determining x' as a critical threshold of an environmental variable could provide unique ecological insight into some species–habitat relationships, although it only makes sense to calculate x' in this context if there is a strong and statistically significant effect of predictor variable X on the response variable. It is worth noting that all logistic regression equations (curves) have an inflection point at $\hat{y} = 0.5$ (unless $\beta_X = 0$), although it may be very gradual (non-abrupt) and occur at a value of X that is <0, and perhaps not empirically possible or observable.

9.3.1 Logistic population growth as a logistic regression

Recall that in Chapters 2 and 4, I briefly mentioned the logistic population growth model. It describes growth in a population in which there is a carrying capacity, K. The logistic population growth model has been prominent in the historical development of ecology and wildlife ecology because it explicitly recognizes that growth of all populations in a natural setting is limited. Resources, typically food, to

support growth are limited. This is true whether the population is currently at carrying capacity or not. All populations have a carrying capacity.

In the form of a logistic regression equation, logistic population growth can be approximated by $N_t/K = e^{\ln(N_0/K) + rt}/(1 + e^{\ln(N_0/K) + rt})$ where K = carrying capacity, N_0 is the initial population size, and r is the intrinsic rate of increase. Thus the term N_t/K gives the size of the population as a proportion of the carrying capacity. This could be based on the current population (assuming t time steps have elapsed) or a projected population size at some time point into the future. Importantly, N_t/K is always ≤ 1 given that the population cannot permanently overshoot the carrying capacity—and recall that logistic regression can only be applied to a response variable constrained as $0 \leq y \leq 1$. Also, to be clear, in formulating logistic population growth as a logistic regression equation, $\beta_0 = \ln(N_0/K) = y$-intercept, $\beta_1 = r$, and $X_1 = $ time. Given that $N_0/K < 1$, $\ln(N_0/K)$ is always a negative number (i.e., the y-intercept is negative as estimated by logistic regression). Logistic population growth as a regression equation is also equivalent to $N_{t-1}/K = -1 \times [(N_t - N_{t-1})/rN_{t-1}] + 1$, which derives from this expression for logistic population growth, $N_t = [rN_{t-1}(1 - (N_{t-1}/K))] + N_{t-1}$ (Fig. 4.1). Hence, logistic regression is an incredibly flexible and insightful way to model many ecological processes and patterns, including population growth and habitat associations.

9.3.2 Multiple logistic regression as the method of choice

To a great extent, multiple logistic regression is the method of choice for conducting a habitat analysis. It makes use of relatively easy to collect data, species presence–absence, rather than species abundance, which requires more effort and hence might often have greater error. The equation for a multiple regression model provides a probability (or at least an index to a probability—see Chapter 11) that a particular location is suitable habitat for a species given the environmental conditions and factors at the location. Probabilities are very useful in obtaining a better understanding of an ecological (or any scientific) pattern or process. Multiple logistic regression is the behind-the-scenes model in several different techniques for assessing resource use by a species and modeling or mapping the spatial distributions of species. These techniques are related to habitat analysis but are a bit more derived and specialized for addressing questions that are more specific than whether a given species tends to associate with a particular set of environmental variables (i.e., habitat analysis as I have presented it). In Chapter 11, I discuss these related techniques in more detail.

9.4 Classification and regression trees

CART (classification and regression trees) was invented by the statistician Leo Breiman and colleagues in the late 1970s. According to Breiman et al. (1984), the idea of classification trees in the statistical sense originated with Morgan and Messenger (1973) and the precursor to regression trees was the automatic interaction detection procedure developed by Morgan and Sonquist (1963). The goal of both types of tree is to sequentially create smaller and smaller subsets of the initial dataset such that the variation in the response variable within each subset is progressively minimized. That is, building the trees consists of creating increasingly homogeneous groups or subsets of observations. Machine-learning algorithms create the groups by identifying particular values of the predictor variables (from the set of [X]) that separate an existing group into two other groups both of which are more internally homogeneous than the group that they split from. Thus, each branching point on the tree is dichotomous and the splitting or branching criterion is always a single value of one of the predictor variables (e.g., canopy cover <60 percent as in Fig. 8.5). The two groups created from the splitting need not be equal in size (number of observations) and typically are not. Further, a given predictor variable can be used at more than one branching point on a tree and some predictor variables in the set of [X] might not be used at all if they are not helpful in creating homogeneous groups. Hypothetically, tree building could continue in this progressive way until there are $N-1$ branch points resulting in N terminal nodes or final groups (where N = sample size or total number of observations). In such a case each final group would consist of just one observation and hence be perfectly homogeneous. However, in practice, extreme and complete tree building such as this does not occur given that there are algorithms that determine when to no longer split a group (and hence define a terminal group) and algorithms for pruning a tree that is too large or overspecified (see below). This general verbal description applies to classification trees and regression trees; the main difference is that classification trees are used for categorical (qualitative) response variables and regression trees are used for quantitative response variables.

The label "regression tree" might seem to be a misnomer, considering that regression analysis is not strictly being used in building a regression tree. However, this label is historical legacy. Breiman et al. (1984) had a very inclusive and general definition of regression. They wrote "Regression analysis is the generic term revolving around the construction of a predictor d(x) starting from a learning sample, L" (Breiman et al. 1984, p. 221). In this broad sense of regression, the label is accurate. Also note that they referred to a "learning sample"; this is the initial dataset (or a portion of it) that is used to train the algorithms to correctly classify observations. Again, CART utilizes some algorithms that can be considered examples of machine-learning, at least as this term was understood back in the 1970s. To fully understand CART, it is important to recognize that each observation in an initial dataset is eventually "assigned" to only one terminal group. Membership in a terminal group is based upon the particular value or identity of the response variable of the given observation; observations in the same terminal group have very similar or identical identities. The predictor variables of an observation are not being classified; rather they are used to guide the classification. By analogy consider dichotomous keys as used to determine the taxonomic identities of given individual organisms. The process of "keying out" specimens involves following a hierarchical taxonomic key wherein each decision point on the key requires an assessment of a particular trait of the specimen. For example, antennae length greater than or equal to body length might be a trait distinguishing long-horned beetles (family Cerambycidae) from other types of beetle. The specimens are being classified, not their traits; the traits are simply used to guide the classification or get the specimen into the correct group. Similarly, when CART is used as a method of habitat analysis on species presence–absence data, the environmental characteristics of a given surveyed location (sampling plot or other spatial unit) are used to classify the location to a group consisting of either presence locations, absence locations, or some mix of both. The locations or plots are being classified.

9.4.1 Splitting of groups

When regression and classification trees are being constructed, algorithms seek to split a group (node) into two smaller groups in such a way that the within-group *homogeneity* of the response variable is *maximized*. Alternatively, some descriptions of CART refer to node *impurity* and the algorithms operating to *minimize* the impurity (in this context, homogeneity and impurity are simply opposites of one another). For regression trees, homogeneity of a node (group) is assessed by calculating either the mean squared error $[H = \Sigma(y_i - \bar{y})^2 / n]$ or the mean absolute deviation $[H = \Sigma |y_i - \bar{y}|/ n]$ of the $i = 1$ to n observations assigned to the group. Lower values of H indicate greater homogeneity in the sense that the sum of squares or sum of the absolute deviations is smaller for the given group size. Note that in both metrics, the comparison of y_i values is to the mean $(\bar{y})$ not to the predicted value $(\hat{y})$ as would be the case in a regression. So the algorithms are based on finding the least-squares or least-absolute deviation estimators for the initial group (that could be split) and the two groups that would result from the split, and then comparing these values. In the jargon of CART, this is referred to as assessing the *information gain* of a split. This is simply a determination of whether the splitting of a group into two smaller groups further increases the homogeneity (or reduces the impurity) of the response variable. Thus, information gain IG $= H - [(H_1 \times n_1)/n] - [(H_2 \times n_2)/n]$, where H is the homogeneity of the initial group and H_1 and H_2 are the homogeneities of the two smaller groups that would result from the split. If H_1 and or H_2 are substantially less than H, then IG is relatively large and hence the split provides a gain in information. In such a case, the initial group should be split; that is, the tree should grow two

more branches each with a node or leaf at the end. In practice, growing a tree in this way requires that the researcher specify a priori a value for IG that must be met in order for the algorithm to perform the split. This critical value can be thought of as a percentage of the homogeneity in the initial group (H). For example, in order for the algorithm to accept a split then perhaps the information gain must be at least 10 percent or greater. The formula for assessing IG also favors more equitable splits ($n_1 \approx n_2$), but again as mentioned previously n_1 need not equal n_2. The phrase "information gain" might seem a bit mysterious. However, consider that ultimately we want a tree to be informative. That is, the classification criteria for the terminal groups should be useful (have high predictive accuracy) in placing observations into the correct group. By analogy, a taxonomic key should be informative and useful for classifying a specimen to its correct taxon and consequently the members (specimens) in a terminal group should be relatively homogeneous particularly as compared with members in other terminal groups and more inclusive groups closer to the proximal portion of the tree.

The calculation of homogeneity proceeds a bit differently for classification trees. Recall that classification (as opposed to regression) trees involve a categorical response variable that also sometimes is qualitative. In ecology, examples of such variables are species identity, ecological guild, diet, behaviors, life stage, and sex of an organism—really any property or characteristic of an individual organism, population, species, ecological community, and even ecosystem that can be defined or delineated as a category. As such, one might be interested in developing a classification tree so as to predict or correctly assign an observation to a category, and thereby gain some ecological knowledge about the categories as well as the predictor (classifying) variables. In the more focused context of a habitat analysis, classification trees are often used for species presence–absence as the response variable. As should be very clear by now, such data are typically treated as a binary variable coded as 1 (species present at the sampling location) or 0 (species absent). In this scenario, the observational units (to be classified) are plots or some type of spatial location that has been surveyed for the species of interest. As an aside, classification trees can be applied to a response variable in which there are more than just two categories. Given that a response variable is categorical (and often non-numeric), homogeneity of a group cannot be assessed by mean squared error or mean absolute deviation. Therefore, classification trees rely on metrics of homogeneity (or its opposite, impurity) that are based on proportional representation of the categories among the observations within a group.

In building classification trees, two of the more common measures of node homogeneity are also well known to some ecologists in a different context. These are the Gini index (or coefficient) and the Shannon entropy coefficient, which are often used as abundance-based metrics of species diversity within ecological communities such as when a researcher wants more detailed information than simply species richness. The Gini index as a measure of node homogeneity is $H = 1 - \Sigma p_i^2$, and the Shannon coefficient is $H = -1 \times \Sigma p_i \ln(p_i)$ where p_i = the proportion of all observations (in the group) representing category i. Again, for species presence–absence as the response variable there are only two categories, hence $p_{pres} + p_{abs} = 1$ where p_{pres} = proportion of observations (or plots) in which the species was present and p_{abs} = proportion in which the species was absent (or at least not found during sampling or surveying). Smaller values of the Gini index and the Shannon coefficient indicate greater homogeneity, which means less equitable representation between the two categories. Information gain is calculated and interpreted in the same way as explained above for regression trees, IG = $H - [(H_1 \times n_1)/n] - [(H_2 \times n_2)/n]$.

9.4.2 Pruning of trees

As previously mentioned, tree building could hypothetically continue until each observation is a node unto itself. However, such a tree would not be very informative or useful in prediction and it would be overfitting the data (i.e., modeling noise or error in the data rather than meaningful pattern) (Breiman et al. 1984; Krzywinski and Altman 2017). Therefore, there must be ways to halt the splitting of nodes. One approach is to have a predefined value for IG that must be met in order to further split a

node (explained above). Although this approach is widely implemented in many different CART software packages and computer codes, there are drawbacks (Breiman et al. 1984; De'ath and Fabricius 2000). If the splitting algorithm is based on an amount of information gain that is too small, then one will still end up with an overbuilt and overfitted tree. If the algorithm is based on an amount of information gain that is too large, then the benchmark for splitting is too conservative. This could prevent the discovery of important splitting variables that only come into play as the tree expands; that is, variables that help define group membership once other variables have reduced overall heterogeneity somewhat. To overcome these procedural issues, Breiman et al. (1984) introduced and suggested the ideas of tree pruning and cross-validation.

Ultimately in conducting CART we want to obtain the best tree, one that is neither too small nor too large (for reasons discussed above) and that ideally has the best predictive accuracy. In the method of pruning, an overly large tree is first allowed to grow (although perhaps not to the very largest size possible). Then sub-trees of it are produced that have fewer terminal nodes and thus overall smaller size, but otherwise the same branching pattern. The best sub-tree is identified as the one having the lowest misclassification error rate (if a classification tree) or the lowest overall sum of squares (if a regression tree). Then sub-trees of this tree are produced and examined accordingly. Finally, the pruning process stops when the given sub-tree does not have any sub-trees that are better with regard to predictive accuracy. In this way, one need not examine every possible tree (of a given dataset), which would be computationally time-consuming and inefficient.

Cross-validation involves randomly dividing the initial dataset into two subsets, one to be used in building the tree and one used to test the predictions of the tree. This process is repeated over and over with different random divisions of the initial dataset and hence different trees being constructed and tested. The tree with the best predictive accuracy is taken as the best tree. Some statisticians consider cross-validation to be a type of pruning. It can be to the extent that an overly large tree is first constructed and then cross-validation used in the process of examining the predictive accuracy of its sub-trees. However, cross-validation need not involve a pruning process. That is, the analyst might decide a priori the maximum size of tree to produce. Subsequently, cross-validation is simply used to determine the predictive accuracy of a tree without examining the accuracy of any sub-trees. In a much broader context, cross-validation is a general approach to model validation even beyond CART. It involves using part of a dataset for building or "training" a model and the other part for validating the model (Section 10.3).

Even the algorithms for pruning and cross-validation involve somewhat arbitrary decisions, such as the relative sizes of the building dataset and the testing dataset. However, because these methods use the initial dataset more completely and have an underlying foundation in random sampling theory, they are considered more robust than simply building trees (splitting nodes) on the basis of information gain alone. The overall goal of CART is to obtain the best tree in the sense of parsimony (smaller trees utilizing fewer splitting variables favored over larger trees) and predictive accuracy. However, it is well known that CART can be sensitive to slight changes in the data (Breiman et al. 1984; Elith et al. 2008; Krzywinski and Altman 2017). That is, the branching pattern and variables used to define splits can be affected to some extent by unknown measurement or sampling error in the predictor and response variables as well as simple natural heterogeneity unaccounted for by any of the included predictor variables. Tree *bagging* has been proposed as a way to overcome this data sensitivity. Bagging involves taking bootstrap samples (resampling with replacement) from the overall dataset and constructing trees from the random samples. Then the output of the trees is combined. For regression trees the output is averaged given that the output is numerical such as the mean of the response variable for observations in the same terminal group. For classification trees, the output is assignment of a given observation to a category, so arithmetic averaging is not possible. Rather the output is combined by placing a given observation in the category that the greatest number of bootstrapped trees assigned it to. *Boosting* (or boosted regression trees) is similar to bagging except that the resampling intentionally targets problematic observa-

tions that are difficult to classify or produce substantial prediction error (Prasad et al. 2006; De'ath 2007; Elith et al. 2008). Again, the idea is to obtain a combined tree that is stable or relatively insensitive to peculiarities in the data and is also useful for prediction or making inferences about the effects of the predictor variables on the response variable.

The construction of *random forests* is another approach to obtain robust trees in the face of data sensitivity. In constructing random forests, the potential predictor (splitting) variables for each node are a random subset of all possible variables, and the random subsets differ among the nodes. Thus, a given tree is constructed based on only some of the predictor variables being eligible as splitting variables at each node. By leaving out some variables, other variables might be used for a given split when otherwise they would not have been chosen by the algorithm. This process is repeated over and over producing thousands of trees (a forest) whose output is then combined as described in the previous paragraph. The overall idea here is that individual trees may be sensitive to the particular structure (including measurement error) in the original data, but the forest is not because it is a combination of all the trees (Prasad et al. 2006; De'ath 2007).

9.4.3 Using CART for habitat analysis

As a method of habitat analysis, CART can be applied when the response variable is species presence–absence (Fig. 8.5), abundance (Fig. 8.6), or any other variable measuring the activity level of individuals or demographic performance of a population at spatial locations where some set of environmental variables has also been measured. Really, any quantitative response variable that could be examined in a regression analysis is also suitable for constructing a regression tree. When the response variable is categorical then a classification tree can be constructed. Upon conducting a CART analysis and obtaining results, meaningful associations between the species-response variable and a given environmental variable can be inferred by the proportion of times the variable is used to split nodes and whether it tends to split the very first node and also define terminal nodes. The first split is important because it then sets the stage for all subsequent branching in the tree and the last splits are important because they typically are identifying the predictor variable and a fairly precise range of values that most accurately determines group membership. For example, canopy cover ≥80.5 percent was the deciding factor in predicting presence of the hypothetical beetle species as well as a relatively high abundance of the beetle (Figs. 8.5 and 8.6).

CART is used routinely in a very wide array of research fields, including the social sciences. It has a very rich and deep history in the statistical field of classification and the computing field of machine-learning. Improvements to CART are still being developed particularly with regard to splitting algorithms and stopping rules. For a discussion of CART in a broad historical context, the interested reader should see Steinberg (2009) or Loh (2014) and companion papers to Loh's review. De'ath and Fabricius (2000) is still one of the best introductory explanations of CART for an ecological audience.

9.5 Multivariate techniques

Multivariate statistical analysis is a huge area of statistics encompassing many different techniques. The common feature of all multivariate techniques is that multiple *response* variables are examined simultaneously. Indeed, this is the definition of "multivariate," as opposed to "multiple" as in multiple regression, which is not a multivariate technique because it involves multiple predictor variables but only a single response variable. (A technique called multivariate multiple regression actually exists, but it is not discussed in this book.) In general, multivariate techniques can involve either ordination or classification. Ordination techniques attempt to produce order in and among the response variables (which might mean finding underlying structure among the variables) and classification techniques seek to classify sampling replicates (e.g., survey plots or observations) into groups based upon the response variables measured for the replicate. Put another way, the goal of ordination is to *separate* sampling units along an environmental gradient (as represented by one or more response variables), whereas the goal of classification is to

group similar units (as based on one or more response variables) *together* in distinct classes (Gotelli and Ellison 2004).

9.5.1 Principal components analysis

Principal components analysis (PCA) is a multivariate ordination technique that can be used to examine the habitat associations of a species. PCA was briefly introduced in Section 7.9 and discussed in Chapter 8 as applied to the beetle data. In this section, I explain the mechanics of PCA a bit more. As with many multivariate techniques, PCA creates "new" variables from linear combinations of multiple response variables. Thus, each new variable has a contribution from each of the original response variables inputted into the PCA. The number of new or outputted variables equals the number of inputted variables, although some of the outputted variables may often be meaningless and hence useless for further analysis or interpretation (see below). The conceptual goal of PCA and other multivariate techniques is to reveal order or pattern in multivariate data that might not otherwise be apparent. The practical goal is to produce new variables that may be more meaningful or appropriate for further analyses than are the original variables (Section 7.9). (As a brief digression, in describing PCA in the context of a habitat analysis, it may seem confusing or misguided to refer to the inputted environmental variables as "response variables." However, by the strict definition of "multivariate analysis," the environmental variables are response (or outcome) variables that have been *measured* and are being *analyzed simultaneously*.)

The variance–covariance matrix is at the very heart of most multivariate techniques. The numbers stored in this matrix are the variances and covariances of the input variables. For example, if we included five environmental variables in our habitat analysis, then we would have the following variance–covariance matrix:

$$C = \begin{bmatrix} var(1) & cov(1,2) & cov(1,3) & cov(1,4) & cov(1,5) \\ cov(2,1) & var(2) & cov(2,3) & cov(2,4) & cov(2,5) \\ cov(3,1) & cov(3,2) & var(3) & cov(3,4) & cov(3,5) \\ cov(4,1) & cov(4,2) & cov(4,3) & var(4) & cov(4,5) \\ cov(5,1) & cov(5,2) & cov(5,3) & cov(5,4) & var(5) \end{bmatrix}$$

where $var(i)$ is the variance of variable i and $cov(i,j)$ is the covariance between variables i and j. Note that $cov(i,j) = cov(j,i)$. In applying a PCA and other multivariate techniques, each variable is first standardized (converted to have mean = 0 and SD = 1) and then the variance–covariance matrix is filled. To briefly return to the hypothetical beetle data, the variance–covariance matrix is filled with these numbers for canopy cover (variable 1), leaf litter depth (2), woody debris volume (3), ratio oak:non-oak (4), and soil type (5):

$$\mathbf{C} = \begin{bmatrix} 1 & 0.899 & 0.264 & 0.055 & -0.046 \\ 0.899 & 1 & 0.253 & 0.033 & -0.029 \\ 0.264 & 0.253 & 1 & -0.085 & -0.049 \\ 0.055 & 0.033 & -0.085 & 1 & 0.138 \\ -0.046 & -0.029 & -0.049 & 0.138 & 1 \end{bmatrix}$$

The variance for each variable equals 1 given that the SD = 1. Note that the covariances (off-diagonal elements) are also the Pearson correlation coefficients of Table 8.1. An important point here is that the variance–covariance matrix (in raw and standardized form) exists apart from doing any multivariate analysis. It is a descriptor of the variation in and among a set of variables.

As with many statistical techniques, PCA proceeds by solving a set or series of equations that can be represented and solved by using matrix algebra. Consider the equation: $\mathbf{CX} = \lambda\mathbf{X}$, where $\mathbf{C}$ is the variance–covariance matrix just described, $\mathbf{X}$ is a vector having n rows (where n = number of response variables), and λ is a scalar (i.e., a single number). This equation in matrix form actually represents n separate equations with the first being: $var(1)x_1 + cov(1,2)x_2 + \ldots cov(1,n)x_n = \lambda x_1$, and the last being: $cov(n,1)x_1 + cov(n,2)x_2 + \ldots var(n)x_n = \lambda x_n$. That is, numbers in the first row of matrix $\mathbf{C}$ are each multiplied by the number in each successive row of $\mathbf{X}$, and these products are then summed and set equal to λx_1. This continues until the last rows in $\mathbf{C}$ and $\mathbf{X}$ are multiplied together and set equal to λx_n. Typically, there are multiple different values of λ that can satisfy the equations and each of those values of λ has a different (unique) set of numbers in $\mathbf{X}$. Each λ is an eigenvalue and there are n eigenvalues for the variance–covariance matrix. Each eigenvalue has an associated eigenvector represented by

the numbers in $\mathbf{X}$. The equation $\mathbf{CX} = \lambda\mathbf{X}$ can also be represented as $\mathbf{CX} - \lambda\mathbf{X} = 0$ or $(\mathbf{C} - \lambda\mathbf{I})\mathbf{X} = 0$ where $\mathbf{I}$ is the identity matrix for $\mathbf{C}$ (i.e., an $n \times n$ matrix wherein the diagonal elements are 1s and the off-diagonal elements are 0s) and $\mathbf{0}$ is an $n \times 1$ vector of zeros. If there are real number solutions for $\mathbf{X}$, then this equation holds: $|\mathbf{C} - \lambda\mathbf{I}| = 0$ where $\mathbf{C} - \lambda\mathbf{I}$ is a vector of n rows. That is, if the n determinants of the vector $\mathbf{C} - \lambda\mathbf{X}$ are equal to zero then the λ values are eigenvalues of the variance–covariance matrix $\mathbf{C}$. All of this matrix algebra, just described, is a way to get eigenvectors from the variance–covariance matrix. The eigenvectors are matrices (vectors) and they are important because they hold the factor loadings.

An eigenvalue and its corresponding eigenvector define a principal component, sometimes referred to as an "axis." As an equation, a principal component can be represented as $Z_{ik} = x_{i1}y_{1k} + x_{i2}y_{2k} + \ldots x_{ij}y_{jk}$, where x_{ij} is a number (corresponding to response variable j) in the eigenvector for principal component i, and y_{jk} is the measured and standardized value of response variable j for observation or sampling unit k. The values x_{ij} for $i = 1$ to n principal components and $j = 1$ to n response variables are called *factor loadings*. That is, in PCA an eigenvector contains the factor loadings for a given principal component. For example, there are five principal components (Z_1, Z_2, Z_3, Z_4, and Z_5) for the hypothetical beetle data (Table 9.4). The equation for each principal component (Z_i) allows one to calculate a PC score for each survey or sampling unit (Section 7.9). This is essentially a new composite variable that can be used in further analyses, as represented by the scatterplot (Fig. 8.8) and t-tests that I applied to the scores of PCs 1 and 2 of the beetle data (Section 8.4.5). Another useful property of a principal component is the proportion (or percentage) of the overall variance within and among response variables that is accounted for by the particular factor loadings of the PC. This proportion is obtained as λ_i divided by $\Sigma\lambda_i$ for $i = 1$ to n. Recall that λ_i is one of the eigenvalues from the variance–covariance matrix. For any principal component, λ_i is also the variance in the PC scores. For example, for the 100 plots of the beetle dataset, PC 1 scores ranged from –2.64 to 3.14 (Fig. 8.8) and had a variance of 2.03. That is, $\lambda_1 = 2.03$, which is

Table 9.4 Results of the principal component analysis applied to the environmental variables measured in 10 × 10 m plots for a study of the habitat associations of a hypothetical beetle species.

	λ	% Cov. explained	Factor loadings (eigenvector X)					Sum of factor loadings
			Canopy cover	Leaf litter depth	Woody debris volume	Oak:non-oak tree ratio	Soil type	
PC 1	2.033	40.7	−0.666	−0.664	−0.334	−0.021	0.061	−1.624
PC 2	1.167	23.3	0.089	0.087	−0.278	0.712	0.632	1.242
PC 3	0.899	18.0	−0.117	−0.100	0.586	−0.369	0.704	0.704
PC 4	0.801	16.0	−0.176	−0.214	0.684	0.596	−0.317	0.573
PC 5	0.101	2.0	−0.709	0.704	0.012	0.021	−0.016	0.012

40.7 percent of $\Sigma\lambda_i = 5$ (Table 9.4). Note that $\Sigma\lambda_i =$ the number of response variables because each response variable was initially standardized (prior to the PCA being applied), which means that the variance was 1.0 for each variable (this is important in the next paragraph). There is substantial debate about how much variance a PC axis needs to account for in order for it to be deemed worthy of further use in an analysis or even worthy of ecological interpretation (Jackson 1993; Franklin et al. 1995; Quinn and Keough 2002; Peres-Neto et al. 2005; Dray 2008). There is no consensus on a requisite minimal percentage. However, many researchers agree that when a given PC axis accounts for relatively small amounts of structure in the multivariate data it should not be used in any further analyses; that is, it is not very informative (Jackson 1993; Jongman et al. 1995; Quinn and Keough 2002; Gotelli and Ellison 2004).

Rather than making arbitrary decisions as to how the output of a PCA is used, it is important to recognize what exactly a PCA is attempting to do. The mathematical underpinnings of a PCA attempt to find a set of factor loadings that will reproduce as much of the variance–covariance structure as possible given the constraint that all λ values sum to the number of variables. Thus, any given λ_i value must be $\leq n$. As explained previously, a variance–covariance matrix has n eigenvalues (λ values), the largest eigenvalue and its eigenvector (factor loadings) represent the first PC axis, and so on until the smallest eigenvalue is taken as the nth PC axis. The sum of the factor loadings for the first PC axis is always greater than the sum for the nth PC axis (Table 9.4). Also, factor loadings for PC 1 will tend to be rela-

tively large in magnitude and either consistently positive or negative, as in the beetle example (Table 9.4). Given the constraint that $\Sigma\lambda_i = n$, the beetle example could have hypothetically had a value of $\lambda_1 \approx 5$. However, such an eigenvalue does not have an eigenvector **X** that would satisfy $\mathbf{CX} = \lambda\mathbf{X}$. If a PCA is applied to a set of response variables that are highly correlated with one another, then λ_1 will tend to be much larger than all the other λ values and the factor loadings will have the same sign and about the same magnitude. If a PCA is applied to a set of random response variables that are completely uncorrelated with one another, then the λ values will not differ much. In such a case, there is no one PC axis that explains a substantial amount of variation among the variables; for example, PC 1 might account for only a little more than $(1/n)$ proportion of the explained variation. This is why PCA is typically unnecessary (or even useless) unless there is some amount of multicollinearity among the variables (Gotelli and Ellison 2004).

Notice that in Table 9.4 the factor loadings for canopy cover and leaf litter depth tend to be about the same on a given PC axis, except for the very last axis (PC 5) in which case they are nearly "polar" opposites. This is because these two variables are highly correlated. This correlation is reflected in their very similar factor loadings on the first four axes, but for PC 5 there is so little unexplained covariance remaining among the five variables that the factor loadings for canopy cover and leaf litter depth offset one another and the factor loadings for the other three variables are all close to zero. Note also that each axis typically has one or two variables that have factor loadings considerably greater in magnitude than the loadings of the other variables. For example, oak:non-oak tree ratio and soil type have the greatest factor loadings for PC 2. The typical interpretation of such situations is that the given PC axis is mostly characterized or defined by factor X. That is, for the beetle survey plots, we would conclude that canopy cover and leaf litter depth characterize the first axis, whereas oak:non-oak tree ratio and soil type characterize the second axis. Further, the combination of woody debris volume and soil type characterize PC 3 and woody debris along with oak:non-oak tree ratio characterize PC 4. The percentage of the covariance accounted for by PC 5 is so low (2 percent) that any interpretation of it is probably unnecessary or even inappropriate. As illustrated by Table 9.4, PCA can produce a wealth of information about the inter-relationships among a set of environmental variables—and potentially a lot of patterns for an analyst to interpret (imagine if there had been 10 variables instead of five). However, PCA as a method of habitat analysis is truly informative only when it is followed by some type of comparison of observations representing species presence–absence or abundance (as was done for the hypothetical beetle species in Section 8.4.5 and Fig. 8.8). That is, conducting a PCA on a set of environmental variables alone (without linking it to species data) does not say anything about how the species might associate with some of those variables.

PCA is a robust and flexible multivariate technique that is widely used for many different purposes in various biological fields (not just ecology) and other disciplines. The explanation that I have given above is focused (for obvious reasons) on its use in conducting a habitat analysis wherein the input data are environmental variables measured on survey or sampling plots or some other type of spatial unit for which the researcher has also recorded some type of species response (e.g., presence–absence or abundance). However, this need not be the case; PCA can be applied to other types of variables measured on other types of replicates. In addition, PCA can serve purposes other than directly being involved in an analysis of habitat. It is widely used to control for multicollinearity among variables and as a variable-reduction technique for creating composite variables that then undergo statistical analysis, as was discussed in Section 7.9.

9.5.2 Discriminant function analysis

Discriminant function analysis (DFA) is another widely used multivariate technique, although its mathematical operation and purposes for application are very different from PCA. There is one conceptual similarity to PCA. In both PCA and DFA, algorithms create linear combinations of the response variables (i.e., input variables) and as such each variable has an associated coefficient (recall the factor loadings of PC axes). A main difference is that in PCA a linear combination of the variables (the PC axis) is intended to coalesce the variables. It unites the variables into a new composite variable that captures some amount of the total covariance among the original variables (e.g., PC axis 1 typically explains a relatively large percentage, say >30 percent, of the covariance). DFA does the opposite. The linear combination of variables, referred to as the discriminant function, seeks to maximize the differences between two or more groups (categories) defined by a grouping variable. PCA involves no such grouping variable. The coefficients for each variable in a discriminant function have values such that the ability of the function to discriminate among the groups and place an observation into the correct group is as good as possible (given the data). DFA does not produce composite variables, as does PCA. Rather DFA uses a discriminant function to classify observations to groups. At its heart, DFA is a classification technique.

DFA also has another interesting property and usage. It is mathematically identical to a one-factor MANOVA; Quinn and Keough (2002) thoroughly discussed this as have other authors. This means that DFA can substitute for a MANOVA and thus be used for the same purposes as a MANOVA (Section 9.1 and Box 9.1). The first discriminant function (LDF 1) obtained in a DFA is equivalent to a MANOVA maximizing the variance among groups relative to the variance within groups. This produces an F-value (ratio of the variances) and associated P-value that is a statistical test of whether the centroids of the groups are different (Box 9.1).

Thus, a MANOVA that has a statistically significant or otherwise relatively small P-value will also very likely correspond to an LDF that is very good at discriminating the groups. This also then entails that the LDF is likely good at classification or predicting the group that an observation should belong to.

But how exactly does an LDF classify an observation? There are two main approaches. One way is to determine the classification function (not the same as the discriminant function) for each group (Tabachnick and Fidell 1996; Quinn and Keough 2002). The classification function (CF) has a coefficient for each variable and a constant. For a given group g, the classification coefficients ($\mathbf{D_g}$) are obtained through matrix algebra as $\mathbf{D_g} = \mathbf{C} \times \mu_g$ where $\mathbf{C}$ is the variance–covariance matrix, μ_g is a vector of the means of the variables for observations in the given group, and $\mathbf{D_g}$ of course is also a vector of the same size as μ_g. It holds the classification coefficients, d_{gi} for $i = 1$ to n variables. The constant term is obtained as $d_{g0} = \mathbf{D_g} \times \mu_g$. For each observation, a classification score is then calculated using the CF for each group. Group membership is based on whichever CF produces the highest score for the observation. To briefly return to the beetle dataset, recall that I applied DFA using species presence–absence as the grouping variable. Thus, for the group composing survey plots where the beetle was captured (presence plots), $\text{CF}_{\text{pres}} = 0.904 \times$ canopy cover + $0.888 \times$ leaf litter depth + $0.308 \times$ woody debris volume + $0.366 \times$ oak:non-oak tree ratio $- 0.014 \times$ soil type + 0.961. The classification for the absence plots is $\text{CF}_{\text{abs}} = -0.603 \times$ canopy cover $-$ $0.592 \times$ leaf litter depth $- 0.206 \times$ woody debris volume $- 0.244 \times$ oak:non-oak tree ratio + $0.009 \times$ soil type + 0.427. The two CFs are very different as should be the case, and each is different from the discriminant function calculated in Section 8.4.5, LDF 1 = $1.061 \times$ canopy cover $- 0.173 \times$ leaf litter depth + $0.014 \times$ woody debris volume + $0.618 \times$ oak:non-oak tree ratio $- 0.082 \times$ soil type + 0.218. Thus, the classification score for a particular observation k if assigned to a given group g is determined as $\text{CF}_g(k) = d_{g1}y_{1k} + d_{g2}y_{2k} + \ldots d_{gn}y_{nk} + d_{g0}$, for $y = 1$ to n response (input) variables. In a similar way, the PC score for a given observation k on a given axis is calculated based on factor loadings and the y_{nk} values of the observation.

The second way to classify an observation is to use Bayes theorem, prior probabilities, and posterior probabilities (James et al. 2013). Bayes theorem states that the probability of a given event A occurring if event B also occurs can be determined from this formula, $P(A \mid B) = [P(B \mid A) \times P(A)] / P(B)$, where $P(B \mid A)$ is the probability of event B occurring given that A occurs, $P(A)$ = probability of A occurring without regard to or independent of B, and $P(B)$ is the probability of B occurring without regard to A. (A brief aside: Bayes theorem has been around for over two centuries. It is the foundation of Bayesian statistics, an approach to statistical analysis and hypothesis testing that is becoming more and more common.) $P(A \mid B)$ and $P(B \mid A)$ are *conditional probabilities*; A is conditioned on B and vice versa. Most importantly, they allow us to determine *posterior probabilities*. They give us a way to calculate an "updated" probability of an event (e.g., A or B) happening *posterior to* (or after) we acquire information on the probability of another event occurring conditioned on the event of interest. For example, we may know ahead of time the values of $P(A)$ and $P(B)$, these are the *prior probabilities*. But if we were to also acquire information about $P(B \mid A)$ then we could get an even better estimate of $P(A)$; or vice versa get an even better estimate of $P(B)$ if we knew $P(A \mid B)$. All of this is important to DFA and classification because we have the prior probabilities and the conditional probabilities of group membership. In the case of a DFA applied to a dataset in which species presence–absence is the grouping variable, the prior probabilities are simply the proportions of observations of species presence and absence (as long as the presence–absence data derive from a random and representative sampling design). Thus for the hypothetical beetle species, the prior probabilities are 0.4 and 0.6, respectively, for presence and absence. Unfortunately, in DFA classification, calculating the conditional probabilities and subsequently the posterior probabilities is not straightforward. The conditional probabilities are assumed to be represented by the probability density function for a multivariate normal distribution of the input variables. Thus for a given observation k, $f(k) = \dfrac{1}{(2p)^{n/2} \, |\mathbf{C}|^{0.5}} \times e^{-0.5(k-\mu)^{\mathrm{T}} \mathbf{C}^{-1}(k-\mu)}$ where p is the prior probability (for a given group), $\mathbf{C}$ is the variance–covariance matrix of the input variables, μ is a vector of their means, and n is the number of input variables. For the interested reader, James et al. (2013, pp. 142–44) give an excellent description of the mathematical derivation of the posterior probabilities, but be aware that some of the symbols they use are different from what I have used. Fortunately, many statistical software programs (including the MASS package in R, Appendix 9.1) provide the posterior probabilities in the output. Classification of an observation to a group is then easy; one need only look at the posterior probabilities for each group. The observation is then classified to (or predicted to be a member of) that group in which it has the greatest posterior probability. This Bayesian method of classification is better than the method that uses CFs (although in many instances they will give nearly identical classifications of a set of observations). It is better because it uses the prior probabilities. For example, a beetle survey plot randomly selected from the set of 100 is slightly more likely to have the species missing (without even taking into consideration the environmental variables) given that there are overall a greater proportion of absence plots (0.6) than presence plots (0.4), and assuming the plots were randomly located in the overall study area. The odds of a plot not having the species is 1.5 to 1 (= 0.6/0.4). Thus, if the prior probabilities actually represent somehow the background or latent abundance of beetles available to occur in plots, then we need to take this information into account when deriving the posterior probabilities.

Why is it necessary to use DFA to classify observations? After all, for each of the observations in the dataset, we already know its class (e.g., presence or absence) because it was recorded during data collection before DFA was even applied to the data. Classification *after* the DFA is not absolutely necessary unless one wants to assess the predictive accuracy of the linear discriminant function. In the context of a habitat analysis, we might want to know how well the environmental variables really do predict presence and absence of the species. However, herein, there is a slight problem with DFA. The discriminant function is constructed from a dataset that is then also subsequently used to test the discriminant function. This is circular—we should expect that the LDF does well at accurately

classifying observations if the same data are used to both build the LDF and test it (Quinn and Keough 2002; Gotelli and Ellison 2004). The solution to this problem is to simply divide the dataset into two portions, one for constructing the LDF (e.g., 80 percent of the observations) and the other (20 percent) for testing its ability to accurately classify observations. Alternatively, one could go collect more data under the same sampling protocol and thus obtain a new dataset for testing the LDF. Quantifying classification accuracy is covered in more detail in Section 10.3.

Another purpose of DFA, mentioned earlier, is to identify the input variables most important in discriminating the groups and perhaps compare among the variables. In this case, one need only examine the coefficients of the LDF (Section 8.4.5); there is no need for testing classification accuracy. In using DFA as a method of habitat analysis, it is useful to examine the relationship between the LDF and the posterior probability of species presence (i.e., the probability of the sampling unit having the species given its particular values for the environmental variables that were measured). If the LDF is truly good at discriminating between the groups then a plot of the posterior probability vs. the LDF will have a sigmoidal shape (Fig. 9.4, top panel). As with the plot of a logistic regression curve, a well-defined S-shape indicates that one or more variables are substantially different between the two groups. However, if the LDF score for an observation (e.g., a beetle survey plot) is calculated from only one of the variables while holding the other variables constant at their mean values, then the sigmoidal form of the relationship erodes (Fig. 9.4, middle panel). This is often true even when the focal variable is the single best predictor (e.g., canopy cover for the hypothetical beetle species). If we then add back in another relatively strong predictor (e.g., oak:non-oak tree ratio), then the relationship between the posterior probability of species presence and the LDF begins to gain back some of its sigmoidal shape (Fig. 9.4, bottom panel). Note that in applying DFA the input variables are typically first standardized and then the LDF is derived. Given that the standardization of a variable entails that the mean = 0, when variables are held constant at their means (Fig. 9.4, middle and bottom panels)

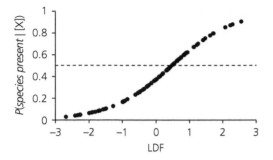

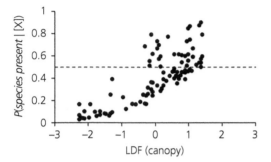

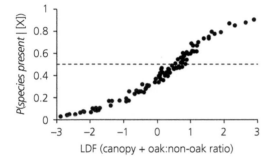

Figure 9.4 For the hypothetical beetle species (Chapter 8), relationship between the linear discriminant function and the posterior probability of the species being present in a survey plot. In the top panel, the LDF score for each survey plot has been calculated based on the full LDF (all five environmental variables), LDF = 1.061 × canopy cover − 0.173 × leaf litter depth + 0.014 × woody debris volume + 0.618 × oak:non-oak tree ratio − 0.082 × soil type + 0.218. In the middle panel, the LDF score is determined by canopy cover only, LDF = 1.061 × canopy cover. In the bottom panel, the LDF score is determined by canopy cover and oak:non-oak tree ratio, LDF = 1.061 × canopy cover + 0.618 × oak:non-oak tree ratio. Each black dot represents a survey plot.

they are effectively removed from the LDF. However, this is not the same as constructing the initial LDF on only one or a subset of the variables. Indeed, one might actually want to conduct DFA on different sets of input variables and then compare

the LDFs and their coefficients in an analogous way that different multiple regression models can be compared (Section 10.2).

Finally, let's briefly return to Bayes' theorem and posterior probabilities. First, I will recast the theorem in terms of species presence at a particular location given the set of environmental conditions or factors at the location. Remember from Chapter 5 that I presented a *general framework of analysis* for uncovering species–habitat associations. This entails conducting any of various statistical analyses (e.g., those presented in this chapter) wherein the strength of the association between a species response (e.g., presence–absence or abundance) and a set of measured environmental variables [X] is quantified and sometimes tested for statistical significance. Following this framework and applying Bayes' theorem, we have:

$$P(species\,present\,|[X_k]) = [P([X_k]\,|\,species\,present) \\ \times P(species\,present)] / P([X_k])$$

Thus, the probability that the species is present at a particular location k given the particular set of environmental conditions at the location depends on (1) the species' background probability of occurrence, $P(species\,present)$, (2) the probability $P([X_k])$ that the location has the particular conditions described by $[X_k]$, and (3) the probability that the species being present also entails that the location must have the environmental conditions $[X_k]$. This latter probability is the strength of the association between species presence–absence and the environmental variables that also presumably are habitat characteristics (if the relationship is strong). This probability specifies how likely the presence of the species at the location indicates that a specific set of environmental conditions must also be present at the location. This is what the LDF of a DFA is assessing. We can gain a lot of insight from the above equation and Fig. 9.5.

First, a large value for $P([X_k]\,|\,species\,present)$ does not necessarily lead to a large value for $P(species\,present\,|[X_k])$; nor are the two probabilities complements of one another. If $P(species\,present)$ is small, then $P(species\,present\,|[X_k])$ will also be small, although it will rapidly increase as the former increases (Fig. 9.5, top panel, solid line). This makes sense—if the species is not very common to begin with (e.g., P(*species present*) < 0.1) then even if there is a really strong association between the species presence and certain environmental conditions (e.g., $P([X_k]|$ *species present* $= 0.9$), the species still isn't predicted to occur with a high probability even when the environmental conditions are met; that is, $P($ *species present* $|[X_k])$ will be <0.5. In a similar way, if the specific environmental conditions are relatively uncommon (e.g., $P([X_k]) < 0.18$) then the species again is not predicted to occur with high probability even though the species–habitat relationship may be strong (Fig. 9.5, top panel, dashed line). Of course, the corollary to this is that the particular environmental conditions $[X_k]$ are a very good indicator of species presence if they and the species are common enough (Fig. 9.5, top panel).

Second, if $P([X_k]|$ *species present*$) = 0.5$ then $P([X_k])$ must = 0.5 also. This is because through probability theory and the algebraic rearrangement of Bayes' theorem, $P([X_k]) = P([X_k]|$ *species present*$) \times P($*species present*$) + (1 - P([X_k]|$ *species present*$)) \times (1 - P($*species present*$))$. This is a somewhat trivial mathematical result. However, it does represent the situation where $P($*species present* $|[X_k]) = P($*species present*$)$ (Fig. 9.5, middle panel). That is, in this situation, the environmental variables do not further help us in predicting species presence (or absence) at a location, beyond what $P($*species present*$)$ predicts.

Third, if the species–habitat relationship is weak or essentially non-existent (e.g., $P([X_k]|$ *species present*$) = 0.1$) then the environmental variables provide little or no further information on the species probability of occurrence; that is, $P($*species present* $|[X_k]) < P($*species present*$)$ (Fig. 9.5, bottom panel). This holds regardless of the frequency of the given set of environmental conditions. In fact, in this situation if we were to use the environmental variables to predict species presence we would drastically underestimate the actual frequency of the species occurring among a set of locations and the probability it occurs in any one location.

I have mostly explained DFA under the scenario where the grouping variable consists of only two categories or groups (e.g., species presence–absence). However, this need not be the case—DFA can be applied when there are many more than just two groups (in which case there will also be more than just one LDF). Bayes' theorem also holds when there are more than two groups. In that situation,

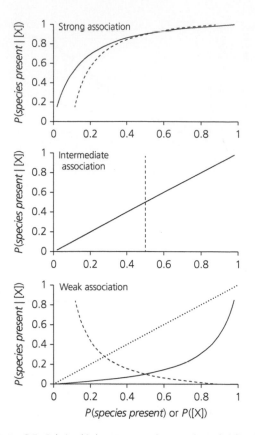

Figure 9.5 Relationship between a species posterior probability of being present at survey location given the set of [X] environmental variables (*y*-axis) versus background probability of species occurrence (*P*(*species present*) = solid black line) and versus probability that the location has the particular set of environmental variables (*P*([X]) = dashed line). Relationships are shown for three different strengths of the association between species presence–absence and the environmental variables: strong, *P*([X]|*species present*) = 0.9; intermediate, *P*([X]|*species present*) = 0.5; and weak, *P*([X]|*species present*) = 0.1. Stippled line in bottom panel is the line of unity representing a one-to-one correspondence between *P*(*species present*|[X]) and either *P*([X]) or *P*(*species*). Note that the two curves lie mostly below this line indicating that both of these prior probabilities are relatively useless in improving the posterior probability.

Bayes' theorem is best conceived as a probabilistic way of dividing the total sample space (James et al. 2013). Each group, g, has a prior probability specified by the proportion of observations in the overall dataset that fall into the group or by some knowledge about the background rate of occurrence. Importantly, $\Sigma P(g) = 1$, such that Bayes' theorem can be written as $P(A_g | B) = \dfrac{P(B | A_g) \times P(A_g)}{\Sigma [P(B | A_G) \times P(A_G)]}$

for $g = 1$ to G groups (James et al. 2013). Bayes' theorem and prior probabilities are a fundamental aspect of using DFA as a classification technique. Classification is prediction, and when observations (survey plots) are classified to a presence, absence, or abundance group based upon their environmental conditions then we can assess model fit and accuracy (Sections 10.1 and 10.3) and infer whether the examined environmental variables truly do characterize the habitat of the species.

9.6 Occupancy modeling

Occupancy modeling came on the scene as a powerful statistical tool for ecologists in the early 2000s (MacKenzie et al. 2002, 2003; Tyre et al. 2003; Bailey et al. 2004). Unlike the other methods discussed so far, this one was invented by ecologists, for a particular analytical purpose. The main intent of occupancy modeling is to estimate a species probability of occupancy (occurrence) at survey or sampling locations while also taking into account (estimating) the *detection probability* of the species. Occupancy modeling was built upon the conceptual foundations of density- and abundance-estimation techniques (including simple mark–recapture methods) that also involve obtaining estimates of detection probability (e.g., Burnham and Overton 1978; Otis et al. 1978; Seber 1982; Huggins 1989; Nichols 1992). In ecological research, estimating detection probability has become a fundamental aspect of many different survey design protocols and statistical methods that share the common feature of being applied to data that essentially record an individual or a species as being present (or not) at a particular location at a particular time. Thus, detection probability is important to estimation of individual survival, population density or abundance, frequency of a species occurrence over various spatial extents (range size), and species richness, as well as habitat occupancy of a species. Species are detected imperfectly; nowadays most ecologists recognize this as a basic fact of data collection and thus when possible they use statistical methods that take this into account. Such methods are not intended solely to estimate detection probability but rather to estimate it and use it in getting a better (more accurate) estimate of another parameter of greater interest such as individual survival rate, population density, and species richness. These methods, including occupancy modeling, require repeat-survey data; that is, repeatedly surveying for the species at the same locations (sampling plots or any other spatial unit) on multiple different occasions. As such, applying occupancy modeling as a method of habitat analysis requires some careful thought and foresight on study design such as optimizing the number of surveyed spatial units and the number of repeat visits to each (MacKenzie 2005; MacKenzie and Royle 2005; Bailey et al. 2007; Guillera-Aroita et al. 2010; MacKenzie et al. 2017). The amount of time between visits might also need to be carefully considered.

One crucial assumption of occupancy modeling (and other similar methods that involve estimating a detection probability) is *population closure*. This simply means that during the course of the survey "season," the time period from the first survey to the last, no individuals are leaving or entering the "population." This assumption makes sense given that one is often trying to get a snapshot estimate of abundance or population size from data collected over some time period. The closure assumption is relevant to occupancy modeling because it is necessary for another more important assumption. Occupancy modeling assumes that if an individual of the focal species was ever detected at a given survey location (spatial sampling unit) during any of the surveying occasions, then it was present and available to be detected during every survey occasion (MacKenzie et al. 2002). This is the *availability assumption*. It is easy to see that this assumption is more likely to be violated as the length of time between survey occasions increases and as the overall duration of the survey season increases. Granted, for some abundance-estimation techniques there are ways of relaxing the closure assumption and there are versions of occupancy modeling that also estimate colonization and extinction rates particularly over time periods longer than a single survey season (MacKenzie et al. 2003). However, in this book, I focus on the most basic form of occupancy modeling, the *single-species, single-season* model (MacKenzie et al. 2002).

It is also important to recognize that just because a particular individual or any given individual of the focal species is present at a given surveying location, it still may not be detected during some (or even all) of the survey occasions. That is, availability does not automatically equal detection; this is the essence of estimating a detection probability. There are some survey occasions (perhaps a majority) where the species may not be detected even though it is present. Our field methods and protocols for finding and counting organisms are not perfect. Moreover, some organisms may be particularly cryptic, elusive, inaccessible, or somehow prone to avoiding capture and/or any form of non-capture counting. Thus, imperfect detectability can be due to certain species traits and behavioral characteristics as well as the limitations of our survey methods including direct observation by a human. Although automated surveying (e.g., camera traps and acoustic recorders) is becoming more common, for many species we still use survey protocols that in some way or another depend on a human's perceptual abilities. Even automated surveying is not perfect.

The requisite data for occupancy modeling are detection/non-detection (coded as 1,0) observations of the focal species at N locations surveyed on T occasions. Sometimes researchers will refer to conducting occupancy modeling on presence–absence data although this is not technically correct; non-detection does not equate (necessarily) to absence. The response variable or data for an occupancy model is a *detection history*; this is a series of 1s and 0s indicating whether the focal species was detected at a given location during each of the survey occasions. Importantly, a detection history (in this context) does not indicate detection/non-detection of any particular individual. Indeed, individuals are not known in any way. They are not marked or otherwise identifiable as unique individuals. This is actually one of the great advantages of occupancy modeling and related abundance-estimation techniques. They do not require the researcher to keep track of individual organisms as would be necessary for the classic mark–recapture methods.

Detection histories enable estimation of a detection probability, p_{it}. This is the probability of detecting at least one individual of the focal species at location (or site) i during survey occasion t, given that the species is available to be detected. Individuals are "accessible" to the survey method

(e.g., a survey method involving above-ground observation or capture is not being used for a focal species that is primarily subterranean). In the description below, I use the same symbolism as MacKenzie et al. (2002). For a given location we can define the *likelihood* of obtaining the observed detection history assuming certain values of p_{it} and the probability of occupancy (ψ_i). Thus for a detection history of 0110, $L_i = \psi_i \times (1 - p_{i1}) \times (p_{i2}) \times (p_{i3}) \times (1 - p_{i4})$.

Note that surveying for a species is a binomial sampling process in which the species either is detected or not detected (there is not a third outcome), therefore non-detection occurs with probability, $1 - p_{it}$ (e.g., during survey occasions 1 and 4). We define an occupied location as one in which there was a detection during at least one survey occasion. The likelihood equation for such locations is $L_{i(1)} = \psi_i \prod [(p_{it}...p_{iT}) \,|\, t^*][(1 - p_{it}...1 - p_{iT}) \,|\, t']$. The opposite of this is a location where the species was never detected, a so-called all-zero detection history, hence $L_{i(0)} = \psi_i \prod (1 - p_{it}...1 - p_{iT}) \,|\, t' + (1 - \psi_i)$. In these equations, t^* represents a survey occasion in which there was a detection and t' represents a survey occasion in which there was not a detection. These equations are modified versions of equation 1 in MacKenzie et al. (2002). Note that for an all-zero detection history the first term in the likelihood equation is the probability that the species was present (location occupied) but never detected and the second term is the probability that the site was never occupied. In these equations for the likelihood of a given detection history, we are not trying to figure out if the location was occupied or not (either it was or was not). Rather, we (or more precisely computer algorithms) are attempting to find values of ψ_i and p_{it} that maximize L_i. This maximization process is being done simultaneously for all survey locations. The product of all L_i for $i = 1$ to N locations is maximized to get a total likelihood, $L_{total} = \prod L_{i(1)} \times \prod L_{i(0)}$. Occupancy modeling is intended to get accurate estimates of ψ_i while taking into account detection probability (i.e., obtaining accurate estimates of p_{it}).

To fully understand occupancy modeling, it is important to see the difference between a likelihood and a probability. Probability pertains to events whereas likelihood pertains to observed outcomes of the events. Unfortunately, the two words are used interchangeably in common parlance and sometimes in scientific communication as well. A likelihood indicates the chance of getting an observed outcome (e.g., a particular detection history) assuming a particular probability of the event leading to the outcome. We observe the outcome and then try to deduce what the probability (of the event leading to the outcome) is most likely to be. On the other hand, we might know a priori the probability of the event happening and thus we can determine or predict the most likely outcome (regardless of whether it is subsequently observed to occur or not) and typically we can also determine the probabilities of each possible outcome of the event. For example, if a detection history is 0110 (or any ordered combination of two 0s and two 1s), then the likelihood estimate is maximized at $L \approx 1$ when $\psi = 1$ with p_{i1} and $p_{i4} \approx 0$ and p_{i2} and $p_{i3} \approx 1$. Even though there are two 0s in the detection history, L cannot be maximized if $\psi \approx 0$ (this makes sense because we know that ψ must be non-zero if there are also two 1s in the detection history). Further, note that if the location is truly occupied, then the chances of obtaining non-detections on occasions 1 and 4 are greatest when detection probability is at or close to zero (for whatever reason) during those two survey occasions. For a detection history of all 0s, the likelihood is maximized at $L \approx 1$ when $\psi = 0$ regardless of the p_{it} values. But this does not necessarily mean that the location actually is unoccupied. The likelihood of an all-zero detection history might be maximized at values of $\psi > 0$ and certain non-zero values of the p_{it}, depending upon the covariates included in the occupancy model (see below). Also, even when a given location is known to be occupied because the species was detected at least once, the predicted probability of occupancy might be $\psi < 1$ depending upon the covariates. Again, ψ is the predicted or expected probability of occupancy, not the observed occupancy state. By the way, in the original description of an occupancy model, MacKenzie et al. (2002) used θ to represent both ψ and p in a generic way—something to be aware of when reading some early papers on occupancy modeling.

So, how is occupancy modeling used as a method of habitat analysis? So far, I have only explained it as a likelihood analysis of a series of 0s and 1s.

However, occupancy modeling entails much more than this. Typically, a researcher using occupancy modeling has collected additional data such as particular characteristics of the survey locations (i.e., environmental variables as I've been referring to throughout the book) that might affect whether the location is occupied by the species. One might also include variables associated with the actual survey process (e.g., time of year or time of day of the surveying, weather conditions, and particular human observer doing the survey) as these latter factors could affect the detection process and hence the probability of detection. MacKenzie et al. (2002) originally used the generic term "covariate" to refer to these variables and so I will retain that usage. Covariates are incorporated into an occupancy model through the use of the logistic equation (or its linear equivalent, the logit-link function). Indeed, logistic regression is the foundation of occupancy modeling. The probability of occupancy and the detection probability can both be modeled as logistic responses to the covariates. A separate logistic regression model is used for detection probability and for occupancy probability, but both utilize the same detection/non-detection data. In addition, some covariates can be modeled (have a potential effect) on both detection and occupancy. For example, the amount of understory vegetation might influence the effectiveness of a visual survey for a small ground-dwelling organism and yet also be an important habitat characteristic and hence affect occupancy. In such a case, one would want to include understory vegetation as a covariate on the detection process as well as occupancy. Covariates included in the logistic regression for occupancy are often called site-level covariates because for a given site they remain constant over the course of the survey season. Covariates in the logistic regression for detection are often called observation- or survey-level covariates because they are unique to a combination of location and survey occasion and for a given location they can vary from one survey occasion to the next. For example, weather conditions may vary, time of day of surveying might vary, and even different human observers might survey the same location on different occasions.

The use of covariates in an occupancy model allows for more accuracy in estimating ψ_i and p_{it}. It allows these parameters to vary among survey locations (for ψ_i and p_{it}) and survey occasions (for p_{it}) inasmuch as the covariates influence either occupancy or detection and values for the covariates themselves vary among survey locations and occasions. Essentially, the logistic regression models within an occupancy model produce values of ψ_i and p_{it} (as functions of the covariates) that are then used in the equation for maximizing total likelihood. Thus, in the context of occupancy modeling, logistic regression is based on maximum likelihood estimation instead of least-squares estimation. That is, the logistic regression equations provide regression coefficients for each covariate that lead to better (more accurate) estimates of ψ_i and p_{it} as assessed through maximizing the total likelihood, which itself is the product of the likelihoods for all the survey locations and their detection histories. Put simply, there is procedural integration between the logistic regression equations and the likelihood equations.

Occupancy modeling can be used to analyze habitat associations of a species in that the covariates for occupancy probability can include environmental variables presumed to influence occupancy and hence define habitat for the species. This is similar to using multiple linear regression and multiple logistic regression to identify the predictor variables that are most important in affecting the species response (abundance or presence–absence). Each covariate in an occupancy model has a coefficient representing the effect of the covariate on either occupancy probability, detection probability, or both. If the covariates are standardized prior to conducting the model then the coefficients can be directly compared with one another. Each coefficient also has a standard error (typically estimated by bootstrapping) and hence the statistical significance of the covariate (testing whether the coefficient $\neq 0$) is obtained based on a z-test. Output from an occupancy model also includes the predicted (or expected) probability of occupancy for each location and the probability of detection for each location for each of its survey occasions, given the detection history and values of the covariates at the location and during the survey occasions. Assessing the fit of an occupancy model is an integral part of the overall procedure of occupancy modeling, more

so than assessing model fit in the context of the other five habitat analysis methods that I have discussed. In addition, in the world of occupancy modeling there is a history or tradition of applying many different models (sometimes dozens), each defined by a different set of covariates, to the detection/non-detection data collected for a single study. Comparing the different models in a formal way (e.g., through the use of Akaike information criterion (AIC)) then further facilitates identifying the environmental variables that might be most important in characterizing the habitat of the species. I further discuss model fit and comparison in Chapter 10.

To briefly illustrate the use of occupancy modeling to analyze habitat associations, I now refer to the study of Warren et al. (2013). They used occupancy modeling to test whether canopy cover and slope of terrain (hillsides) were important characteristics of the habitats of black-and-white warblers and golden-cheeked warblers in central Texas, United States (Fig. 9.6). They conducted 5-min. point counts for each species at each of 36 survey locations at each of six study sites. Each point count was repeated four times during the breeding season in March and April 2009. Thus, for this study, $N = 216$ and $T = 4$. The area surveyed was assumed to be a 100-m radius circle centered on the location of the person conducting the point count. In addition to canopy cover and slope of each location, other covariates were aspect (portion of survey circle with NE or NNW orientation and hence relatively mesic), time of day of the point count, "season" of the point count (week 1, 2, 3, or 4), and study site (actually five different dummy variables coded as 0,1). Covariates for occupancy included only canopy cover, slope, and aspect. Covariates for detection probability included all the variables as well as (time of day)^2 to account for the possibility that singing (and hence detection) is mostly likely mid-morning. Canopy cover and slope were included as covariates for detection probability in that these two variables might presumably affect acoustic and visual detection of the birds. Note that it would be inappropriate to model time of day and season as covariates on occupancy as these variables are not constant over all the survey occasions, and moreover, if either actually does affect occupancy, then

Figure 9.6 Photos of a black-and-white warbler (top), golden-cheeked warbler (middle), and their juniper–oak woodland habitat (bottom). The habitat for both species is characterized by relatively steep hillsides with dense canopy cover. In order, photo credits are unknown photographer, Alan Schmierer, and the author.

obviously the closure assumption has been severely violated and one should then not be applying an occupancy model in the first place.

Warren et al. (2013) used various combinations of covariates in a few dozen models for each species. Given the number of covariates for occupancy and detection, approximately 500 models were possible—good statistical practice always puts an upper limit on the number of models that are conducted (Section 10.2). Further, they used an approach in which they first attempted to find the best model for detection and then for occupancy, followed by models that used covariates for both detection and occupancy, rather than running every possible model from the very start. In the end, the best models (lowest AIC_c values and relatively good fit, Section 10.2) for both species included an effect of canopy cover, slope, and aspect although canopy cover appeared to have a substantially greater effect on occupancy by black-and-white warblers than occupancy by golden-cheeked warblers, as evidenced by model-averaged regression coefficients. For each species there was some effect of each covariate on detection probability. More importantly, estimated detection probabilities for black-and-white warblers ranged from 0.02–0.22 and for golden-cheeked warblers, 0.16–0.69. These probabilities are well below 1 (perfect detection) further supporting the need to use occupancy models to take into account imperfect detection and consequently get better estimates of occupancy and how it might be affected by the tested environmental variables. This is a clear example of how useful occupancy modeling can be in identifying the habitat associations of a species.

It should be clear that occupancy modeling is more than just multiple logistic regression. The two methods do not give identical results (same values for the coefficients) because occupancy modeling is taking into account that some of the non-detections may not represent true absences (or put differently, for each non-detection there is an estimated probability that the non-detection represents imperfect detection rather than absence of the species). For a logistic regression model applied to presence–absence data, an implicit assumption is that the recorded zeros truly do represent species absence.

As with any statistical technique, occupancy modeling comes with assumptions about the survey design and data. In addition to the closure assumption, locations should be independent of one another and survey occasions at a particular location should be independent. This means that during a given survey occasion, detecting (or not detecting) the species at location i should not depend on whether the species was detected (or not detected) at any other location. Similarly, within a given survey location, detecting the species during occasion t should not depend on whether the species was detected during occasion $t - 1$. For any given study design intended for an occupancy model, assessing whether locations and occasions are independent can be somewhat nuanced and context-specific, hence I refer the reader to MacKenzie et al. (2017, pp. 152–4 and 439–76) for more in-depth discussion. Another important assumption of many versions of occupancy modeling is that there is no or very minimal chance of a false positive during the surveying. That is, the recorded detections truly do represent the occurrence of the species of interest; it has not been confused with another species, no misidentification. However, even this fundamental requirement of occupancy modeling can be loosened by modifying some models to allow for false positive detections (Royle and Link 2006; MacKenzie et al. 2017). As with any type of empirical research, proper study design and forethought can go a long way toward assuring that the collected data are amenable to whatever statistical analysis is planned.

Newer versions of occupancy models relax some of these assumptions and expand the applicability of occupancy models beyond simply estimating species occupancy under the single-species, single-season model (MacKenzie et al. 2017). Any reader interested in using occupancy modeling as a method of habitat analysis is advised to learn more about the technique beyond the basic introduction given in this book. I recommend the following references: MacKenzie et al. (2002) and MacKenzie et al. (either the 2006 or 2017 edition). Also, there is stand-alone software, PRESENCE, developed by MacKenzie and colleagues for occupancy modeling; the user's manual is another good basic introduction. Occupancy modeling can also be conducted with the "unmarked" R package (Appendix 9.1).

9.7 Conclusion

In this chapter I discussed six methods of habitat analysis. There is some commonality among them as evidenced by the analysis of the hypothetical beetle data (here and Chapter 8) often revealing canopy cover and oak:non-oak tree ratio as the two variables that most influence presence–absence and abundance. Also, it is somewhat revealing that the variance–covariance matrix of the predictor variables is an integral part of the mathematical foundation of many of the methods, and hence we should expect that the methods give very similar results. However, I created the hypothetical beetle data to be intentionally simple with at least two of the five environmental variables having strong effects on beetle abundance. When applied to more complex and realistic datasets, the six methods of habitat analysis would likely be seen to have greater differences and more definitive strengths and weaknesses. A thorough method-by-method comparison of performance on complex datasets is beyond the introductory scope of this book. However, I recommend the following studies that have directly compared performance and applicability of two or more of the methods: Segurado and Araújo (2004), Ahmadi-Nedushan et al. (2006), Meynard and Quinn (2007), Leclere et al. (2011), Sharma et al. (2012), and Gorosito et al. (2016). Although each method may have its particular strengths and weaknesses, it probably is not appropriate or useful to declare one method better than the others, particularly not as a blanket statement. The appropriate method(s) to use will depend on the study design, structure of the data, purpose of the analysis, and perhaps even the particular ecology/biology of the focal species (Chapter 6). In Chapter 12, I give a bit more guidance on selecting a method.

Appendix 9.1 R code and packages useful in conducting habitat analyses

In this appendix, R code is shown in italics. Some of the functions may require downloading and installing particular R packages. I assume the reader has some prior knowledge and experience in using R. Most of the functions described below are contained within the "stats" R package (and other packages). If you have previously conducted statistical analyses in R then you probably already have the "stats" package installed (loaded) on your computer. It includes a wide variety of statistical procedures. To ensure proper use of any of the functions listed below, the user should further consult the R documentation for the function and package.

Analysis of variance—An ANOVA can be conducted in R using the function *aov()*. Of course, ANOVA is such a common statistical procedure that nearly every statistical software package can conduct ANOVAs, even EXCEL® has the capacity to perform some types of ANOVA.

Multiple linear regression—These models can be constructed in R as generalized linear models using the *glm()* function. To perform an ordinary least squares (OLS) regression, use *family = gaussian*.

Multiple logistic regression—Same as above, but use *family = binomial*.

Negative binomial regression—The function *glm.nb()* will conduct a negative binomial regression model. It requires installation of the "MASS" package.

Zero-inflated regression—The function *zeroinfl()* will conduct a zero-inflated regression model. Use *dist = "negbin"* if the abundance data are to be modeled as having a negative binomial distribution (clumped spatial dispersion among survey units), *dist = "poisson"* if abundance data are assumed to have a random spatial dispersion. The *zeroinfl()* function requires installation of the "pscl" package.

Classification and regression trees (CART)—Both types of tree can be constructed using the function *rpart()*. If the modeled response variable is binary (e.g., presence–absence) or categorical then a classification tree is outputted. If the modeled response variable is quantitative (e.g., species abundance) then a regression tree is outputted. There is no need for the user to explicitly specify which type of tree to produce. The *rpart()* function requires installation of the "rpart" package.

Principal components analysis (PCA)—The function *prcomp()* will conduct a PCA. This function is included in the "stats" package.

Discriminant function analysis (DFA)—The function *lda()* will conduct a linear discriminant analysis. It requires installation of the "MASS" package.

Occupancy modeling—The "unmarked" package can be used to conduct various versions of occupancy modeling. Data entry is a bit different from that required for most R functions and packages. The function *unmarkedFrameOccu()* is used to organize the species detection data and covariate data into a single data frame that can then be analyzed with the single-season, single-species occupancy model of MacKenzie et al. (2002). The function for conducting the occupancy model is *occu()*.

In addition to the functions and packages listed above, the following functions may be useful: *qqnorm()* produces a q-q plot that is a diagnostic for normality; *logLik()* gives the log-likelihood for nearly any model that is estimated using maximum likelihood; *deviance()* gives the model deviance, and *AIC()* gives the Akaike information criterion. All of these functions are included in the "stats" package.

References

Ahmadi-Nedushan, B., St-Hilaire, A., Bérubé, M., Robichaud, E., Thiémonge, N., and Bobée, B. (2006). A review of statistical methods for the evaluation of aquatic habitat suitability for instream flow assessment. *River Research and Applications*, 22, 503–23.

Anscombe, F.J. (1949). The statistical analysis of insect counts based on the negative binomial distribution. *Biometrics*, 5, 165–73.

Bailey, L.L., Simons, T.R., and Pollock, K.H. (2004). Estimating site occupancy and species detection probability parameters for terrestrial salamanders. *Ecological Applications*, 14, 692–702.

Bailey, L.L., Hines, J.E., Nichols, J.D., and MacKenzie, D.I. (2007). Sampling design trade-offs in occupancy studies with imperfect detection: examples and software. *Ecological Applications*, 17, 281–90.

Beck, H., Gaines, M.S., Hines, J.E., and Nichols, J.D. (2004). Comparative dynamics of small mammal populations in treefall gaps and surrounding understorey within Amazonian rainforest. *Oikos*, 106, 27–38.

Blasco-Moreno, A., Pérez-Casany, M., Puig, P., Morante, M., and Castells, E. (2019). What does a zero mean? Understanding false, random and structural zeros in ecology. *Methods in Ecology and Evolution*, 10, 949–59.

Bliss, C.I. and Fisher, R.A. (1953). Fitting the negative binomial distribution to biological data. *Biometrics*, 9, 176–200.

Bonar, M., Manseau, M., Geisheimer, J., Bannatyne, T., and Lingle, S. (2016). The effect of terrain and female density on survival of neonatal white-tailed deer and mule deer fawns. *Ecology and Evolution*, 6, 4387–402.

Breiman, L., Friedman, J., Stone, C.J., and Olshen, R.A. (1984). *Classification and Regression Trees*. Taylor and Francis, Boca Raton, Fla.

Brown, J.H., Mehlman, D.W., and Stevens, G.C. (1995). Spatial variation in abundance. *Ecology*, 76, 2028–43.

Burnham, K.P. and Overton, W.S. (1978). Estimation of the size of a closed population when capture probabilities vary among animals. *Biometrika*, 65, 625–33.

Cameron, A.C. and Trivedi, P.K. (2013). *Regression Analysis of Count Data*. Cambridge University Press, Cambridge.

Condit, R., Ashton, P.S., Baker, P., et al. (2000). Spatial patterns in the distribution of tropical tree species. *Science*, 288, 1414–18.

Crawley, M.J. (2015). *Statistics: An Introduction Using R*. Wiley Publishing, Chichester.

Dalley, K.L., Taylor, P.D., and Shutler, D. (2008). Nest-site characteristics and breeding success of three species of boreal songbirds in western Newfoundland, Canada. *Canadian Journal of Zoology*, 86, 1203–11.

De'ath, G. (2007). Boosted trees for ecological modeling and prediction. *Ecology*, 88, 243–51.

De'ath, G. and Fabricius, K.E. (2000). Classification and regression trees: a powerful yet simple technique for ecological data analysis. *Ecology*, 81, 3178–92.

Dray, S. (2008). On the number of principal components: a test of dimensionality based on measurements of similarity between matrices. *Computational Statistics and Data Analysis*, 52, 2228–37.

Elith, J., Leathwick, J.R., and Hastie, T. (2008). A working guide to boosted regression trees. *Journal of Animal Ecology*, 77, 802–13.

Franklin, S.B., Gibson, D.J., Robertson, P.A., Pohlmann, J.T., and Fralish, J.S. (1995). Parallel analysis: a method for determining significant principal components. *Journal of Vegetation Science*, 6, 99–106.

Lawless, J.F. (1987). Negative binomial and mixed Poisson regression. *Canadian Journal of Statistics*, 15, 209–25.

Lewin, W.-C., Freyhof, J., Huckstorf, V., Mehner, T., and Wolter, C. (2010). When no catches matter: coping with zeros in environmental assessments. *Ecological Indicators*, 10, 572–83.

Lloyd, M. (1967). Mean crowding. *Journal of Animal Ecology*, 36, 1–30.

Glover, T. and Mitchell, K. (2002). *An Introduction to Biostatistics*. McGraw Hill, Boston, Mass.

Gorosito, I.L., Bermúdez, M.M., Douglass, R.J., and Busch, M. (2016). Evaluation of statistical methods and sampling designs for the assessment of microhabitat selection based on point data. *Methods in Ecology and Evolution*, 7, 1316–24.

Gotelli, N.J. and Ellison, A.M. (2004). *A Primer of Ecological Statistics*. Sinauer Associates, Sunderland, Mass.

Guillera-Aroita, G., Ridout, M.S., and Morgan, B.J.T. (2010). Design of occupancy studies with imperfect detection. *Methods in Ecology and Evolution*, 1, 131–39.

He, F., Legendre, P. and LaFrankie, J.V. (1997). Distribution patterns of tree species in a Malaysian tropical rain forest. *Journal of Vegetation Science*, 8, 105–14.

Hilbe, J. (2011). *Negative Binomial Regression*. Cambridge University Press, Cambridge.

Holoubek, N.S. and Jensen, W.E. (2015). Avian occupancy varies with habitat structure in oak savanna of the south-central United States. *Journal of Wildlife Management*, 79, 458–68.

Huggins, R.M. (1989). On the statistical analysis of capture experiments. *Biometrika*, 76, 133–40.

Inselman, W.M., Datta, S., Jenks, J.A., Jensen, K.C., and Grovenburg, T.W. (2015). *Buteo* nesting ecology: evaluating nesting Swainson's hawks in the northern Great Plains. *PLoS ONE*, 10, e0137045.

Jackson, D.A. (1993). Stopping rules in principal components analysis: a comparison of heuristical and statistical approaches. *Ecology*, 74, 2204–14.

James, G., Witten, D., Hastie, T., and Tibshirani, R. (2013). *An Introduction to Statistical Learning, with Applications in R*. Springer, New York.

Johnson, D., Longshore, K., Lowrey, C., and Thompson, D.B. (2015). Habitat selection and survival of pronghorn fawns at the Carrizo Plain National Monument, California. *California Fish and Game*, 101, 267–79.

Jongman, R.H.G., Ter Braak, C.J.F., and Van Tongeren, O.F.R. (1995). *Data Analysis in Community and Landscape Ecology*. Cambridge University Press, Cambridge.

Krzywinski, M. and Altman, N. (2017). Classification and regression trees. *Nature Methods*, 14, 757–58.

Leclere, J., Oberdorff, T., Belliard, J., and Leprieur, F. (2011). A comparison of modeling techniques to predict juvenile 0+ fish species occurrences in a large river system. *Ecological Informatics*, 6, 276–85.

Loh, W.-Y. (2014). Fifty years of classification and regression trees. *International Statistical Review*, 82, 329–48.

MacKenzie, D.I. (2005). What are the issues with presence-absence data for wildlife managers? *Journal of Wildlife Management*, 69, 849–60.

MacKenzie, D.I. and Royle, J.A. (2005). Designing occupancy studies: general advice and allocating survey effort. *Journal of Applied Ecology*, 42, 1105–14.

MacKenzie, D.I., Nichols, J.D., Lachman, G.B., Droege, S., Royle, J.A., and Langtimm, C.A. (2002). Estimating site occupancy rates when detection probabilities are less than one. *Ecology*, 83, 2248–55.

MacKenzie, D.I., Nichols, J.D., Hines, J.E., Knutson, M.G., and Franklin, A.B. (2003). Estimating site occupancy, colonization, and local extinction when a species is detected imperfectly. *Ecology*, 84, 2200–207.

MacKenzie, D.I., Nichols, J.D., Royle, J.A., Pollock, K.H., Bailey, L.L. and Hines, J.E. (2017). Basic presence/absence situation. Pages 115–207 in *Occupancy Estimation and Modeling*. Second edition. Academic Press, London.

Martin, T.G., Wintle, B.A., Rhodes, J.R., et al. (2005). Zero tolerance ecology: improving ecological inference by modelling the source of zero observations. *Ecology Letters*, 8, 1235–46.

McCullagh, P. and Nelder, J.A. (1983). *Generalized Linear Models*. Second edition. Chapman and Hall, New York.

McMahon, L.A., Rachlow, J.L., Shipley, L.A., Forbey, J.S., and Johnson, T.R. (2017). Habitat selection differs across hierarchical behaviors: selection of patches and intensity of patch use. *Ecosphere*, 8, e01993.

Meynard, C.N. and Quinn, J.F. (2007). Predicting species distributions: a critical comparison of the most common statistical models using artificial species. *Journal of Biogeography*, 34, 1455–69.

Morgan, J.N. and Sonquist, J.A. (1963). Problems in the analysis of survey data, and a proposal. *Journal of the American Statistical Association*, 58, 415–34.

Morgan, J.N. and Messenger, R.C. (1973). *THAID, a Sequential Analysis Program for the Analysis of Nominal Scale Dependent Variables*. Survey Research Center Institute for Social Research, University of Michigan, Ann Arbor, Mich.

Myers, R.H. and Montgomery, D.C. (1997). A tutorial on generalized linear models. *Journal of Quality Technology*, 29, 274–91.

Myers, J.A., Chase, J.M., Jiménez, I., et al. (2013). Beta-diversity in temperate and tropical forests reflects dissimilar mechanisms of community assembly. *Ecology Letters*, 16, 151–57.

Nichols, J.D. (1992). Capture-recapture models. *Bioscience*, 42, 94–102.

Otis, D.L., Burnham, K.P., White, G.C., and Anderson, D.R. (1978). Statistical inference from capture data on closed animal populations. *Wildlife Monographs*, 62, 3–135.

Paterson, J.E., Steinberg, B.D., and Litzgus, J.D. (2014). Effects of body size, habitat selection and exposure on hatchling turtle survival. *Journal of Zoology*, 294, 278–85.

Peak, R.G. and Thompson, F.R. (2014). Seasonal productivity and nest survival of golden-cheeked warblers vary with forest type and edge density. *Condor*, 116, 546–59.

Peres-Neto, P.R., Jackson, D.A., and Somers, K.M. (2005). How many principal components? stopping rules for determining the number of non-trivial axes revisited. *Computational Statistics and Data Analysis*, 49, 974–97.

Prasad, A., Iverson, L.R., and Liaw, A. (2006). Newer classification and regression tree techniques: bagging and random forests for ecological prediction. *Ecosystems*, 9, 181–99.

Quinn, G.P. and Keough, M.J. (2002). *Experimental Design and Data Analysis for Biologists*. Cambridge University Press, Cambridge.

Reidy, J.L., Thompson, F.R., and Kendrick, S.W. (2014). Breeding bird response to habitat and landscape factors across a gradient of savanna, woodland, and forest in the Missouri Ozarks. *Forest Ecology and Management*, 313, 34–46.

Roach, M.C., Thompson, F.R., and Jones-Farrand, T. (2019). Effects of pine-oak woodland restoration on breeding bird densities in the Ozark-Ouachita Interior Highlands. *Forest Ecology and Management*, 437, 443–59.

Ross, G.J.S. and Preece, D.A. (1985). The negative binomial distribution. *Journal of the Royal Statistical Society, Series D, (The Statistician)*, 34, 323–35.

Royle, J.A. and Link, W.A. (2006). Generalized site occupancy models allowing for false positive and false negative errors. *Ecology*, 87, 835–41.

Seber, G.A.F. (1982). *The Estimation of Animal Abundance and Related Parameters*. Second edition. Macmillan Publishing, New York.

Segurado, P. and Araújo, M.B. (2004). An evaluation of methods for modelling species distributions. *Journal of Biogeography*, 31, 1555–68.

Sharma, S., Legendre, P., Boisclair, D., and Gauthier, S. (2012). Effects of spatial scale and choice of statistical model (linear versus tree-based) on determining species-habitat relationships. *Canadian Journal of Fisheries and Aquatic Sciences*, 69, 2095–11.

Shorrocks, B. and Sevenster, J.G. (1995). Explaining local species diversity. *Proceedings of the Royal Society of London B*, 260, 305–309.

Sileshi, G. (2006). Selecting the right statistical model for analysis of insect count data by using information theoretic measures. *Bulletin of Entomological Research*, 96, 479–88.

Smith, M.J., Boland, C.R.J., Maple, D., and Tiernan, B. (2009). The Christmas Island blue-tailed skink (*Cryptoblepharus egeriae*): a survey protocol and an assessment of factors that relate to occupancy and detection. *Records of the Western Australian Museum*, 27, 40–44.

Snedecor, G.W. and Cochran, W.G. (1989). *Statistical Methods*. Eighth edition. Iowa State University Press, Ames, Iowa.

Sokal, R.R. and Rohlf, F.J. (1995). *Biometry*. Third edition. W.H. Freeman Publishing, New York.

Steinberg, D. (2009). CART: Classification and regression trees. Pages 180–201 in *The Top Ten Algorithms in Data*

Mining. Wu, X. and Kumar, V. (editors). CRC Press, Chapman and Hall, Boca Raton, Fla.

Student (1907). On the error of counting with a haemacytometer. *Biometrika*, 5, 351–60.

Tabachnick, B.G. and Fidell, L.S. (1996). *Using Multivariate Statistics*. Third edition. Harper Collins, New York.

Taylor, L.R., Woiwod, I.P., and Perry, J.N. (1978). The density-dependence of spatial behaviour and the rarity of randomness. *Journal of Animal Ecology*, 47, 383–406.

Trexler, J.C., Travis, J., and McManus, M. (1992). Effects of habitat and body size on mortality rates of *Poecilia latipinna*. *Ecology*, 73, 2224–36.

Tyre, A.J., Tenhumberg, B., Field, S.A., Niejalke, D., Parris, K., and Possingham, H.P. (2003). Improving precision and reducing bias in biological surveys: estimating false-negative error rates. *Ecological Applications*, 13, 1790–801.

Veech, J.A., Crist, T.O., and Summerville, K.S. (2003). Intraspecific aggregation decreases local species diversity of arthropods. *Ecology*, 84, 3376–83.

Vik, P. (2014). *Regression, ANOVA, and the General Linear Model, a Statistics Primer*. Sage Publishing, Thousand Oaks, Calif.

Warren, C.C., Ott, J.R., and Veech, J.A. (2013). Comparative occupancy and habitat associations of black-and-white (*Mniotilta varia*) and golden-cheeked warblers (*Setophaga crysoparia*) in the juniper-oak woodlands of central Texas. *American Midland Naturalist*, 169, 382–97.

Young, L.J., Campbell, N.L., and Capuano, G.A. (1999). Analysis of overdispersed count data from single-factor experiments: a comparative study. *Journal of Agricultural, Biological, and Environmental Statistics*, 4, 258–75.

Zeileis, A., Kleiber, C., and Jackman, S. (2008). Regression models for count data in R. *Journal of Statistical Software*, 27, e8.

Zuur, A.F., Ieno, E.N., and Elphick, C.S. (2009). A protocol for data exploration to avoid common statistical problems. *Methods in Ecology and Evolution*, 1, 3–14.

CHAPTER 10

Post-analysis Procedures

In this chapter, I discuss additional procedures that can be conducted after the habitat analysis. These are assessing model fit, comparing models, and assessing predictive accuracy. In general, the output of a habitat analysis can be viewed as predictions about species presence–absence, abundance, activity level, or whatever was the measured response variable. As such, the habitat analysis produces a statistical model relating the species response to one or more environmental variables. As a piece of knowledge, we want the model to be a parsimonious, complete, and truthful depiction of a species' habitat associations. As a predictive tool, we want the model to be accurate, robust, reliable, and transferable.

10.1 Assessing model fit

In the most general terms, model fit describes how well the variation in the response variable is explained by or due to variation in the predictor variable(s). Assessing model fit often involves a comparison of the overall amount of variance in the response variable that is explained by the predictor variables versus the amount that is due to random noise or error in the data. We can represent this relationship as $\sigma^2_{total} = \sigma^2_{[x]} + \sigma^2_{error}$. In a model that fits the data well, the explained variance in the response variable due to the effect of the predictor variables ($\sigma^2_{[x]}$) is large relative to the error variance, regardless of their absolute values. Imagine a set of N values for X that range between 1 and some upper limit. A second set of values meant to represent the response variable is generated by multiplying X by a constant c, such that $Y = Xc$. The pure effect of X on Y is indicated by c, and the variance in Y is equal to the variance in X multiplied by c^2. The function

$Y = Xc$ represents a model with perfect fit; there is no error variance because there is no source of error. However, nature, data, and the analysis of species–habitat associations are not this clean and straightforward. Rather there is always some error owing to various causes such as design and measurement inadequacies or other important environmental variables not included in the model. So, to make the data more realistic, we will add a random number between $-a$ and a, such that $Y = Xc + rnd[-a, a]$ with a variance indicated by σ^2_{total}. A habitat analysis on these data might utilize least-squares or maximum likelihood estimation to obtain a function relating Y to X, represented as $\hat{Y} = X\beta_1 + \beta_0$. The estimate for β_1 will be very close to c, unless $-a$ to a covers a wide interval. The variance in $\hat{Y}$ is the amount of variance in the observed values of the response variable Y that is explained by variance in X (this explained variance was denoted earlier as $\sigma^2_{[x]}$). Thus, the error variance is $\sigma^2_{error} = \sigma^2_{total} - \sigma^2_{[x]}$. Again, if $rnd[-a, a]$ is greatly constrained, then $\sigma^2_{[x]}$ will be just slightly less than σ^2_{total}, and σ^2_{error} will be minimal indicating good model fit. However, if $rnd[-a, a]$ is large relative to c and the values of X, then σ^2_{error} will be large and the model (or function) will have poor fit due to the excessive error. See Table 10.1 for numerical examples of this explanation of model fit. This way of explaining model fit may seem to resemble linear regression and therefore be applicable to that method of analysis only. This is not the case; this explanation is a general description valid for most statistical techniques that rely on estimating some type of function relating Y to X or estimating differences in Y among groups that are defined on the basis of X (e.g., ANOVA).

For some methods of habitat analysis (Chapter 9), the assessment of model fit may be "automatically" incorporated into the results obtained from the statistical procedure. For example, coefficients of determination (R^2 values) are metrics of model fit for *linear* regression models and most statistical software and computer code for conducting regression provide these values in the output. Typically, this output also includes the so-called "adjusted" R^2 values that take into account the inclusion of multiple predictor variables in a regression model, $R^2_{adj} = 1 - (1 - R^2) \times (N - 1)/(N - k - 1)$, where N = sample size and k = number of predictor variables in the regression model. Mathematically, an increasingly better fit is obtained as more predictor variables are added (even if they contribute very little to explaining

Table 10.1 Numerical examples of model fit defined as the percentage of variance of the response variable (Y) explained by a function or model relating $\hat{Y}$ to X. Dataset and model 1 exhibit excellent fit because there is not much random error (quantified by a) in Y. Model 3 has very poor fit to the data because there is substantial random error relative to c. For all datasets, $N = 24$.

	(1) $c = 3$, $a = 5$, Model: $\hat{Y}= 3.14X - 2.75$			(2) $c = 3$, $a = 25$ Model: $\hat{Y}= 1.88X +12.39$			(3) $c = 2$, $a = 40$ Model: $\hat{Y}= 1.01X + 19.55$		
	X	Y	$\hat{Y}$	X	Y	$\hat{Y}$	X	Y	$\hat{Y}$
1	20	55	60.0	17	48	44.4	13	31	33.9
2	1	0	0.4	15	21	40.6	12	2	32.8
3	15	50	44.3	1	13	14.3	4	46	24.0
4	4	17	9.8	15	40	40.6	11	43	31.7
5	10	25	28.6	10	46	31.2	9	40	29.5
6	7	17	19.2	4	20	19.9	10	9	30.6
7	2	2	3.5	18	59	46.2	14	41	35.0
8	18	58	53.8	14	60	38.7	13	65	33.9
9	13	37	38.1	3	1	18.0	20	53	41.6
10	5	15	12.9	5	38	21.8	9	18	29.5
11	1	0	0.4	1	4	14.3	13	2	33.9
12	11	36	31.8	17	26	44.4	15	37	36.1
13	14	38	41.2	7	35	25.6	3	23	22.8
14	10	27	28.6	18	46	46.2	9	1	29.5
15	20	62	60.0	18	30	46.2	8	14	28.4
16	19	57	56.9	20	69	50.0	17	58	38.3
17	18	49	53.8	16	40	42.5	18	48	39.4
18	7	27	19.2	19	41	48.1	22	43	31.7
19	13	39	38.1	8	25	27.4	5	43	25.1
20	16	47	47.5	14	54	38.7	5	4	25.1
21	1	1	0.4	6	22	23.7	13	41	33.9
22	14	40	41.2	12	28	35.0	15	51	36.1
23	19	61	56.9	15	22	40.6	10	57	30.6
24	6	13	16.1	6	34	23.7	19	3	40.5
Variance		412.8	402.4		284.2	128.5		407.3	24.8

Model = (402.4/412.8) × 100 = 97.5%   45.2%   6.1%
Error = 100 − 97.5 = 2.5%   54.8%   93.9%

variation in the response variable) to the model and hence this unwanted effect must be controlled. By the way, this is usually what is being referred to when authors mention that a statistical model is "overfitted." For *logistic* regression, there are several metrics available for calculating R^2 and adjusted R^2 values (Menard 2000), but none are actual R^2 values in that they are not based on the residual sum of squares (they are sometimes labeled as "pseudo R^2" values or "R^2 analogs") (Hosmer and Lemeshow 1989; Menard 2000). Multiple logistic regression and other forms of generalized linear models (GLMs) do not use least-squares calculations to obtain the coefficient estimates; rather they are based on maximum likelihood estimation (Boxes 9.2 and 9.4). Nonetheless, for any type of regression, examination of residuals is a good initial way of assessing model fit; relatively small standardized residuals indicate good model fit. For ANOVA, the comparison of variance (of the response variable) among groups versus within groups is essentially

an assessment of model fit. The F-statistic is based upon this ratio and hence assessing model fit is implicitly incorporated into significance testing. For any method of analysis, statistical significance typically coincides with good model fit (although see Box 10.1).

The chi-square test and other goodness-of-fit (GOF) tests also simultaneously test for statistical significance as well as model fit. These tests compare observed values of a response variable to expected values. For example, the often-used chi-square test computes $\chi^2 = \Sigma[(O_j - E_j)^2/E_j]$ for $j = 1$ to J categories or classes. The observed and expected values are frequencies or counts of items in a class. The expected values may be from either a theoretical or an empirical distribution. For a given number of classes (degrees of freedom), increasing values of χ^2 indicate poorer fit of the observed to the expected values in that the difference between O and E for at least one class is increasing. For a given calculated value of the χ^2 test statistic, the portion of

---

Box 10.1 Sample size, statistical significance, and model fit

It is easy to demonstrate analytically that statistical significance of a model is more likely to be achieved as sample size increases while model fit may remain unaffected and relatively poor (low). Statistical significance does not guarantee good model fit. As a demonstration, consider a simple linear regression model in which fit is assessed by the R^2 value calculated as $R^2 = 1 - (SS_{res}/SS_{tot})$ where $SS_{res} = \Sigma(y_i - \hat{y}_i)^2$ and $SS_{tot} = \Sigma(y_i - \bar{y})^2$ for $i = 1$ to n observations (i.e., the sample size). Statistical significance is based upon the F-statistic, $F_{N, n-1} = MS_{reg}/MS_{res}$, where $MS_{reg} = SS_{reg}/N = \Sigma(\hat{y}_i - \bar{y})^2/N$ and $MS_{res} = SS_{res}/(n - N - 1)$ where $N = $ total number of predictor variables (Section 9.2). Assume that $n = 30$, $N = 1$, and $SS_{reg} = 150$, such that each observation on average contributes a value of 5 to SS_{reg}. In a similar way, let's assume that $SS_{res} = 15,000$ such that each observation on average contributes a value of 500 to SS_{res}. Therefore, for this dataset of 30 observations, the simple one-factor linear regression has $F_{1,28} = 0.28$, $P = 0.60$, and $R^2 = 0.01$. If the per-observation amount of explained variance remains constant (at 5) and the per-observation amount of residual error remains constant (at 500), then $n = 386$ is the minimum sample size that would provide statistical significance ($F_{1,384} = 3.84$, $P = 0.05$) but R^2 remains the same at 0.01 because the ratio of SS_{res}/SS_{tot} (or SS_{reg}/SS_{res}) has not changed. Extending this

example, when $n = 1,000$, $F_{1,998} = 10$ and $P = 0.0016$; when $n = 2,000$, $F_{1,1998} = 20$ and P is highly significant at 0.000008. In both cases, R^2 remains equal to 0.01. The regression model explains only 1% of the variance in the response variable and yet it is statistically significant given such large sample sizes. Here is what is occurring: as sample size increases, MS_{reg} increases substantially while MS_{res} actually decreases slightly leading to ever-increasing F-values. Given a large enough sample size, statistical significance of a model can be achieved even though the model has poor fit to the data and likely does not have any real ecological meaning. The effect size of the predictor variable, measured by the regression coefficient β_x, will be very small even though significant. The opposite result can attain when sample size is too small. For example, if each observation on average contributes a value of 10 to SS_{reg} and only 40 to SS_{res} then the predictor variable explains a relatively large amount of the variance in the response variable ($R^2 = 0.20$) regardless of sample size. However, if $n \leq 19$ then the model will not be statistically significant at alpha $\alpha = 0.05$. This exercise highlights how the power of a statistical test (its ability to reveal a true significant difference or effect) typically increases with increasing sample size. This exercise also suggests that R^2 may not always be an inappropriate metric for model fit.

the chi-square distribution to the right gives the probability of obtaining by chance a χ^2 value greater than the calculated value, whereas the portion to the left is the probability of obtaining by chance a lower χ^2 value. Thus, a chi-square "model" fits the data well when the observed frequencies are close to expected frequencies, which means that the χ^2 test statistic is relatively small, left-tailed P-value is small and right-tailed P-value is large. Note that this concept of model fit is different from that described in the previous paragraph in that there is no comparison of model-explained variance to error variance. Also, chi-square and other GOF tests were not presented as methods of habitat analysis in Chapter 9. Again, these methods compare counts of items (e.g., survey plots or individual organisms) among categories. Perhaps, a habitat analysis could define categories on the basis of environmental (habitat) variables and then use a chi-square test to examine the difference in observed counts of individuals (or occupied plots) in the categories versus an expectation of random, even, or some specified distribution of individuals among categories. This approach could lead to identification of important habitat requirements of a species, but in a very indirect way compared with most of the more direct methods presented in Chapter 9. Nonetheless, some preference indices are based on chi-square and GOF tests (Section 11.2) thus indicating their utility and value in studying processes such as habitat selection. Further, even if chi-square and GOF tests are not used as the method of habitat analysis, such tests in modified form may still be useful in assessing model fit.

In particular, chi-square tests can be used to examine the fit of logistic regression models. Recall that the predicted y-values from a logistic regression model are confined to range between 0 and 1 (Section 9.3). The Hosmer–Lemeshow test is based on dividing that output range into a specified number of groups (often 10) (Hosmer and Lemeshow 1980; Hosmer et al. 1989). Then the *observed* frequencies of $y = 1$ and $y = 0$ values are calculated for each group. The *expected* frequency of $y = 1$ for each group is determined by a summation of the predicted y-values of the replicates in each group. The expected frequency of $y = 0$ for a group is simply the total number of replicates in the group minus the expected number of $y = 1$. In the case of a habitat analysis, the replicates are usually survey plots or other spatial sampling units where the species was recorded as either present ($y = 1$) or absent ($y = 0$). Also, recognize that the predicted y-values from a logistic regression model can be interpreted as "probabilities"; hence for a given group the summation of predicted y-values for each replicate is equivalent to taking the average predicted y-value for the group and multiplying it by the number of replicates in order to get the expected value of $y = 1$ for the group (i.e., this is a common way that expected values are obtained for any chi-square test). Table 10.2 illustrates the Hosmer–Lemeshow test for model fit as applied to a multiple logistic regression model for the hypothetical beetle species (Section 8.4.3 and Table 8.4). There are many other ways of assessing model fit for logistic regression models and GLMs (Hosmer et al. 1997; Menard 2000; Fagerland et al. 2008; Nakagawa and Schielzeth 2013) and sometimes these metrics of fit are referred to as R^2. Authors typically are clear that such R^2 analogs are not the true R^2 values derived from least-squares estimation of regression coefficients. As such, the R^2 analogs and other GOF metrics do not provide an estimate of the model-explained variance in the response variable, as a true least-squares R^2 value does.

Chi-square tests are also used to examine the fit of occupancy models (Section 9.6). Recall that occupancy models require temporally repeated surveying of the spatial sampling units. As such, at the end of the study, each unit has a detection history. This is a series of 1s and 0s indicating whether the focal species was detected within the unit during each of the survey occasions. Each unique detection history (e.g., 01101 in a design with five repeat surveys) can serve as a category or class for the chi-square test. If there are T repeat surveys then there are 2^T possible detection histories (e.g., $2^5 = 32$ detection histories). For each detection history, there is an *observed* number of survey units that had the particular detection history. The *expected* number of survey units having the given detection history (e.g., 01101) is simply the summation of *every* survey unit's probability of having that particular detection history. As explained in Section 9.6, the probability of a given detection history for a unit i surveyed T times

Table 10.2 The Hosmer–Lemeshow test for model fit applied to the multiple logistic regression model of beetle presence–absence in survey plots as a function of percentage canopy cover (CC), depth of leaf litter (LL), volume of woody debris (WD), ratio of oak to non-oak trees (RO), and soil type (ST): $\hat{Y} = e^{-4.79 + 0.058CC - 0.052LL + 0.076WD + 0.231RO - 0.066ST} / (1 + e^{-4.79 + 0.058CC - 0.052LL + 0.076WD + 0.231RO - 0.066ST})$.[1]

Category $\hat{Y}$	Plots	Observed values		Expected values		$(O - E)^2/E$	
		Absence Y = 0	Presence Y = 1	Absence	Presence $\Sigma\hat{Y}$of category	Absence	Presence
0–0.099	17	17	0	16.1	0.9	0.052	0.913
0.1–0.199	8	7	1	6.8	1.2	0.004	0.025
0.2–0.299	11	7	4	8.3	2.7	0.204	0.629
0.3–0.399	12	7	5	7.8	4.2	0.080	0.147
0.4–0.499	19	11	8	10.4	8.6	0.041	0.049
0.5–0.599	12	5	7	5.3	6.7	0.021	0.017
0.6–0.699	11	3	8	4.1	6.9	0.278	0.163
≥0.7	10	3	7	1.9	8.1	0.615	0.146

$$\chi^2 = \Sigma(O - E)^2/E$$
$$\chi^2 = 3.38$$
$$P = 0.759$$

[1] Each plot ($N = 100$) is assigned to a category based upon its $\hat{Y}$ value. Then for each category the numbers of absence and presence plots are counted to get the observed values. The expected value for presence for a given category is obtained by summing the $\hat{Y}$ values for the plots in that category. For example, there are eight plots in the 0.1–0.199 category with $\hat{Y} = 0.117, 0.120, 0.141, 0.146, 0.150, 0.151, 0.157$, and 0.190, which sum to 1.2. Alternatively, the expected value could be obtained by multiplying the average $\hat{Y}$ for the category by the number of plots in the category, $0.147 \times 8 = 1.2$. The χ^2 test statistic is then obtained by summing the last two columns in the table. With 6 degrees of freedom, $\chi^2 = 3.38$ is not statistically significant. The logistic regression model has relatively good fit to the data in that the observed and expected values are not significantly different.

depends upon the species detection probability during each survey (p_{it}) and the probability of occupancy (ψ_i). These probabilities are estimated by the occupancy model. Once we have observed and expected values for each category (detection history) then a chi-square test can be conducted. As always, if the χ^2 test statistic is relatively small then the model fits the data well. In the context of an occupancy model, this simply means that the observed number of survey units having each particular detection history is about what we would expect if the parameters of the model (e.g., estimated detection probabilities and occupancy probabilities as affected by the covariates or predictor variables) are correct. Applying the chi-square test as an assessment of fit for an occupancy model actually involves a few more steps than I have presented here. MacKenzie and Bailey (2004) provide a very lucid description of the entire process. One final point: if T is large relative to the number of survey

units then there may be some detection histories that have observed frequency = 0 (no survey units had the particular detection history) and expected frequency < 2. Zeros as observed values do not present a problem for chi-square testing; however, the general and longstanding rule of thumb is that expected frequencies should be > 5. This speaks to the importance of careful planning, such as deciding the number of survey units to use relative to number of survey occasions, prior to collecting data for use in an occupancy model. It may also be worth consulting the literature to see if other researchers have developed and implemented remedies for the situation when very low expected frequencies are unavoidable.

Returning to the topic of maximum likelihood, calculating *model deviance* is another way of assessing model fit (or more appropriately, lack of fit) for statistical models whose parameters are estimated by maximum likelihood. The deviance is calculated as $D = -2[\ln(L_M) - \ln(L_F)]$, where L_M is the maximum likelihood for the model of interest and L_F is the maximum likelihood for a hypothetical model that fits the data perfectly (sometimes also called a saturated model), thus L_F is always $> L_M$. Recall that likelihoods were discussed briefly in Chapter 9 with regard to occupancy modeling. The maximum likelihood for a fitted model of interest identifies the parameters, such as regression coefficients, that give the greatest likelihood value (L_M) for the given observed set of data expressed as a relationship between predictor variables and response variable. Those parameter estimates are then taken as the best for explaining the relationship between the response variable and the predictor variables. In other words, given the observed data, *what values of the model parameters give the greatest likelihood*? The perfectly fit model entails having a number of predictor variables equal to N, the sample size. Essentially, L_F pertains to an overfitted model that is simply a construct which the fitted model of interest is compared with—it is needed to calculate model deviance. The smaller the model deviance, then the better fit of the model.

Each observation in the dataset can also be thought of as having a deviance, essentially the difference between the y-value predicted by the fitted model and that predicted by the perfectly fitted model. However, recognize that the latter value is simply the observed y-value, given that the perfectly fitted model completely specifies ("predicts") the observed y-value. In this way, the deviance of each observation is equivalent to a residual. Further, the model deviance can be calculated as the sum of these observation-level deviance values. Hence, model deviance is analogous to the residual sum of squares from a least-squares estimated model. (By the way, this insight is nothing new. Most statistical references that discuss model deviance also explain this connection between model deviance and least-squares residuals.) Given that deviance can be determined for each observation, a model has a distribution of deviance values; some computer code (e.g., code for conducting GLMs in R) and software return characteristics of the distribution such as minimum, median, and maximum values. Also, the model deviance, D, has a theoretical distribution that is often approximated by the

chi-square distribution with degrees of freedom equal to $N - k$, where k is the number of parameters estimated by the model. Thus, one can test whether a given value of D is statistically significant (large) and thus indicative of significantly poor fit of the model. However, model deviance is typically used in comparing models (see next section) and not in determining whether any particular model has a significantly poor (or good) fit to the data.

10.2 Comparing models

In modern-day ecology, model comparison has its roots in the so-called information-theoretic approach to scientific inference (Burnham and Anderson 1998). This approach or framework is much deeper, philosophically, conceptually, and computationally than just comparing models based upon likelihood. Nonetheless, Burnham and Anderson (1998) popularized the use of the Akaike information criterion (AIC) (Akaike 1973, 1974) as a way to compare and rank statistical models. AIC is based on maximum likelihood, $AIC = 2k - 2\ln(L_M)$, where again k = number of model parameters (predictor variables and any intercept term). As explained in the previous paragraph, there is a connection between maximum likelihood estimation and least-squares estimation. More formally, $AIC = 2k + N \times \ln(SS_{res})$; see Burnham and Anderson (2002, p. 63) for the derivation of this equation. Although AIC can be expressed as a function of the residual sum-of-squares, AIC alone is not really a measure of model fit. This is because neither of the two equations for AIC (given above) compare the model likelihood or residual sum of squares to anything. There is no way to know if the "error" represented by either quantity is comparatively large or not. Therefore, AIC should be used strictly for model comparison. When comparing two or more models, the one with the lower AIC value has the *better* fit to the data, but this does not necessarily mean that its fit is good (i.e., that the model is accurate). Note that both equations above include the number of model parameters (the "$2k$" term) in an additive way; all other things being equal, an increase in k leads to a higher AIC value. This is important because we might often be comparing models that differ in the number of predictor variables. As mentioned previously, as the number of parameters (predictor variables) increases, the model will become better fitted to the data to the point of being unnecessarily overfitted. Some authors refer to the $2k$ term as a way of "penalizing" a model for having extra predictor variables. Even with the penalty, when overall sample size N is small (e.g., <50), the model with the lowest AIC might still represent an overfitted model. To overcome this, statisticians recommend using a corrected form of AIC, that is AICc = AIC + $(2k^2 + 2k)/(N - k - 1)$. In the past two decades, AIC-based model comparison has become very popular and widely used, not just in ecological research and comparing species–habitat models, but in all of science. One reason for its popularity is that it departs from the traditional frequentist statistical approach of applying tests of significance to null hypotheses. Model comparison through AIC can still involve hypothesis testing but it need not involve significance testing.

To further illustrate model comparison, I return to the regression models constructed to analyze the habitat associations of the hypothetical beetle species (Chapter 8). Recall that there were five predictor variables: percentage canopy cover, depth of leaf litter, volume of woody debris, ratio of oak:non-oak trees, and soil type, all of them measured in each of 100 survey plots. For the model comparison, I included the full model (all five predictor variables) along with a model that included canopy cover and ratio of oak:non-oak trees, and each of the two single-factor models, canopy cover only, and oak:non-oak ratio only. In the various habitat analyses of Chapter 8, canopy cover and oak:non-oak ratio were revealed to have the greatest influence on the response variable, beetle presence–absence or abundance, and hence their inclusion in this model comparison exercise. With five predictor variables there are 31 possible models (or combinations of the predictor variables, including the full model and the five single-factor models); this number is determined as $2^k - 1$, although it can also be determined through combinatorics. In a technical sense, there is no limit to the number of models to include in a model comparison. If I had implemented all 31 models then all could have been included in the model comparison.

Table 10.3 presents a comparison of the models constructed using multiple linear regression (Section 8.4.2), negative binomial regression (Section

Table 10.3 Comparison of statistical models and assessment of model fit for the analyses conducted on the hypothetical beetle species. AIC and ΔAIC values are useful for model comparison but do not assess model fit. Model deviance (D) and percentage reduction in D can be used to assess model fit and compare models. Model comparison should only be conducted for models using the same statistical method and applied to the same dataset, hence the divisions in the table. Note that some models were applied only to the non-zero abundance data ($N = 40$). All models, except the zero-inflated negative binomial (ZINB) model, were conducted as generalized linear models (see Chapter 9).

Method	N	Model	k^1	AIC	ΔAIC	D	%D red.[2]
Linear regression	100	Full	6	529.4	5.3	1,014.0	28.4
		Canopy + ratio oak:non-oak	3	524.1	0	1,020.8	27.9
		Canopy only	2	541.1	17.0	1,233.7	12.8
		Ratio oak:non-oak only	2	536.7	12.6	1,180.9	16.6
	40	Full	6	220.5	5.1	408.6	28.7
		Canopy + ratio oak:non-oak	3	215.4	0	418.1	27.0
		Canopy only	2	223.2	7.8	534.0	6.8
		Ratio oak:non-oak only	2	217.1	1.7	458.5	20.0
Negative binomial regression	40	Full	7	215.2	4.6	41.2	27.1
		Canopy + ratio oak:non-oak	4	210.6	0	41.4	24.5
		Canopy only	3	217.6	7.0	41.7	6.2
		Ratio oak:non-oak only	3	211.8	1.2	41.3	18.6

(Continued)

Table 10.3 (Continued)

Method	N	Model	k^1	AIC	ΔAIC	D	%D red.[2]
Zero-inflated NB regression	100	Full	7	320.5	4.2	170.4	−118.0
		Canopy + ratio oak:non-oak	4	316.3	0	172.3	−120.3
		Canopy only	3	331.0	14.7	188.9	−141.7
		Ratio oak:non-oak only	3	336.1	19.8	194.0	−148.2
Logistic regression	100	Full	6	120.0	5.8	108.0	19.8
		Canopy + ratio oak:non-oak	3	114.2	0	108.3	19.6
		Canopy only	2	120.5	0.5	116.5	13.4
		Ratio oak:non-oak only	2	130.0	10.0	127.0	5.7

[1] The number of model parameters (k) for each model includes the y-intercept. The models involving a negative binomial distribution have the extra parameter (θ) that estimates intraspecific clumping.

[2] This refers to the reduction in model deviance when the predictor variables are added to the model. It is the percentage difference between the deviance of the null or intercept-only model (not shown) and the model of interest.

9.2.1), zero-inflated negative binomial regression (Section 9.2.2), and logistic regression (Section 8.4.3). Note that for each set of four models, comparison is limited to the models constructed using the same regression method and applied to the same dataset ($N = 40$ or 100 survey plots). The table gives ΔAIC values for each model within its set of four. The change (increase) of an AIC value is calculated as the difference between the model with the lowest AIC value and each of the other models. ΔAIC is a widely used metric for comparing models; nearly all published model comparison tables include it. Table 10.3 is not strictly formatted as a model comparison table. A more formal and orthodox model comparison table would include other information (such as the log-likelihood values) and arrange the models from top to bottom in order of increasing AIC and ΔAIC values. Such a table also would certainly not include models implemented through different statistical techniques. ΔAIC provides a direct comparison of all models to the model taken to be the "best," which is the one with the greatest $\ln(L)$ value (remember maximum likelihood is involved in estimating parameters of the model) and hence the lowest AIC value. Given that models are usually arranged in order of increasing ΔAIC value, they are essentially ranked in the table. The ordering also makes it visually easier to see what is sometimes referred to as a "top" or "candidate" model set. Over the years, various authors have proposed cutoff values for ΔAIC that define the candidate model set. One common threshold is ΔAIC < 2, meaning that all models with such low ΔAIC values are included in the candidate set. Some authors will even use language such as "equally good" or "equally competitive" to describe models in the candidate set. This implies that all the models in the set are reasonable representations of the relationship between the response variable and predictor variables. Unfortunately, no one has ever provided a rigorous and objective rationale for the seemingly arbitrary cutoff values and qualitative word choice used in interpreting AIC values. Rather than dig deeper into this issue, I refer the reader to the literature on AIC and the information-theoretic approach; Burnham and Anderson (2002) might be a good starting point.

Now back to Table 10.3. It clearly shows that the model including canopy cover and oak:non-oak tree ratio is the best (of the four models examined) for explaining beetle presence–absence and abundance. AIC values are lowest for this model as implemented by all four regression techniques. However, this does not necessarily mean that this model has a good fit to the data. To assess model fit, we can examine the deviance values. Interestingly, the full model has the lowest D values and hence also the greatest percentage reduction in D, although the canopy + oak tree ratio model is a close second (Table 10.3). The likely reason for this is that the full model has more predictor variables and the three additional variables (depth of leaf litter, volume of woody debris, and soil type) explain a little bit more of the variance in the response variable beyond that explained by canopy cover and oak:non-oak tree ratio. A particularly unusual result shown in the table is that the D values for the zero-inflated negative binomial (ZINB) models are actually greater than the D value for the intercept-only model. That is, the %D reduction values are negative indicating that, as predictor variables are added, the models actually take on a much poorer fit to the data. This suggests that the unique feature of the ZINB regression technique in estimating the probability of a zero count (Section 9.2.2) is not really needed or useful for the beetle data. Lastly, as previously mentioned, it is not appropriate to compare AIC or D values among models implemented by different statistical techniques. However, I take the liberty of pointing out that the logistic regression models had the lowest AIC values (Table 10.3). At a few places in the book, I have shown favor to logistic regression as the method of choice for conducting a habitat analysis—see also Chapter 12.

Back to the question of how many models to run. In the analyses presented in Chapter 8, I only used the full regression model. However, when a researcher has measured multiple environmental variables, each of which could be potentially important in characterizing the species' habitat, an important decision must be made with regard to how many models to conduct and which ones. There is no universally accepted strategy for this. Typically, there is no need to conduct every possible model (every possible combination of predictor variables), particularly when more than four predictor variables have been measured. Consider that when a dataset has as few as six predictor variables, 63 unique models are possible. The danger in conducting an excess number of models is that it leads to an increase in the study-wide rate of Type I error (Chapter 12). Good statistical and scientific practice is to examine only a subset of all possible models. Further, these obviously should not be randomly chosen, but rather some type of logical rationale should be employed in deciding which models to

examine. The set of models could be completely specified prior to running any models, or perhaps models are conducted based on a particular strategy that uses the output of current models to identify the next model(s) to conduct (e.g., stepwise regression).

Assessing model fit and comparing models are not required components of a habitat analysis. They should be considered additional examinations that one could use to further support the inferences made from the habitat analysis. If habitat is viewed as a set of multiple integrated environmental factors that a species requires for living, foraging, and reproducing (Chapter 12), then identifying that set may be very important. From this perspective, it would be very useful to examine and compare different statistical models and perhaps identify the single best one or the top few. This is essentially the statistical procedure of *model selection*. Models are compared with the intent of identifying (i.e., selecting) the one whose predictor variables best explain variation in the response variable. Finally, the words "prediction" and "predicted" are often used when discussing model fit. We refer to "predicted y-values" or values of the response variable predicted by the statistical model, and even use (as I have done throughout this book) the label of "predictor variable" for the independent (X) variables of a statistical model. This is informal and rather loose use of the word. Prediction as a word and property of a statistical model is used in a much more exact context when assessing predictive accuracy (next section).

10.3 Assessing predictive accuracy

Statistical models of a species' habitat associations can sometimes be used to predict where the species could occur and where it is expected to be absent. This is a common and often required step in habitat suitability modeling and species distribution modeling (Sections 11.4 and 11.5); however, it is also a necessary step if one is to use their statistical model for tasks such as identifying areas that might be most suitable for species reintroduction and areas that may be most in need of formal protection. That is, even apart from developing a sophisticated model to predict species presence–absence or habitat suitability over a broad geographic region, there may be great utility in examining the ability of the model to accurately predict presence–absence at particular locations on a landscape. After conducting a habitat analysis, the need for assessing the predictive capacity of the model (developed during the habitat analysis) depends to some extent on the purpose of having conducted the initial habitat analysis (Chapter 4). At the very least, an assessment of predictive accuracy allows for a further independent test of the validity of the identified habitat associations (i.e., the statistical model) beyond just examining model fit.

A key feature of assessing predictive accuracy that distinguishes it from model fit is that a "new" independent dataset is used. This additional dataset can represent collection or compilation of original data or it could be a previously collected set of data from a source other than the study that was used to develop the statistical model. Importantly, this is another key feature, the dataset used for assessing predictive accuracy must be different from that used in constructing the model (although see "cross-fold validation" below). The procedure for assessing predictive accuracy and quantifying error rates is sometimes referred to as *model validation*. That is, one is attempting to validate their statistical model as correct and useful for prediction, although the phrase "model validation" is used somewhat broadly and can sometimes refer to a procedure that is essentially an assessment of model fit rather than predictive accuracy.

The predictive accuracy of a model can be assessed whether the model predicts binary outcomes (e.g., species presence–absence) or non-binary outcomes represented by discrete or continuous variables (e.g., abundance and activity level). For a binary response variable such as presence–absence there are four possible outcomes that can be predicted and realized: true positive, true negative, false positive, and false negative (Fig. 10.1). (Alternatively, we could think of these outcomes as true presence, true absence, false presence, and false absence; however, the "positive-negative" labeling is more general.) A true positive occurs when the species is predicted to be present at a particular location (given the set of environmental factors at that location) and actually is observed at

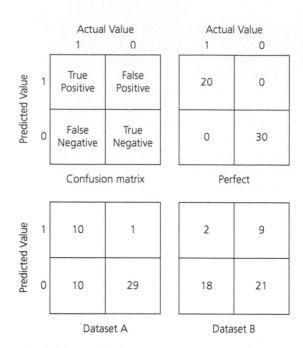

Figure 10.1 The four possible outcomes for a statistical model or diagnostic test that has a binary response such as species presence–absence. These outcomes are sometimes presented in a confusion matrix or table (this label derives from the literature on machine-learning). A matrix that perfectly validates the model has actual (observed) values equal to the predicted values. As an example, for the hypothetical beetle species surveyed at 50 locations, a dataset that perfectly validates the model would have 20 locations representing true positives and 30 representing true negatives. That is, each location predicted to have the species actually does have it and the species is absent from each location where it was predicted to be absent. The confusion matrices for Datasets A and B differ from one another and from the "perfect" dataset. For both datasets, the model being assessed is $\hat{Y} = e^{-4.89 + 0.051CC + 0.226RO}/(1 + e^{-4.89 + 0.051CC + 0.226RO})$ where CC = percentage canopy cover and RO = the ratio of oak:non-oak trees and predicted values are based on $\hat{Y}_{crit} = 0.5$.

the location. A true negative occurs when the prediction is absence and indeed the species is not detected at the location (given enough survey effort and sufficiently high detection probability). A false positive occurs when the prediction is presence but the species is not detected (again assuming enough survey effort and high detection probability). A false negative occurs when the prediction is absence yet the species is subsequently found at the location.

As previously discussed, the output of a logistic regression model is a number between 0 and 1 (Section 9.3). Although this number is not really a probability in the strict sense, we can use it to predict whether a given location will have the species of interest or not. To do this, we select a cutoff value such as $\hat{Y}_{crit}$ = 0.5. Assuming we have data on the environmental variables (predictor variables in the model) for the given location, we use the logistic regression equation to determine $\hat{Y}_i$ for location i. If $\hat{Y}_i > \hat{Y}_{crit}$ then we predict a species presence (1) for location i, otherwise we predict species absence (0). To illustrate this procedure, imagine that we collected an independent set of species–environment data for the hypothetical beetle species surveyed at 50 new locations (the survey method could be the same plot-based approach as initially used or it could be a different method) (Appendix 10.1). We tabulate the number of true positives (TP), true negatives (TN), false positives (FP), and false negatives (FN) for Datasets A and B (Fig. 10.1). First, note that the number of observed (actual) species presences for both datasets is 20; this is equal to TP + FN (recall that an "FN" is a species presence at a location where it was predicted not to occur). The observed number of species absences is TN + FP, and equal to 30 for both datasets (Fig. 10.1). For these two hypothetical validation datasets, I intentionally set the ratio of observed absence:presence (3:2) equal to that in the hypothetical dataset used to build the model (Appendix 8.1). Also, for ease of comparison, Datasets A and B have the same values for the predictor variables, although this obviously would not be the case for two *real* and independent validation datasets. Second, the numbers of TP, TN, FP, and FN for both datasets depart from those of a dataset that would perfectly validate the model. Third, Datasets A and B differ from one another, and Dataset A appears to be the better match to the "perfect" dataset (Fig. 10.1). There are many different metrics for assessing predictive accuracy that derive from different combinations of TP, TN, FP, and FN (Fielding and Bell 1997; Allouche et al. 2006). Two of the most fundamental are the true positive rate, TPR = TP/(TP + FN), and the true negative rate, TNR = TN/(TN + FP). In the context of species presence–absence data, TPR is the number of locations (or sampling units) where the species was *predicted to be present and was recorded as such*, as a proportion of the total number of locations having the species (this latter quantity typically

includes some locations that were FN). If there are not any FN locations then TPR = 1 (or 100 percent) and the model is perfect at predicting or identifying locations where the species is present. TNR is the number of locations where the species was *predicted to be absent and was recorded as such*, as a proportion of the total number of locations lacking the species. Some of the absence locations might actually be FP, these locations were predicted to have the species, but even with sufficient survey effort it was not detected. If there are no FP locations, then all recorded absences were predicted as such and TNR = 1 (or 100 percent); the model is perfect at predicting absence. The TPR is also known as the *sensitivity* of a model or diagnostic test and the TNR is known as *specificity*. A model with high (or good) predictive accuracy has high sensitivity and high specificity; it is very sensitive and very specific. High sensitivity means that the model (or test) rarely commits errors of omission when identifying positive outcomes (species presence), whereas high specificity means that the model rarely commits errors of commission in identifying negative outcomes (i.e., predicting a species absence at a location where it is subsequently found).

For the hypothetical beetle species, I examined the predictive accuracy of the logistic regression model that included only percentage canopy cover and oak:non-oak tree ratio as this model appeared to have the best fit as indicated by the lowest AIC value (Table 10.3). As shown in Table 10.4, using Dataset A for validation reveals the model to have greater sensitivity and specificity than does using Dataset B, regardless of the $\hat{Y}_{crit}$ value that is used. With either dataset and any $\hat{Y}_{crit}$ value, specificity is greater than sensitivity indicating that the model is better at identifying locations where the species should be absent and is found to be absent than it is at identifying locations that have suitable habitat (appropriate amounts of canopy cover and ratio of oak:non-oak trees) and hence are likely to have the beetle species. The positive predictive value (PPV) expresses the number of observed positives (1s or species presences) as a proportion of predicted positives (the latter includes TP and FP). A high PPV value indicates that the model does not overpredict positives or species presence. Dataset A reveals the logistic model to have a greater PPV than does Dataset B (Table 10.4). The negative predictive value (NPV) does the opposite; it indicates whether the test overpredicts negatives or species absence. The model has a higher NPV value (less overprediction) when examined with Dataset A than B (Table 10.4). The correct classification rate is simply the number of true positives and true negatives as a proportion of all outcomes or locations. When examined with Dataset A, the model has a higher correct classification rate than when examined with Dataset B regardless of $\hat{Y}_{crit}$ value (Table 10.4).

The last two rows of Table 10.4 give metrics of prediction *error*. The false discovery rate (FDR) indicates the extent to which the model incorrectly predicts positive outcomes or species presence. FDR values are substantially greater when examining model performance with Dataset B than with Dataset A. The misclassification rate is the complement of the correct classification rate. Dataset B reveals the model to have greater misclassification rates than does Dataset A. Importantly, the model has a substantially higher rate of correct classification than rate of misclassification, at least when validated with Dataset A. This alone suggests that the logistic regression model with canopy cover and

oak:non-oak tree ratio as predictor variables would be a useful tool in predicting beetle presence–absence in some situations (granted the beetle species is imaginary).

Table 10.4 also shows that the prediction accuracy and error metrics depend on the choice of $\hat{Y}_{crit}$. This is because the $\hat{Y}_{crit}$ value affects the number of TP, TN, FP, and FN outcomes. One common technique for assessing the performance of statistical models and diagnostic tests over the entire range of $\hat{Y}_{crit}$ is to produce a receiver operating characteristic (ROC) curve and calculate its area under the curve (AUC) value (Fielding and Bell 1997; Bewick et al. 2004; Allouche et al. 2006; Brown and Davis 2006; Jiménez-Valverde 2012). The ROC curve has its historic beginnings in signal detection and information retrieval theory that emerged in the 1950s (Peterson and Birdsall 1953; Peterson et al. 1954). The name initially referred to the ability of a human observer (the operator) to use equipment (a receiver) to detect some type of signal (e.g., acoustic or radar) and distinguish it from noise or clutter. "Characteristic" refers to properties of the receiver and operator as these properties affect the process of detecting and distinguishing the signal. A false positive occurs when noise or clutter is mistaken for

Table 10.4 Metrics of predictive accuracy and error for the logistic regression model, $\hat{Y} = e^{-4.89 + 0.051CC + 0.226RO}/(1 + e^{-4.89 + 0.051CC + 0.226RO})$, CC = percentage canopy cover and RO = ratio of oak:non-oak trees, applied to Datasets A and B. Results are shown for three different $\hat{Y}_{crit}$ values. Note that all of the metrics are constrained to range between 0 and 1.

Metric	Equation[1]	Dataset A			Dataset B		
		$\hat{Y}_{crit} = 0.4$	$\hat{Y}_{crit} = 0.5$	$\hat{Y}_{crit} = 0.6$	$\hat{Y}_{crit} = 0.4$	$\hat{Y}_{crit} = 0.5$	$\hat{Y}_{crit} = 0.6$
Sensitivity	TPR = TP/(TP + FN)	0.600	0.500	0.350	0.200	0.100	0.100
Specificity	TNR = TN/(TN + FP)	0.933	0.967	1	0.667	0.700	0.833
Positive predictive value	PPV = TP/(TP + FP)	0.857	0.909	1	0.286	0.182	0.286
Negative predictive value	NPV = TN/(TN + FN)	0.778	0.744	0.698	0.556	0.538	0.581
Correct classification rate	= (TP + TN)/n	0.800	0.780	0.740	0.480	0.460	0.540
FDR	= 1 − PPV	0.143	0.091	0	0.714	0.818	0.714
Misclassification rate	= (FP + FN)/	0.200	0.220	0.260	0.520	0.540	0.460

[1] See text for abbreviations; n = total number of locations, survey plots, or sample size.

a real signal and a false negative occurs when the settings of the receiver are set in a way in which the human observer does not detect a real signal. Thus, an ROC curve is a graph of the rate of true positives versus false positives over the entire range of $\hat{Y}_{crit}$ from approximately 0 to 1. It shows how true and false positives both increase as $\hat{Y}_{crit}$ decreases. Remember that a given test, location, or survey plot i is classified as a TP or FP based on whether $\hat{Y}_i > \hat{Y}_{crit}$, so if $\hat{Y}_{crit}$ is really small (e.g., 0.1) then most $\hat{Y}_i$ will be greater than $\hat{Y}_{crit}$. An ideal ROC curve, representing a model with excellent predictive performance, increases and reaches a plateau very quickly such that AUC → 1. This would indicate a

model that has a much higher TPR (or sensitivity) than FPR over all values of $\hat{Y}_{crit}$.

Note the difference in the ROC curves and AUC values for the logistic regression model applied to Datasets A and B (Fig. 10.2). Also, the ROC curve for the model applied to Dataset B traces the line of no discrimination and has an AUC value <0.5 (Fig. 10.2); both of these characteristics indicate a model that has very poor predictive performance. Indeed, the model's classification of each unit i to TP, TN, FP, and FN is essentially random. I also constructed graphs of sensitivity and specificity versus $\hat{Y}_{crit}$ (Fig. 10.3). These graphs show how there is an inherent tradeoff between sensitivity and specificity. Ideally, a statistical model or test maximizes both. The intersection of the sensitivity and specificity curves gives the $\hat{Y}_{crit}$ value where both are simultaneously maximized. As applied to Dataset

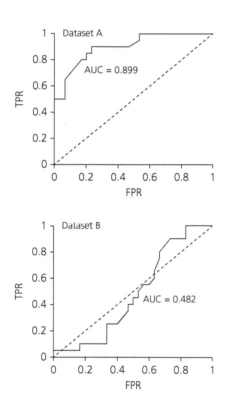

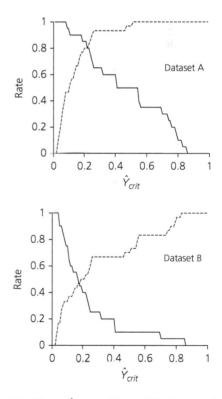

Figure 10.2 Receiver operating characteristic (ROC) curves for the logistic regression model, $\hat{Y} = e^{-4.89 + 0.051CC + 0.226RO}/(1 + e^{-4.89 + 0.051CC + 0.226RO})$ (CC = percentage canopy cover and RO = the ratio of oak:non-oak trees) applied to Datasets A and B. Although not explicitly depicted, $\hat{Y}_{crit}$ decreases from left to right. The line of no discrimination (stippled) represents a model, as applied to a particular validation dataset, that cannot reliably discriminate true positives from false positives. TPR = FPR along the entire line.

Figure 10.3 Effect of $\hat{Y}_{crit}$ on sensitivity (solid line) and specificity (stippled line) of the logistic regression model, $\hat{Y} = e^{-4.89 + 0.051CC + 0.226RO}/(1 + e^{-4.89 + 0.051CC + 0.226RO})$ (CC = percentage canopy cover and RO = the ratio of oak:non-oak trees) applied to Datasets A and B.

A, the model has a maximum sensitivity and speci-ficity of 0.8 occurring at $\hat{Y}_{crit}$ = 0.22, whereas for Dataset B these values are 0.47 and 0.18, respect-ively (Fig. 10.3). Presumably, the intersection of the sensitivity and specificity curves identifies a $\hat{Y}_{crit}$ value that could be used in other applications of the model in predicting positive and negative outcomes (i.e., species presences and absences) as long as sen-sitivity and specificity are sufficiently high (e.g., 0.7 or better).

The use of Datasets A and B to illustrate the pre-dictive performance or accuracy of a statistical model is not intended as a comparison of whether the logistic regression model applies better to Dataset A or B, or which is the better (more valid) dataset. Note that I did not assess model fit. In gen-eral, if a model has good (high) predictive accuracy then it probably also fits the data well. However, a model could fit the data well and yet still not be very good at prediction. Assessing the predictive accur-acy of a model is an on-going task that should be conducted on as many independent datasets as pos-sible, particularly if the model is to be put to some important use in conservation or natural resources management (Chapter 4). Also, the validation data-sets should have large sample sizes (much greater than just n = 50) so as to estimate the performance metrics (Table 10.4) with sufficient precision. A sam-ple size of 50 allows precision only in increments of 0.02.

Often, independent datasets for model validation are not easily available. As a result, cross-fold valid-ation is used. In this technique, a large dataset (per-haps, n = 500 or even much greater) is randomly divided into two portions typically with a 90:10 or 80:20 split. The larger portion is then used to con-struct the model (i.e., used in the actual habitat analysis) and the smaller portion is used to assess predictive performance. Further, the process is typ-ically repeated many times representing many dif-ferent unique divisions of the dataset. There are various types of cross-fold validation that differ in the particular way in which the dataset is split or subsampled, although these variants are not important to a basic description of the technique. One particularly common method is k-fold valid-ation in which the dataset is divided into k sub-samples of equal size. Then each subsample is in

turn used as the validation dataset for a model con-structed on the other $k-1$ subsamples that are com-bined into a single larger model-building dataset. Cross-fold validation has its beginnings in machine-learning, hence the model-building portion of the dataset is often referred to as the "training" data and the validation portion is referred to as the "testing" data.

In explaining predictive accuracy, so far I have only discussed logistic regression models as applied to binary response variables such as species presence–absence. Some of the other methods presented in Chapter 9 (e.g., classification and regression trees (CART), occupancy modeling) also directly or indir-ectly produce results that can be subsequently treated as binary in an assessment of predictive accuracy. However, for methods that are applied to non-binary and non-categorical response variables, either discrete or continuous, the assessment of predictive accuracy cannot be based on 0s and 1s (TP, TN, FP, and FN). For example, with regard to habitat analysis, multiple linear regression models are typically applied to data in which the response variable is species abundance or activity level. As such, their predictions are also in the form of abun-dance or activity. A multiple linear regression model can be considered a predictive equation for the abundance (or activity) expected at a particular cur-rently unsurveyed location given the values of the various environmental (habitat) variables at the location. Once the relevant environmental variables are measured, we can predict species abundance at locations constituting a validation dataset. We then measure the abundance of the species at the loca-tions and then compare these observed values to predicted (expected) values using any of the many statistical tests available for this purpose (e.g., chi-square and other GOF tests, paired t-tests, regression of observed vs. predicted values, Kolmogorov–Smirnov test). The predictive performance of the model is then assessed (and perhaps quantified in some way) based on how well predicted values match the observed values.

The predictive accuracy of a statistical model is a reliable indicator of how well the predictor vari-ables of the model truly do represent real effects on the response variable. If a model is good at predic-tion then it is probably based on some reality or genuine knowledge. Thus, with regard to habitat analysis, a good predictive model indicates that the quantitative effects of the environmental variables have been accurately quantified. The habitat associations of the species have been correctly identified and described. Predictive accuracy is also about generality and transferability of the model. Ideally, the model performs well in prediction beyond the spatial and temporal scope of the data used to construct the model. In the example of the hypothetical beetle species, the logistic regression model as applied to Dataset B did not perform well in predicting presence–absence. Such a failure could arise if the locations of Dataset B are geographically too far removed from the four forests and the initial set of 100 survey plots used to develop the model (Chapter 8). Habitat associations can sometimes change over the geographic range of a species (e.g., Revilla et al. 2000; Oliver et al. 2009; Dumyahn et al. 2015) weakening transferability of a model over space. Also, the model might have poor predictive performance for Dataset B if there is some other unmeasured and excluded variable(s) that is most important in determining beetle presence–absence at the 50 locations comprising the dataset. Another possible explanation is survey or sampling error in the

species presence–absence data. False positives can easily occur if the model predicts presence but survey effort is insufficient and/or species detection probability is too low to reveal the species. False negatives are probably less likely, but they also can occur when the model predicts absence a similar species is mistakenly recorded as the species of interest. As previously stated, predictive performance is best assessed using many different validation datasets; however, there will always be constraints on the applicability of a model due to limited transferability and generality.

Assessing model fit and predictive performance is an active area of research at the nexus of ecology and statistics (e.g., Lawson et al. 2014; Fieberg et al. 2018). In particular, predictive performance is very important in species distribution modeling (Section 11.5) in that the validity of a species distribution model lies in how well it predicts species presence–absence over relatively large geographical areas (Meynard and Quinn 2007; Merow et al. 2014; Guillera-Arroita et al. 2015). Many of the techniques that I discussed in this section (e.g., AIC model comparison, ROC and AUC to assess specificity and sensitivity) are used in a wide variety of academic disciplines, from the many different fields of science (broadly defined) and even non-science research arenas such as economics, sociology, and education.

Appendix 10.1 Two hypothetical datasets for testing the accuracy of the logistic regression model, $\hat{Y} = e^{-4.89 + 0.051CC + 0.226RO}/(1 + e^{-4.89 + 0.051CC + 0.226RO})$, in predicting the probability of the hypothetical beetle species being present at a given location. The model includes canopy cover (CC) and oak:non-oak tree ratio (RO) as predictor variables.

Location	Canopy cover	Oak:non-oak tree ratio	Predicted probability ($\hat{Y}$)	Observed P–A[1] Dataset A	Observed P–A[1] Dataset B
1	20	1.33	0.027	0	0
2	20	1.67	0.030	0	0
3	24	0.83	0.030	0	0
4	21	1.71	0.031	0	0
5	21	2.75	0.039	0	0
6	25	2.00	0.041	0	1
7	23	3.00	0.046	0	1
8	31	1.67	0.051	0	0
9	34	1.13	0.052	0	0
10	34	1.29	0.054	0	0
11	34	2.00	0.063	0	1
12	33	2.67	0.069	0	0

(Continued)

Appendix 10.1 (Continued)

Location	Canopy cover	Oak:non-oak tree ratio	Predicted probability ($\hat{Y}$)	Observed P–A[1] Dataset A	Observed P–A[1] Dataset B
13	42	0.83	0.072	0	0
14	33	3.00	0.074	0	1
15	42	1.50	0.082	1	1
16	41	2.67	0.100	1	1
17	31	5.00	0.102	0	1
18	48	1.25	0.103	0	0
19	30	6.00	0.119	0	1
20	46	3.00	0.134	0	0
21	40	4.50	0.138	0	1
22	46	3.33	0.143	0	0
23	58	1.25	0.161	0	0
24	51	3.00	0.166	0	1
25	58	1.60	0.172	0	1
26	62	1.14	0.187	1	0
27	57	2.50	0.195	0	1
28	40	7.00	0.220	1	0
29	27	10.00	0.222	0	1
30	64	2.00	0.236	0	1
31	33	9.00	0.236	1	0
32	66	1.67	0.241	0	0
33	71	0.63	0.245	1	1
34	57	4.00	0.254	0	0
35	55	4.50	0.256	1	0
36	74	1.40	0.310	1	1
37	83	1.13	0.401	1	1
38	64	5.50	0.405	1	1
39	58	8.00	0.469	0	0
40	70	6.00	0.509	0	0
41	73	6.00	0.547	1	0
42	90	2.20	0.549	1	0
43	56	10.00	0.556	1	0
44	72	9.00	0.693	1	1
45	72	10.00	0.739	1	0
46	82	8.00	0.750	1	0
47	77	10.00	0.785	1	0
48	83	9.00	0.799	1	0
49	86	9.00	0.822	1	0
50	86	10.00	0.853	1	1

[1] Observed presence–absence refers to the datasets used to assess predictive accuracy of the logistic regression model. The model itself was constructed using the observed data in Appendix 8.1.

References

Akaike, H. (1973). Information theory and an extension of the maximum likelihood principle. Pages 267–81 in *Second International Symposium on Information Theory, Tsahkadsor, Armenia, USSR, September 2–8, 1971*. Petrov, B.N. and Csáki, F. (editors). Akadémiai Kiadó, Budapest, Hungary.

Akaike, H. (1974). A new look at the statistical model identification. *IEEE Transactions on Automatic Control*, 19, 716–23.

Allouche, O., Tsoar, A., and Kadmon, R. (2006). Assessing the accuracy of species distribution models: prevalence, kappa, and the true skill statistic (TSS). *Journal of Applied Ecology*, 43, 1223–32.

Bewick, V., Cheek, L., and Ball, J. (2004). Statistics review 13: receiver operating characteristic curves. *Critical Care*, 8, 508–12.

Brown, C.D. and Davis, H.T. (2006). Receiver operating characteristics curves and related decision measures: a tutorial. *Chemometrics and Intelligent Laboratory Systems*, 80, 24–38.

Burnham, K.P. and Anderson, D.R. (1998). *Model Selection and Inference: a Practical Information-Theoretic Approach*. Springer-Verlag, New York.

Burnham, K.P. and Anderson, D.R. (2002). *Model Selection and Inference: a Practical Information-Theoretic Approach*. Second edition. Springer-Verlag, New York.

Dumyahn, J.B., Zollner, P.A., and Smith, W.P. (2015). Microhabitat comparison of swamp rabbit sites between periphery and core of the species range. *Journal of Wildlife Management*, 79, 1199–206.

Fagerland, M.W., Hosmer, D.W., and Bofin, A.M. (2008). Multinomial goodness-of-fit tests for logistic regression models. *Statistics in Medicine*, 27, 4238–53.

Fieberg, J.R., Forester, J.D., Street, G.M., Johnson, D.H., ArchMiller, A.A., and Matthiopoulos, J. (2018). Used-habitat calibration plots: a new procedure for validating species distribution, resource selection, and step-selection models. *Ecography*, 41, 737–52.

Fielding, A.H. and Bell, J.F. (1997). A review of methods for the assessment of prediction errors in conservation presence-absence models. *Environmental Conservation*, 24, 38–49.

Guillera-Arroita, G., Lahoz-Monfort, J., Elith, J., et al. (2015). Is my species distribution model fit for purpose? Matching data and models to applications. *Global Ecology and Biogeography*, 24, 276–92.

Hosmer, D.W. and Lemeshow, S. (1980). A goodness-of-fit test for the multiple logistic regression model. *Communications in Statistics*, 9, 1043–69.

Hosmer, D.W. and Lemeshow, S. (1989). *Applied Logistic Regression*. Wiley Publishing, New York.

Hosmer, D.W., Jovanovic, B., and Lemeshow, S. (1989). Best subsets logistic regression. *Biometrics*, 45, 1265–70.

Hosmer, D.W., Hosmer, T., Le Cessie, S., and Lemeshow, S. (1997). A comparison of goodness-of-fit tests for the logistic regression model. *Statistics in Medicine*, 16, 965–80.

Jiménez-Valverde, A. (2012). Insights into the area under the receiver operating characteristic curve (AUC) as a discrimination measure in species distribution modeling. *Global Ecology and Biogeography*, 21, 498–07.

Lawson, C.R., Hodgson, J.A., Wilson, R.J., and Richards, S.A. (2014). Prevalence, thresholds and the performance of presence-absence models. *Methods in Ecology and Evolution*, 5, 54–64.

MacKenzie, D.R. and Bailey, L.L. (2004). Assessing the fit of site-occupancy models. *Journal of Agricultural, Biological, and Environmental Statistics*, 9, 300–18.

Menard, S. (2000). Coefficients of determination for multiple logistic regression analysis. *American Statistician*, 54, 17–24.

Merow, C., Smith, M.J., Edwards, T.C., et al. (2014). What do we gain from simplicity versus complexity in species distribution models? *Ecography*, 37, 1267–81.

Meynard, C.N. and Quinn, J.F. (2007). Predicting species distributions: a critical comparison of the most common statistical models using artificial species. *Journal of Biogeography*, 34, 1455–69.

Nakagawa, S. and Schielzeth, H. (2013). A general and simple method for obtaining R^2 from generalized linear mixed-effects models. *Methods in Ecology and Evolution*, 4, 133–42.

Oliver, T., Hill, J.K., Thomas, C.D., Brereton, T., and Roy, D.B. (2009). Changes in habitat specificity of species at their climatic range boundaries. *Ecology Letters*, 12, 1091–102.

Peterson, W.W. and Birdsall, T.G. (1953). *The Theory of Signal Detectability, Part I. The General Theory*. Engineering Research Institute, University of Michigan, Ann Arbor, Mich.

Peterson, W.W., Birdsall, T.G., and Fox, W.C. (1954). The theory of signal detectability. *Transactions of the IRE Professional Group on Information Theory*, 4, 171–12.

Revilla, E., Palomares, F., and Delibes, M. (2000). Defining key habitats for low density populations of Eurasian badgers in Mediterranean environments. *Biological Conservation*, 95, 269–77.

CHAPTER 11

Other Techniques Related to Habitat Analysis

In this chapter, I briefly discuss several statistical modeling and analysis techniques that are broadly related to the general framework of a habitat analysis (Chapter 5). The common feature is that all these techniques involve some type of statistical examination of a species–environment relationship. However, these techniques differ from a habitat analysis in that the main purpose is not necessarily to determine how well a species response variable correlates with a set of environmental predictor variables. Rather, these techniques go a bit further and try to connect pattern to process, even if just implicitly. Recall that the basic habitat analysis discussed so far in this book does not try to infer or identify the process(es) that leads to particular habitat associations of a species. Neither does it attempt to predict a species distribution in nature or occurrence at certain locations. The five techniques discussed below, (1) resource selection functions, (2) selectivity and preference indices, (3) compositional data analysis, (4) habitat suitability modeling, and (5) species distribution modeling, either refine the idea of a habitat analysis to clarify how individuals select and use resources or they extend the idea to use it in predictive and probabilistic modeling of a species geographic distribution. As such, at their cores, each of these five techniques implicitly relies on pre-existing knowledge or further derives new knowledge about the habitat associations of a species. Each of these techniques is the toolbox for a much larger topic area and there is substantial supporting literature for all of them. It is beyond the scope of this book to cover all of that literature. Instead, my intention is to provide a brief introduction so that the reader can see the conceptual and operational linkages between these techniques and a basic analysis of habitat.

11.1 Resource selection functions

A resource selection function (RSF) is a general term for any statistical model or analysis that relates a species selection (and subsequent use) of a resource to properties of the resource. Often, the resource is conceived of as being a unit of land in which there are measured properties some of which could be environmental variables defining the habitat of the species. Further, occurrence of individuals of the focal species within the unit constitutes use of the unit. When the concept of an RSF was initially being developed, it was suggested that many different types of statistical models could be used to estimate an RSF (Manly et al. 1993; Alldredge et al. 1998; Boyce and McDonald 1999), although at the time this was not demonstrated on empirical data and thus no practical method was suggested to fit the data to those models (Lele 2009). The original statistical description of an RSF and demonstration on empirical data relied solely upon logistic regression or the logit function (Manly et al. 1993; Boyce and McDonald 1999)—see the basic description of multiple logistic regression in Section 9.3. Since then, statistical estimation of RSFs has extended beyond logistic regression (Lele and Keim 2006; Lele 2009; Lele et al. 2013). Monte Carlo (randomization) algorithms for maximum likelihood estimation of an RSF have been suggested but they do not always converge to a single definitive set of values (coefficients) for the predictor variables (Lele 2009). To date, logistic regression has remained as

Habitat Ecology and Analysis. Joseph A. Veech, Oxford University Press (2021). © Joseph A. Veech. DOI: 10.1093/oso/9780198829287.003.0011

the favored and most often used method for esti-mating an RSF. But therein lies a problem.

In their initial descriptions of the concept and statistical foundation of an RSF, Manly et al. (1993, 2002), Boyce and McDonald (1999), and Boyce et al. (2002) emphasized that an RSF estimated as a logis-tic regression does not provide the actual probabil-ity that a *given* resource unit is selected, but rather an RSF gives a value that is *proportional* to the prob-ability of selection. Although the logistic regression equation always produces a number bounded between 0 and 1 (recall that the equation is $\hat{y} = e^{\beta_0 + \beta_1 x_1 + \ldots \beta_N x_N}/(1 + e^{\beta_0 + \beta_1 x_1 + \ldots \beta_N x_N})$), in a strict statistical sense this number cannot be interpreted as a probability. The species response variable (presence–absence) can be measured under three different sampling designs: (1) used vs. unused resource units, (2) used vs. available resource units, and (3) unused vs. available resource units (Manly et al. 2002). "Used" entails that the resource unit was surveyed and the species revealed; "unused" entails that the resource unit was surveyed but the species was not found and hence recorded as absent. "Available" represents resource units that were not surveyed and hence the species was assumed to be absent or pseudo-absent (Section 7.4) for the purposes of coding the data as "0" for a logistic regression. These categories are represented by an underlying sampling distribution such that P_a, P_u, and P_{nu} represent the probabilities that a randomly selected unit is either within the available, used, or unused categories.

In order to legitimately use logistic regression to estimate the *probability* that resource unit i was selected by the species given the properties of i, the researcher needs to know P_a, P_u, and P_{nu} and use them to modify the logistic regression equation (see pp. 83–111 in Manly et al. 2002). Unfortunately, the researcher rarely knows the values for P_a, P_u, and P_{nu}. Note that these probabilities are not derived from the actual sample sizes (number of resource units in) of each of the categories. For example, assume we were using Design 2 in a 1 km² study area arbitrarily divided into 1,600 25 × 25 m plots (resource units). We know the locations of 100 indi-viduals of a hypothetical species (perhaps a small and relatively sedentary rodent, lizard, or insect) and use those locations to compose a sample of

used resource units. We also randomly select 100 plots to represent a sample of available resource units. We do not survey for the species in these plots, but we do collect (or already have data for) a set of environmental variables for these plots and the used plots. P_a and P_u do not necessarily each equal 0.5, and they do not each equal 0.0625 (=100/1,600). There may be other used resource units that we are not aware of and that are not included in the study. Also, some of the available units might actually have an individual of the spe-cies that we are not aware of. This latter source of error is known as the contamination rate (Lancaster and Imbens 1996) and its value is rarely known for resource selection studies (Keating and Cherry 2004; Johnson et al. 2006). The same applies for unused resource units if we were using one of the other designs. Johnson et al. (2006) suggested that for most previous studies employing a use-availability design, contamination rate was probably low enough to not adversely affect the parameters of the RSF as estimated by logistic regression.

There is another even more profound reason not to interpret the output of a logistic regression equation as the probability of species presence, resource selec-tion or use. Logistic regression is based on a theoret-ical binomial sampling process in which there are N trials (e.g., resource units) each of which is known with certainty to represent either a success (e.g., spe-cies recorded as present) or failure (e.g., species recorded as absent). This is never the case with a spe-cies response variable; false negatives (recording species as absent) occur due to low detection prob-ability and even false positives can sometimes occur due to misidentification. Thus, interpreting the out-put of a logistic regression equation as an *absolute* probability is a bit inexact and misleading, although the output can be interpreted in a relative sense and used to compare resource units.

Despite all these issues, RSFs are useful and the technique of estimating them, particularly via logis-tic regression, is widely practiced. Because an RSF does not produce a probability estimate nor does it always give numbers bounded by 0 and 1, Manly et al. (1993, 2002) were led to introduce the idea of a resource selection probability function (RSPF). An RSPF was defined as "a function that gives the probability of a resource unit being used as a function of the values that the unit possesses for certain variables $X_1, X_2, \ldots X_p$" (Manly et al. 2002, p. 44). Clearly, RSPFs were intended to only take on values bounded by 0 and 1. Unfortunately, Manly et al. (2002, pp. 100 and 105) used the equation, $w(x) = \exp\{\beta_0 - \ln[(1 - P_a)P_u/P_a] + \beta_1 X_1 + \ldots \beta_N X_N\}$ to represent the RSPF and yet this equation can sometimes take on values much greater than 1 (Johnson et al. 2006). Therefore, it does not give a probability even if P_a and P_u are known with certainty. Keating and Cherry (2004) thoroughly criticized and exposed the flaws in using the logistic and exponential functions as an RSPF and RSF. However, they also pointed out that logistic regression can be used to compute odds ratios and thus used to compare resource units even without determining absolute probabilities of use. For example, assume for resource unit 1 that $w_1(x = 1) = 1.8$ and for unit 2 $w_2(x = 1) = 0.9$, then computing the odds ratio (1.8/0.9) we find that unit 1 is twice as likely to be used as is unit 2. Since Manly et al. (2002) and the critique of Keating and Cherry (2004), other authors have derived legitimate functions and a more meaningful definition for the RSPF (Johnson et al. 2006; Lele and Keim 2006; Lele 2009; Lele et al. 2013).

The original intent of Manly et al. (1993, 2002) was to develop the idea of an RSF (and RSPF) as a unifying statistical theory and technique for examining resource selection (broadly defined). Previously, Manly (1974, 1985) had proposed selectivity indices for quantifying preferential use of food items and then McDonald et al. (1990) built upon those and may have been the first to suggest these indices (or functions) for use in examining habitat selection. Even further back, the general concept of and statistical methods for analyzing habitat selection gradually emerged in the ecological literature beginning in the 1970s. Neu et al. (1974) and Johnson (1980) described the use-availability design in the context of habitat as the selected resource, but neither used logistic regression; also see Martin (1985) and Alldredge and Ratti (1986, 1992). Ben-David et al. (1996) seem to have independently developed the idea of using logistic regression in a use-availability design in that they did not cite Manly et al. (1993). They also did not use the phrase "resource selection function."

On the one hand, RSFs are nothing special, they are just a dressed up version of using multiple logistic regression as a method of habitat analysis. On the other hand, RSFs and their explicit ties to sampling designs do force the researcher to think carefully about hidden assumptions and the limitations on what can be inferred from the results of their study. RSFs are also sometimes valuable in predicting species presence (or absence) and mapping potential spatial distributions of species (Boyce and McDonald 1999; Meyer and Thuiller 2006). The application of RSFs to address various ecological questions and conservation tasks continues to expand. The role of daily or seasonal movement as it allows an individual animal to encounter (and potentially select and use) different resource units has been examined to refine the concept and measurement of availability (Fieberg et al. 2010; Avgar et al. 2016). RSFs have recently been incorporated into the study of landscape-level gene flow and spatial genetic differentiation; the ingenious idea here is that the inverse of a habitat (land cover) based RSF serves as a measure of resistance to dispersal (for a review and empirical example, see Roffler et al. 2016). The scale-dependency of RSFs has been examined (Boyce 2006; Meyer and Thuiller 2006) and scale-dependent RSFs have been used to better understand the effects of population density on broad-scale habitat selection and fine-scale resource use (van Beest et al. 2014). Some authors have suggested that viewing an RSF as a dynamic property of an individual and a species can lend insight into the study of various organism–environment processes, and in so doing these same authors plea for a greater role of ecological theory in applying and interpreting RSFs (McLoughlin et al. 2010). The concept and quantification of an RSF can have many extended uses beyond just examining species–habitat relationships.

Closely related to RSFs, utilization distributions (UD) and corresponding resource utilization functions (RUF) are based upon species locational data that is more time intensive. Rather than simply modeling species' use and non-use of resource units as a binary response variable (as RSFs do), RUFs model the relationship between the frequency of use of resource units (the UD) and the environmental properties of the units (Marzluff et al. 2004;

Millspaugh et al. 2006; Long et al. 2009). As such, RUFs present a more precise and detailed account of resource selection and habitat use, both as measured properties of an individual or species, and as an ecological concept (Box 11.1). Nonetheless, there is no need to force the RSF-UD-RUF framework onto a study that is simply using multiple logistic regression (as a method of habitat analysis; Section 9.3) to uncover meaningful habitat associations of a species.

11.2 Selectivity and preference indices

This is another vast area of statistical-ecological methodology and research. Selection and preference is studied for a wide array of ecological topics, including prey and diet selection, mate choice, as well as habitat selection and use (Chesson 1978; Johnson 1980; Alldredge and Ratti 1986; Kirkpatrick and Ryan 1991; Wagner 1998; Morris 2003; Beyer et al. 2010). In some studies, the distinction between

---

Box 11.1 Selection vs. use

Although there is a real distinction between selection and use of a resource (or habitat), both terms are often used synonymously or in a non-exact way by authors. For the resource selection literature (i.e., studies citing Manly et al. 1993, 2002), Lele et al. (2013) attempted to clarify terminology and concepts. In particular, they sought to distinguish the probability of resource *selection* from the probability of resource *use*. Further, they also separately defined probabilities of *occupancy*, *occurrence*, and *choice*. Perhaps some of this clarification of terminology has been useful. On the other hand, some of it seems arbitrary. For example, Lele et al. (2013, p. 1185) stated "We denote selection of a resource unit by an animal as the act of using a resource unit if it is encountered." This sentence is just as logical if it is stated as "We denote *use* of a resource unit…as the act of *selecting*." In a revisionist way, Lele et al. (2013) also redefined Manly et al.'s (1993, 2002) probability of selection (and use) as a probability of occupancy.

Many studies of resource selection and use involve recording (detecting) individual animals at spatially explicit locations or within pre-defined spatial sampling units, both referred to as "resource units." When an individual is recorded in a resource unit then it obviously has encountered the unit and its presence is taken as the individual occurring, occupying, selecting, choosing, and using the unit. Therefore, the presence data alone do not allow us to distinguish among the five different probabilities. What matters in distinguishing the different probabilities is what the "absence" data actually represent, such as availability, presumed absence without surveying, or absence upon surveying. The particular study design and function used to derive the probability also determines which of the five probabilities is being estimated (Lele et al. 2013). All of this is important, but it should not overly restrict how we use terms

and envision the conceptual context of resource (including habitat) selection and use.

Consider the behavioral sequence of events for a given animal. Selection is temporally first, followed by use. By logic, it does not make sense to think of an animal as using a resource or habitat without first having selected it (i.e., selection as the event of entering into the resource unit), and before that it must have *encountered* the resource unit. Also, once selected, it does not make sense to think of an animal as not subsequently using the resource or habitat (remaining in it for some period of time) even if the use is brief or only temporary, or else why would the animal have selected the resource? Thus, selection and use may be distinct events, but they are inseparable with regard to an animal's behavioral sequence of action. However, selection and use may be separable given an appropriate study design, and in such cases it may be appropriate to separately estimate probabilities of selection and use. Lele et al. (2013) note the possibility that a resource unit with a highly desirable *type* of resource might have a high probability of selection but low probability of actual use if it is not easily encountered. Their definitions of selection and use are based on the conceptual distinction between a resource type and an actual resource unit. Selection pertains to resource type and use pertains to a physical resource unit. This distinction may not always be necessary in studies of habitat selection and use, particularly when authors are clear that they intend "selection and use" as a combined event. Even in those studies, a particular resource unit may not be selected and used if it is difficult to find or access (encounter probability is low). As discussed in Chapter 3, "habitat selection" has a broad and varied usage in the ecological and evolutionary literature—this simply means that the term is polysemous not ambiguous (Hodges 2008, see also Chapter 1 of this book).

habitat preference and selection is not always clearly stated. Again, this is likely due to the close conceptual and empirical coupling of preference and selection. If an animal has a preference then it will select non-randomly from a set of habitat types. If an animal is seen to select non-randomly then it presumably has innate preferences. Selection is the actual act of choosing and using; preference is a trait or characteristic of the individual and species describing what type of habitat it would disproportionately select.

Many different metrics have been developed to measure selection and preference (Jacobs 1974; Cock 1978; Strauss 1979; Lechowicz 1982; Elston et al. 1996; De Caceres and Legendre 2009). Nearly all of the metrics yield an estimate (of selection or preference) for a resource unit, habitat type, or prey item that is *relative* to other units or to the general availability of the resource in the immediate environment (see table 1.1 in Manly et al. 2002). Also, the metrics generally require that each resource unit (or prey item) can be assigned to a unique category. Ideally, selection and preference are measured and studied in a controlled experimental field setting wherein the study subjects (focal animals) clearly have immediate access to the different types of resources being examined, and the researcher can subsequently record their choices. An example is the eloquent and classic experiment by Wecker (1964) on habitat selection and preference of deer mice. Because selectivity and preference indices are relative, I do not consider them a method of habitat analysis as I have defined it throughout this book, particularly Chapter 7. They are not intended to quantify the relationship between a species response variable and a set of environmental factors. Nonetheless, selectivity and preference indices can be useful in studies directly examining a species' habitat preferences. In addition, a habitat analysis conducted prior to a study of habitat preference and selection might help inform on the experimental design of the latter.

11.3 Compositional data analysis

Compositional data analysis (CoDA) is essentially a form of selectivity analysis in that it is based on examining relative, not absolute, measures of

resource use. However, CoDA is more closely tied to habitat analysis than are selectivity and preference indices. CoDA has its beginnings in geology inasmuch as the original description of it (Aitchison 1982) used geological examples and was partly directed at an audience of geologists. Compositional data are those that fully determine the entire composition of a sample. For a given sample or observation, such data sum to 1 (as proportions) or 100 (as percentages) or more generally sum to a finite upper limit. In geology, the mineral (elemental) composition of a rock or sediment is an example of compositional data. In ecology, the proportions of an individual or a species diet composed of different prey or forage items is compositional. The proportions of different habitat types in a species' home range is compositional (e.g., Nikula et al. 2004; Gehring et al. 2019; Shonfield et al. 2019). CoDA examines the proportions represented by the constituent parts *relative* to one another. It is essentially an examination of ratios and whether a set of ratios (or a single ratio) is significantly large, small, or non-significant.

Aitchison (1982) recognized that previous statistical tests for compositional data were founded upon testing whether the compositional data (as a whole) departed from a multivariate normal distribution. Also, as others before him had, Aitchison (1982) also recognized that a fundamental challenge with compositional data is that the constituent parts are not independent of one another (again because their sum is limited to and must meet a finite constant value). As part of his exposition on CoDA, Aitchison also noted that (1) a series of random "perturbations" can lead to a composition having an additive logistic-normal distribution, and (2) structuring factors or processes can lead to some parts given "priority" over other parts such that the former compose a greater proportion of the composition. These two "principles" of CoDA suggested to Aitchison that parametric statistical tests (e.g., MANOVA, regression analyses, chi-square tests) could be employed and that comparison of parts to one another (as in ratios) was more meaningful than simultaneous comparison of all parts to a theoretical multivariate distribution (Aitchison 1986; Aitchison and Egozcue 2005). Indeed, compositions only provide information about the relative value of the components not absolute values, and thus ratios are the entities to statistically test and study—this is the property of *scale invariance* (Aitchison and Egozcue 2005). CoDA also has the very useful property of *sub-compositional coherence*. This means that a ratio (of proportions of two components) derived from a partial composition of the full composition is identical to the ratio based on the full composition (Aitchison and Egozcue 2005). For example, assume we have five components (A–E) with the following values: A 18, B 20, C 25, D 30, and E 40, which as a composition correspond to the following proportions respectively: 0.135, 0.150, 0.188, 0.226, and 0.301. The ratio of E to A is $0.301/0.135 = 2.22$. Now consider that we only have the subcomposition consisting of A, B, and E, in which case the proportions are 0.231, 0.256, and 0.513. The ratio of E to A remains unchanged, $0.513/0.231 = 2.22$. Again, a key feature of CoDA is that ratios of proportions are analyzed, not the proportions themselves.

Aebischer et al. (1993) may have been the first study to apply CoDA to ecological data, and it was a study of habitat use. They also presented a statistical description that was much less technical than that of Aitchison (1982) and thus more accessible for ecologists. According to Aebischer et al. (1993), Aitchison (1986) demonstrated that the components of a composition (represented as proportions x_i to x_j) could be made independent of one another through the log-ratio transformation, $y_i = \ln(x_i/x_j)$. However this is not correct, as was demonstrated by Jackson (1997) and also emphasized by Elston et al. (1996) and Fattorini et al. (2014). The ratios are multivariate normal (Aitchison 1982) but this does not ensure that they will also be independent of one another, where "independence" is defined and measured as a relatively low correlation coefficient (e.g., $-0.2 < r < 0.2$). Other authors, Pendleton et al. (1998), Manly et al. (2002), and Nikula et al. (2004) also mistakenly claimed that the log ratios $[\ln(x_i/x_j)]$ of a composition are independent of one another. Obviously, the log ratios do not have the unit sum constraint; however, if they are not statistically independent of one another then significance testing with parametric techniques may not be appropriate. That is, the *P*-values generated from such tests may not be trustworthy.

CoDA also has other procedural issues. The data for a composition may sometimes include values of zero. With regard to species–habitat data, this would occur when some individuals do not have a particular habitat type in their home range or conversely some resource units of the given habitat type were never used by individuals of the species. Also, CoDA is often employed in use-availability designs and a particular habitat type may be widely available but rarely or never used, resulting in a lot of zeros in the use data (Aebischer et al. 1993). The problem here is that log values of zero are mathematically undefined. As a remedy, Aitchison (1982) suggested replacing the zeros with very small values (e.g., 0.0005). Although the choice of replacement value affects the test statistic (e.g., t, F, or χ^2), it might not always affect statistical significance or the ranking of preferred habitat types (Aitchison 1982; Aebischer et al. 1993). Thus, CoDA may be relatively robust to zero-value replacements particularly if there are not very many zeros to replace. Other authors have concluded that the need to replace zero values is a serious weakness of CoDA (Elston et al. 1996; Jackson 1997; Thomas and Taylor 2006; Fattorini et al. 2014; Tsilimigras and Fodor 2016). Further, there is evidence that CoDA as applied to some datasets may have excessive Type I error rates; that is finding a significant effect when one does not actually exist (Pendleton et al. 1998; Dasgupta and Alldredge 2002; Bingham and Brennan 2004).

As with other methods of assessing habitat preference and selection, CoDA requires that the behavior of individual animals is not influenced by others and hence each animal represents an independent observation. Further, if the compositional data derive from tracking (GPS or radio-telemetry) individual animals, then the relocation points for an animal should be independent of one another although temporal autocorrelation in such data is difficult to completely control when using CoDA or any other analysis method. For further discussion of statistical design issues relevant to CoDA and other analytical techniques, see Jackson (1997), Alldredge et al. (1998), Otis and White (1999), and Thomas and Taylor (2006). CoDA continues to be widely used in assessing habitat preference and selection as well as the study of other ecological processes (see assessment in Thomas and Taylor 2006). The ISI Web of Science database has 800 citations of Aebischer et al. (1993) between 2007 and 2019, and 738 are related to "habitat" in some way. In recent years there have been improvements to CoDA and these include better handling of zero values (de Valpine and Harmon-Threatt 2013; Fattorini et al. 2014; Pierotti et al. 2017). These newer techniques depart somewhat from Aitchison's original framework, but they could prove very useful in analyzing habitat selection and preference when the habitat data are compositional.

11.4 Habitat suitability models

Habitat suitability modeling has gained substantial prominence in ecology and conservation in the past two decades (Schadt et al. 2002; Brotons et al. 2004; Rondinini et al. 2005; Hirzel and Le Lay 2008; Bradter et al. 2018). "Habitat suitability model" (HSM) is a generic term for a statistical model that uses spatially referenced environmental data to quantify the spatial distribution of a species' habitat, typically within a landscape (10s – 1,000s km²) or region (>10,000 km²). Often, it involves mapping a species probability of occurrence at a specified resolution (e.g., 200 × 200 m pixels) for the entire landscape or region. In a more technical and proper sense, the quantity being mapped is some function (e.g., number between 0 and 1) that indicates the extent to which environmental conditions within the pixel (or other spatially explicit part of the map) represent suitable habitat for the species *relative to other pixels*. Suitable habitat is defined and quantified by a statistical model depicting the species–habitat (or –environment) relationship, such as a RSF. HSMs do not literally predict the probability that the species is present in the area represented by the pixel.

HSMs are most appropriately developed *after* the species–habitat relationship has been established. That is, the researcher has previously conducted habitat analyses (as described in Chapters 7 and 9) to identify important environmental variables that define the habitat of the species—the habitat has been characterized. Then landscape-wide or region-wide data on those variables are obtained, often in the form of spatially explicit data that can be ana-

lyzed and mapped in a GIS computing environment. The main product or output of the HSM is then a map depicting a probability surface, with probability interpreted in a relative sense as previously described. One could then survey for the species at locations that have high suitability and locations that have low suitability so as to obtain data to validate (test) the model and estimate accuracy (Section 10.3). However, construction and use of HSMs typically does not play out in this way. Rather the step of conducting the habitat analysis is integrated with the actual mapping of suitable habitat. That is, the HSM is used to discover the important habitat variables as well as map them, and then also test the accuracy of those variables in identifying suitable and occupied habitat.

Troy et al. (2014) developed an HSM for Newell's shearwater (*Puffinus newelli*) nesting on Kauai, an island in the Hawaiian archipelago (Fig. 11.1). The species data consisted of 35 locations known to regularly have the seabird species between 2007 and 2010, and a set of randomly selected locations considered to represent pseudo-absence. Thus, the study had a use-availability (or presence-availability) design with a binary response variable coded as 1 for species presence and 0 for pseudo-absence. Troy et al. (2014) consulted previous studies on burrowing seabirds to compile a list of 11 environmental variables that could characterize the breeding habitat of Newell's shearwaters. These variables mostly related to factors that could facilitate the bird's burrowing (nesting) behavior, take-off behavior (the birds require a slope and headwind to get airborne), and cover from predators. Troy et al. (2014) then used a series of logistic regression models to identify a subset of the environmental variables that seemed most important in influencing species presence (Fig. 11.1). The best models as indicated by AIC_c values were combined into a single final weighted-average model that became the function (i.e., species–habitat relationship) used to map habitat suitability for the entire island. This step used remote-sensing data and produced a map at 50 m pixel resolution (Fig. 11.1). As with any logistic regression equation, the final model provided a "probability" or number between 0 and 1 for each pixel indicating the relative suitability of that pixel as nesting habitat given the values of the

Figure 11.1 Photo of Newell's shearwater (*Puffinus newelli*). Middle photo shows habitat for Newell's shearwaters in the Upper Limahuli Preserve (National Tropical Botanical Garden). Steeply sloped terrain with native vegetation characterizes the habitat for this species. In particular, the burrows of Newell's shearwaters typically exist within thickets of native fern as seen in the right-side foreground of the photo. Map depicts habitat suitability at 50 m resolution for the entire island of Kauai, Hawaii. Darker shading indicates greater suitability. Lightest shading represents suitability of 0–0.1; darkest shading represents 0.9–1. For scale, Kauai is about 48 km from west to east through the middle of the island. Photo credits are Brenda Zaun (USFWS) for the bird and Jeff Troy for the habitat. Map is adapted from Troy et al. (2014).

environmental variables for that pixel. The overall fit of the final model to the presence-available data was relatively good as indicated by AUC values in the range of 0.84–0.88 (see Section 10.1 for discussion of model fit).

Importantly, the model (map) was also tested or validated with an *independent* set of 29 Newell's shearwater nesting locations (colonies) recorded from surveying during the 2011 and 2012 nesting seasons. On the map, these 29 colonies were depicted as polygons typically consisting of a few to as many as 141 50 × 50 m pixels. For each polygon, habitat suitability was calculated as a mean over all the pixels of the polygon. Although the model was not perfect it did perform relatively well. Nineteen of the 29 polygons had a mean suitability value >0.5 and 25 polygons had at least some pixels with suitability values >0.7 (Troy et al. 2014). Areas occupied by colonies were not internally homogeneous with regard to habitat suitability (environmental factors), but nearly all colonies contained some smaller portions (2,500 m² areas) that had relatively high suitability. The study also incorporated information on anthropogenic threats to the birds (e.g., feral cats, human habitation, agricultural fields, artificial light) and land ownership so as to produce habitat suitability maps that could provide guidance to on-going and future conservation efforts for the species.

It is important to recognize that HSMs and maps depict *potential* habitat, not occupied or even probable habitat. An HSM, or any other type of model, does not give an absolute probability of species presence. In fact, such a probability is essentially undefined; it does not exist in a logical sense. A species either is or is not present at a given location at a given time, and this cannot and need not be quantified by a probability. For example, assume the area represented by a given pixel on a map has a habitat suitability value of 0.8. This does not mean that we could go there and have an 80 percent chance of finding the species and it does not mean that there is an 80 percent chance that the species is actually present there (either it is present or it is not, there's no reason to define a probability for that). It also does not mean that there is an 80 percent chance that a dispersing individual would settle in at that location upon encountering the location. It also certainly does not mean that the location has an 80 percent chance of having suitable habitat (when "suitable habitat" is defined strictly as completely suitable or completely unsuitable rather than as a relative measure). A habitat suitability value of 0.8, as derived from a logistic regression model or perhaps any other species–habitat function, is simply an index that indicates the relative usefulness or suitability of the location to the species. An area that has a suitability value of 0.8 has better habitat conditions for the species than does an area that has a suitability value <0.8. This is why HSMs are useful; their depiction of potential habitat is relative.

With regard to terminology, some previous authors have criticized use of the phrase "suitable habitat" and have even gone so far as to claim that the phrase is redundant and illogical (Hall et al. 1997; Mathewson and Morrison 2015; Krausman and Morrison 2016; Kirk et al. 2018). Their perspective is best illustrated by this quote from Hall et al. 1997, p. 178: "The term, 'suitable' habitat should not be used because if an organism occupies an area that supports at least some of its needs, then it is habitat. So, by definition then, habitat is suitable. Thus there is no such thing as unsuitable habitat, because it is the quality that changes, not the suitability per se." The problem with this perspective is that it ignores the fact that "habitat suitability" is applied as a descriptive term to an entire type of habitat modeling, and has been for several decades. Further there is nothing inherently illogical or redundant in the phrase "suitable habitat," particularly when used in the context of habitat suitability modeling. "Suitability" when expressed as a number between 0 and 1 simply indicates how appropriate the habitat is for the species. The base definition of "suitable" is *appropriate for a particular purpose or situation*. There are different degrees to which a habitat could be appropriate. Despite the argument of Mathewson and Morrison (2015) and Krausman and Morrison (2016), "unsuitable habitat" is not a misnomer. They suggest that if a habitat is unsuitable then it is not habitat for the particular species of interest. This is correct but it ignores the fact that the particular location or set of environmental variables could function as habitat for some other species, thus it is perfectly appropriate to refer to "unsuitable habitat." Again, as with other habitat-related terms, "habitat suitability" is an effective term for communication and is easily understood in context (Hodges 2008). See Section 1.2 for more discussion of this issue.

Habitat suitability modeling is also sometimes referred to as "environmental niche modeling" in that the fundamental niche or habitat requirements (broadly defined) are used to model the potential spatial distribution of the species. However, Kearney (2006) recognized a distinction between habitat suitability modeling and environmental niche modeling. The former is essentially what I have described, whereas the latter according to Kearney (2006) uses basic knowledge of a species, such as its environmental tolerances and requirements, to model and map the potential distribution of the species. Notably, environmental niche modeling defined in this way does not utilize or build upon actual distributional data of the species (Kearney 2006). As an example, Kearney (2006) described an environmental niche model for Bynoe's gecko (*Heteronotia binoei*) based on soil temperature and the rate of physiological development of eggs. He contrasted this with an HSM for the species and showed how the two models gave somewhat different predicted spatial distributions for the gecko in Australia.

Lastly, the term "habitat suitability" has been in the wildlife ecology literature for some time, predating the emergence of HSMs in the general ecology literature. In the former, habitat suitability indices and habitat suitability index models date back to the 1980s (Cole and Smith 1983; Cook and Irwin 1985; Irwin and Cook 1985; Verner et al. 1986 and papers cited within). Even apart from research in wildlife ecology, the term "habitat suitability index" was used as early as 1975 in a study of the coral species *Muricea californica* (Grigg 1975).

11.5 Species distribution models

Species distribution modeling is yet another area of ecological research and methodology that has greatly expanded over the past 10–15 years (Elith et al. 2006; Elith and Leathwick 2009; Franklin, 2010, 2013; Stohlgren et al. 2011; Guisan et al. 2017). A species distribution model (SDM) is similar to an HSM except the goal typically is to model potential species distribution over a much larger spatial extent, from regions to continents. Also, SDMs tend to include climate data more often than do HSMs, and an SDM rarely is restricted to using only habitat data. As with HSMs, development of an SDM involves several steps of (1) selecting the environmental (predictor) variables and corresponding data (often in GIS format), (2) deriving a species–environment function, (3) modeling or mapping the output of the function at a specified resolution over the region of interest, (4) testing the fit of the model output to the environmental data, and (5) assessing the accuracy of the model output often using an independent source of species data or by splitting the original dataset into a model-building portion and a model validation portion (Section 10.1). Similar to HSMs, a logistic regression equation is often used as the species–environment function and hence the output (for each pixel of a map) is a number between 0 and 1. Again, this number is not exactly a true probability, rather it can be interpreted as I explained above for HSMs. Often, Steps 1 and 2 represent an actual analysis of habitat (Chapters 7 and 9) that can be considered a separate stand-alone study whose results are then used in the later steps that involve spatial modeling and map-making. Other times, Steps 1 and 2 are more formally integrated with the later steps. It is also important to recognize that Steps 2–4 are often an iterative process in which different sets of environmental variables are examined so as to find the set that best fits the data. As such and given that SDMs typically involve a substantial amount of species occurrence and environmental data, the modeling process can be computationally intensive and time-consuming.

In many studies involving development of either SDMs or HSMs, the species occurrence data are already on hand prior to building the models. The data might come from previous surveys conducted by the same researcher(s), museum records, or citizen science databases (Section 7.6). The environmental data also might be pre-existing in the form of GIS-formatted remote-sensing data or climate data. Thus, constructing an SDM, and to a lesser extent an HSM, might not require collecting any original data. However, sometimes either or both types of data are field-collected as an integral part of the model-building and map-making phases. In such cases, Steps 1–4 above can be carried out through an iterative process that essentially refines and improves the species–environment function and hence the SDM (or HSM) as more data are obtained. Bliss et al. (2019) and Ott et al. (2019) implemented such an approach in modeling the regional-level spatial distribution of two kangaroo rat species.

There is not a huge difference between SDMs and HSMs, except for the greater spatial scale and greater amounts of data used in SDMs; these are not strict defining differences. Perhaps, SDMs are a bit more likely to rely solely on previously collected data simply because they cover a greater spatial area. Indeed, predicting habitat suitability is sometimes the main goal of an SDM (e.g., Gogul-Prokurat 2011; Li et al 2013; Gama et al. 2016; Ahsani et al. 2018; Nneji et al. 2020). This indicates that the two modeling approaches are really not different from one another, and that word usage is sometimes interchangeable. The key feature is that both HSMs and SDMs involve spatial prediction and mapping; some kind of probabilistic response surface is the output of both.

Lastly, species distribution modeling is often based on a technique called maximum entropy (Maxent). Maxent modeling involves estimating an unknown statistical probability distribution (Phillips et al. 2006). In the context of an SDM, the unknown distribution is the ratio of $f_1(\mathbf{z})/f(\mathbf{z})$, where $\mathbf{z}$ is a vector of environmental variables for locations across the landscape or region of interest, $f_1(\mathbf{z})$ is the set of probabilities of having certain values of the environmental variables at locations *where the species is known to be present*, and $f(\mathbf{z})$ is the set of probabilities for *all locations* (Elith et al. 2011). The locations on the landscape or region can be thought of as pixels given that species distribution modeling typically involves the use of environmental data that is pixelated (i.e., GIS-formatted). (Importantly, Maxent is based upon a presence-availability sampling design, or what Elith et al. (2011) call "presence-only" data in that actual absence data are not being used. Also, Elith et al. (2011) refer to the availability locations as "background data.") The first step in Maxent is to estimate $f_1(\mathbf{z})$ from the presence locations but in a way

that *minimizes* the differences or distances between $f_1(\mathbf{z})$ and $f(\mathbf{z})$ for each of the environmental variables. That is, Maxent tries to get the ratio, $f_1(\mathbf{z})/f(\mathbf{z})$, close to 1. Combined over all variables, the difference between $f_1(\mathbf{z})$ and $f(\mathbf{z})$ is the entropy of $f_1(\mathbf{z})$ *relative* to $f(\mathbf{z})$ (Elith et al. 2011). Recall from physics that entropy is a measure of disorder or lack of information and structure in a system, or a set of numbers. Thus, if the environmental variables truly do have some influence on species presence, there should be less inherent entropy in $f_1(\mathbf{z})$ than in $f(\mathbf{z})$. At the very least, there is structure in $f_1(\mathbf{z})$ represented by the mean values of the environmental variables and the covariances among them. In a similar way, there is inherent structure in $f(\mathbf{z})$, but not as much. Also, recognize that the presence locations (pixels) are a non-random subset of all the locations, hence $f_1(\mathbf{z})$ has less entropy than $f(\mathbf{z})$. Therefore, when the algorithms of Maxent attempt to match up $f_1(\mathbf{z})$ with $f(\mathbf{z})$ (i.e., minimize the differences), they must do so by maximizing the entropy of $f_1(\mathbf{z})$. This is why the technique is called "maximum entropy."

There are many additional steps in Maxent modeling, some of which involve machine-learning algorithms and are based on Bayesian logic. Elith et al. (2011) and Merow et al. (2013) are excellent detailed explanations of Maxent as used for species distribution modeling. Maxent is not perfect. When applying Maxent (and other types of SDMs) to actual data, the researcher sometimes has to make somewhat arbitrary decisions with regard to model parameters. In using Maxent, the researcher has to decide on up to 12 settings prior to model building, more so than with other SDM methods, and model output is sensitive to the settings (Hallgren et al. 2019). The output of Maxent, as used in SDM, are numbers scaled between 0 and 1, although again these should not be strictly interpreted as the probability of occurrence of the species (for reasons discussed earlier). Although widely applied in the past decade, Maxent does have its critics (Peterson et al. 2007; Rota et al. 2011; Pineda and Lobo 2012; Fitzpatrick et al. 2013; Yackulic et al. 2013; Guillera-Arroita et al. 2015) and there are certainly other statistical techniques available for modeling and mapping species distributions (Skidmore et al. 1996; Segurado and

Araújo 2004; Elith et al. 2006; Austin 2007; Meynard and Quinn 2007; Merow et al. 2014). Maxent and SDMs in general have been used to forecast potential changes in species distribution under different climate change scenarios (Araújo and New 2007; Aitken et al. 2008; Kearney et al. 2010; Austin and Van Niel 2011; Bond et al. 2011; Luo et al. 2015; Willis et al. 2015). As with HSMs, SDMs go beyond a simple analysis of species–habitat associations. Indeed, both types of modeling are informed (made better) by careful and thorough analyses of habitat associations (using any of the methods described in Chapter 9) to a priori identify meaningful habitat characteristics or environmental variables to include in the HSM or SDM.

References

Aebischer, N.J., Robertson, P.A., and Kenward, R.E. (1993). Compositional analysis of habitat use from animal radio-tracking data. *Ecology*, 74, 1313–25.

Ahsani, N., Kaboli, M., Rastegar-Pouyani, E., Karami, M., and Kamangar, B.B. (2018). Habitat suitability prediction for *Salamandra infraimmaculata* (Caudata: Amphibia) in western Iran based on species distribution modeling. *Journal of Asia-Pacific Biodiversity*, 11, 203–205.

Aitchison, J. (1982). The statistical analysis of compositional data. *Journal of the Royal Statistical Society Series B*, 44, 139–77.

Aitchison, J. (1986). *The Statistical Analysis of Compositional Data*. Chapman and Hall, London.

Aitchison, J. and Egozcue, J.J. (2005). Compositional data analysis: where are we and where should we be heading? *Mathematical Geology*, 37, 829–50.

Aitken, S.N., Yeaman, S., Holliday, J.A., Wang, T., and Curtis-McLane, S. (2008). Adaptation, migration or extirpation: climate change outcomes for tree populations. *Evolutionary Applications*, 1, 95–111.

Alldredge, J.R. and Ratti, J.T. (1986). Comparison of some statistical techniques for analysis of resource selection. *Journal of Wildlife Management*, 50, 157–65.

Alldredge, J.R. and Ratti, J.T. (1992). Further comparison of some statistical techniques for analysis of resource selection. *Journal of Wildlife Management*, 56, 1–9.

Alldredge, J.R., Thomas, D.L., and McDonald, L.L. (1998). Survey and comparison of methods for study of resource selection. *Journal of Agricultural, Biological, and Environmental Statistics*, 3, 237–53.

Araújo, M.B. and New, M. (2007). Ensemble forecasting of species distributions. *Trends in Ecology and Evolution*, 22, 42–47.

Austin, M. (2007). Species distribution models and ecological theory: a critical assessment and some possible new approaches. *Ecological Modelling*, 200, 1–19.

Austin, M.P. and Van Niel, K.P. (2011). Improving species distribution models for climate change studies: variable selection and scale. *Journal of Biogeography*, 38, 1–8.

Avgar, T., Potts, J.R., Lewis, M.A., and Boyce, M.S. (2016). Integrated step selection analysis: bridging the gap between resource selection and animal movement. *Methods in Ecology and Evolution*, 7, 619–30.

Ben-David, M., Bowyer, R.T., and Faro, J.B. (1996). Niche separation by mink and river otters: coexistence in a marine environment. *Oikos*, 75, 41–48.

Beyer, H.L., Haydon, D.T., Morales, J.M., et al. (2010). The interpretation of habitat preference metrics under use-availability designs. *Philosophical Transactions of the Royal Society B*, 365, 2245–54.

Bingham, R.L. and Brennan, L.A. (2004). Comparison of Type I error rates for statistical analyses of resource selection. *Journal of Wildlife Management*, 68, 206–12.

Bliss, L.M., Veech, J.A., Castro-Arellano, I., and Simpson, T.R. (2019). GIS-based habitat mapping and population estimation for the Gulf Coast kangaroo rat (*Dipodomys compactus*) in the Carrizo Sands Region of Texas, USA. *Mammalian Biology*, 98, 17–27.

Bond, N., Thomson, J., Reich, P., and Stein, J. (2011). Using species distribution models to infer potential climate change-induced range shifts of freshwater fish in southeastern Australia. *Marine and Freshwater Research*, 62, 1043–61.

Boyce, M.S. (2006). Scale for resource selection functions. *Diversity and Distributions*, 12, 269–76.

Boyce, M.S. and McDonald, L.L. (1999). Relating populations to habitats using resource selection functions. *Trends in Ecology and Evolution*, 14, 268–72.

Boyce, M.S., Vernier, P.R., Nielsen, S.E., and Schmiegelow, F.K.A. (2002). Evaluating resource selection functions. *Ecological Modelling* 157, 281–300.

Bradter, U., Mair, L. Jönsson, M., Knape, J., Singer, S., and Snäll, T. (2018). Can opportunistically collected Citizen Science data fill a data gap for habitat suitability models of less common species? *Methods in Ecology and Evolution*, 9, 1667–78.

Brotons, L., Thuiller, W., Araújo, M.B., and Hirzel, A.H. (2004). Presence-absence versus presence-only modelling methods for predicting bird habitat suitability. *Ecography*, 27, 437–48.

Chesson, J. (1978). Measuring preference in selective predation. *Ecology*, 59, 211–15.

Cock, M.J.W. (1978). The assessment of preference. *Journal of Animal Ecology*, 47, 805–16.

Cole, C.A. and Smith, R.L. (1983). Habitat suitability indices for monitoring wildlife populations—an evaluation.

Transactions of the North American Wildlife and Natural Resources Conference, 48, 367–75.

Cook, J.G. and Irwin, L.L. (1985). Validation and modification of a habitat suitability model for pronghorns. *Wildlife Society Bulletin*, 13, 440–48.

Dasgupta, N. and Alldredge, J.R. (2002). A single-step method for identifying individual resources. *Journal of Agricultural, Biological, and Environmental Statistics*, 7, 208–21.

De Caceres, M. and Legendre, P. (2009). Associations between species and groups of sites: indices and statistical inference. *Ecology*, 90. 3566–74.

de Valpine, P. and Harmon-Threatt, A.N. (2013). General models for resource use or other compositional count data using the Dirichlet-multinomial distribution. *Ecology*, 94, 2678–87.

Elith, J. and Leathwick, J.R. (2009). Species distribution models: ecological explanation and prediction across space and time. *Annual Review of Ecology, Evolution, and Systematics*, 40, 677–97.

Elith, J., Graham, C., Anderson, R.P., et al. (2006). Novel methods improve prediction of species' distributions from occurrence data. *Ecography*, 29, 129–51.

Elith, J., Phillips, S.J., Hastie, T., Dudík, M., Chee, Y.E., and Yates, C.J. (2011). A statistical explanation of MaxEnt for ecologists. *Diversity and Distributions*, 17, 43–57.

Elston, D.A., Illius, A.W., and Gordon, I.J. (1996). Assessment of preference among a range of options using log ratio analysis. *Ecology*, 77, 2538–48.

Fattorini, L., Pisani, C., Riga, F., and Zaccaroni, M. (2014). A permutation-based combination of sign tests for assessing habitat selection. *Environmental and Ecological Statistics*, 21, 161–87.

Fieberg, J., Matthiopoulos, J., Hebblewhite, M., Boyce, M.S., and Frair, J.L. (2010). Correlation and studies of habitat selection: problem, red herring or opportunity? *Philosophical Transactions Royal Society B*, 365, 2233–44.

Fitzpatrick, M.C., Gotelli, N.J., and Ellison, A.M. (2013). MaxEnt versus MaxLike: empirical comparisons with ant species distributions. *Ecosphere*, 4, e55.

Franklin, J. (2010). *Mapping Species Distributions: Spatial Inference and Prediction*. Cambridge University Press, Cambridge.

Franklin, J. (2013). Species distribution models in conservation biogeography: developments and challenges. *Diversity and Distributions*, 19, 1217–23.

Gama, M., Crespo, D., Dolbeth, M., and Anastácio, P. (2016). Predicting global habitat suitability for *Corbicula fluminea* using species distribution models: the importance of different environmental datasets. *Ecological Modelling*, 319, 163–69.

Gehring, T.M., McFadden, L.M., Prussing, S.A., et al. (2019). Spatial ecology of re-introduced American martens in the northern Lower Peninsula of Michigan. *American Midland Naturalist*, 182, 239–51.

Gogul-Prokurat, M. (2011). Predicting habitat suitability for rare plants at local spatial scales using a species distribution model. *Ecological Applications*, 21, 33–47.

Grigg, R.W. (1975). Age structure of a longevous coral: a relative index of habitat suitability and stability. *American Naturalist*, 109, 647–57.

Guillera-Arroita, G., Lahoz-Monfort, J., Elith, J., et al. (2015). Is my species distribution model fit for purpose? Matching data and models to applications. *Global Ecology and Biogeography*, 24, 276–92.

Guisan, A., Thuiller, W., and Zimmerman, N.E. (2017). *Habitat Suitability and Distribution Models*. Cambridge University Press, Cambridge.

Hall, L.S., Krausman, P.R., and Morrison, M.L. (1997). The habitat concept and a plea for standard terminology. *Wildlife Society Bulletin*, 25, 173–82.

Hallgren, W., Santana, F., Low-Choy, S., Zhao, B., and Mackey, B. (2019). Species distribution models can be highly sensitive to algorithm configuration. *Ecological Modelling*, 408, e108719.

Hirzel, A.H. and Le Lay, G. (2008). Habitat suitability modelling and niche theory. *Journal of Applied Ecology*, 45, 1372–81.

Hodges, K.E. (2008). Defining the problem: terminology and progress in ecology. *Frontiers in Ecology and the Environment*, 6, 35–42.

Irwin, L.L. and Cook, J.G. (1985). Determining appropriate variables for a habitat suitability model for pronghorns. *Wildlife Society Bulletin*, 13, 434–40.

Jackson, D.A. (1997). Compositional data in community ecology: the paradigm or peril of proportions? *Ecology*, 78, 929–40.

Jacobs, J. (1974). Quantitative measurement of food selection: a modification of the forage ratio and Ivlev's electivity index. *Oecologia*, 14, 412–17.

Johnson, C.J., Nielsen, S.E., Merrill, E.H., McDonald, T.L., and Boyce, M.S. (2006). Resource selection functions based on use-availability data: theoretical motivation and evaluation methods. *Journal of Wildlife Management*, 70, 347–57.

Johnson, D.H. (1980). The comparison of usage and availability measurements for evaluating resource preference. *Ecology*, 61, 65–71.

Kearney, M. (2006). Habitat, environment and niche: what are we modeling? *Oikos*, 115, 186–91.

Kearney, M.R., Wintle, B.A., and Porter, W.P. (2010). Correlative and mechanistic models of species distribution provide congruent forecasts under climate change. *Conservation Letters*, 3, 203–13.

Keating, K.A. and Cherry, S. (2004). Use and interpretation of logistic regression in habitat-selection studies. *Journal of Wildlife Management*, 68, 774–89.

Kirk, D.A., Park, A.C., Smith, A.C., et al. (2018). Our use, misuse, and abandonment of a concept: whither habitat? *Ecology and Evolution*, 8, 4197–208.

Kirkpatrick, M. and Ryan, M.J. (1991). The evolution of mating preferences and the paradox of the lek. *Nature*, 350, 33–38.

Krausman, P.R. and Morrison, M.L. (2016). Another plea for standard terminology. *Journal of Wildlife Management*, 80, 1143–44.

Lancaster, T. and Imbens, G. (1996). Case-control studies with contaminated controls. *Journal of Econometrics*, 71, 145–60.

Lechowicz, M.J. (1982). The sampling characteristics of electivity indices. *Oecologia*, 52, 22–30.

Lele, S.R. (2009). A new method for estimation of resource probability function. *Journal of Wildlife Management*, 73, 122–27.

Lele, S.R., Merrill, E.H., Keim, J., and Boyce, M.S. (2013). Selection, use, choice and occupancy: clarifying concepts in resource selection studies. *Journal of Animal Ecology*, 82, 1183–91.

Lele, S.R. and Keim, J.L. (2006). Weighted distributions and estimation of resource selection probability functions. *Ecology*, 87, 3021–28.

Li, H.-F., Fujisaki, I., and Su, N.-Y. (2013). Predicting habitat suitability of *Coptotermes gestroi* (Isoptera: Rhinotermitidae) with species distribution models. *Journal of Economic Entomology*, 106, 311–21.

Long, R.A., Muir, J.D., Rachlow, J.L., and Kie, J.G. (2009). A comparison of two modeling approaches for evaluating wildlife-habitat relationships. *Journal of Wildlife Management*, 73, 294–302.

Luo, Z., Jiang, Z., and Tang, S. (2015). Impacts of climate change on distributions and diversity of ungulates on the Tibetan plateau. *Ecological Applications*, 25, 24–38.

Manly, B.F.J. (1974). A model for certain types of selection experiments. *Biometrics*, 30, 281–94.

Manly, B.F.J. (1985). *The Statistics of Natural Selection on Animal Populations*. Chapman and Hall, London.

Manly, B.F.J., McDonald, L.L., and Thomas, D.L. (1993). *Resource Selection by Animals, Statistical Design and Analysis for Field Studies*. Chapman and Hall, London.

Manly, B.F.J., McDonald, L.L., Thomas, D.L., McDonald, T.L., and Erickson, W.P. (2002). *Resource Selection by Animals, Statistical Design and Analysis for Field Studies*. Second edition. Kluwer Academic Publishers, Boston, Mass.

Martin, T.E. (1985). Resource selection by tropical frugivorous birds: integrating multiple interactions. *Oecologia*, 66, 563–73.

Marzluff, J.M., Millspaugh, J.J., Hurvitz, P., and Handcock, M.S. (2004). Relating resources to a probabilistic measure of space use: forest fragments and Steller's jays. *Ecology*, 85, 1411–27.

Mathewson, H.A. and Morrison, M.L. (2015). The misunderstanding of habitat. Pages 3–8 in *Wildlife Habitat Conservation*, Morrison, M.L. and Mathewson, H.A. (editors). Johns Hopkins University Press, Baltimore, MD.

McDonald, L.L., Manly, B.F.J., and Raley, C.M. (1990). Analyzing foraging and habitat use through selection functions. *Studies in Avian Biology*, 13, 325–31.

McLoughlin, P.D., Morris, D.W., Fortin, D., Vander Wal, E., and Contasti, A.L. (2010). Considering ecological dynamics in resource selection functions. *Journal of Animal Ecology*, 79, 4–12.

Merow, C., Smith, M.J., and Silander, J.A. (2013). A practical guide to MaxEnt for modeling species' distributions: what it does, and why inputs and settings matter. *Ecography*, 36, 1058–69.

Merow, C., Smith, M.J., Edwards, T.C., et al. (2014). What do we gain from simplicity versus complexity in species distribution models? *Ecography*, 37, 1267–81.

Meyer, C.B. and Thuiller, W. (2006). Accuracy of resource selection functions across spatial scales. *Diversity and Distributions*, 12, 288–97.

Meynard, C.N. and Quinn, J.F. (2007). Predicting species distributions: a critical comparison of the most common statistical models using artificial species. *Journal of Biogeography*, 34, 1455–69.

Millspaugh, J.J., Nielson, R.M., McDonald, L., et al. (2006). Analysis of resource selection using utilization distributions. *Journal of Wildlife Management*, 70, 384–95.

Morris, D.W. (2003). Toward an ecological synthesis: a case for habitat selection. *Oecologia*, 136, 1–13.

Neu, C.W., Byers, C.R., and Peek, J.M. (1974). A technique for analysis of utilization-availability data. *Journal of Wildlife Management*, 38, 541–45.

Nikula, A., Heikkinen, S., and Helle, E. (2004). Habitat selection of adult moose (*Alces alces*) at two spatial scales in central Finland. *Wildlife Biology*, 10, 121–35.

Nneji, L.M., Salako, G., Oladipo, S.O., et al. (2020). Species distribution modelling predicts habitat suitability and reduction of suitable habitat under future climatic scenario for *Sclerophrys perreti*: a critically endangered Nigerian endemic toad. *African Journal of Ecology*, 58, 481–91.

Otis, D.L. and White, G.C. (1999). Autocorrelation of location estimates and the analysis of radiotracking data. *Journal of Wildlife Management*, 63, 1039–44.

Ott, S.L., Veech, J.A., Simpson, T.R., Castro-Arellano, I., and Evans, J. (2019). Mapping potential habitat and range-wide surveying for the Texas kangaroo rat. *Journal of Fish and Wildlife Management*, 10, 619–30.

Pendleton, G.W., Titus, K., DeGayner, E., Flatten, C.J., and Lowell, R.E. (1998). Compositional analysis and GIS for study of habitat selection by goshawks in southeast

Alaska. *Journal of Agricultural, Biological, and Environmental Statistics*, 3, 280–95.

Peterson, A.T., Papeş, M., and Eaton, M. (2007). Transferability and model evaluation in ecological niche modeling: a comparison of GARP and Maxent. *Ecography*, 30, 550–60.

Phillips, S.J., Anderson, R.P., and Schapire, R.E. (2006). Maximum entropy modeling of species geographic distributions. *Ecological Modelling*, 190, 231–59.

Pierotti, M.E.R., Martín-Fernández, J.A., and Barceló-Vidal (2017). The peril of proportions: robust niche indices for categorical data. *Methods in Ecology and Evolution*, 8, 223–31.

Pineda, E. and Lobo, J.M. (2012). The performance of range maps and species distribution models representing the geographic variation of species richness at different resolutions. *Global Ecology and Biogeography*, 21, 935–44.

Roffler, G.H., Schwartz, M.K., Pilgrim, K.L., et al. (2016). Identification of landscape features influencing gene flow: how useful are habitat selection models? *Evolutionary Applications*, 9, 805–17.

Rondinini, C., Stuart, S., and Boitani, L. (2005). Habitat suitability models and the shortfall in conservation planning for African vertebrates. *Conservation Biology*, 19, 1488–97.

Rota, C.T., Fletcher, R.J., Evans, J.M., and Hutto, R.L. (2011). Does accounting for imperfect detection improve species distribution models? *Ecography*, 34, 659–70.

Schadt, S., Revilla, E., Wiegand, T., et al. (2002). Assessing the suitability of central European landscapes for the reintroduction of Eurasian lynx. *Journal of Applied Ecology*, 39, 189–203.

Segurado, P. and Araújo, M.B. (2004). An evaluation of methods for modelling species distributions. *Journal of Biogeography*, 31, 1555–68.

Shonfield, J., King, W., and Koski, W.R. (2019). Habitat use and movement patterns of Butler's gartersnake (*Thamnophis butleri*) in southwestern Ontario, Canada. *Herpetological Conservation and Biology*, 14, 680–90.

Skidmore, A.K., Gauld, A., and Walker, P. (1996). Classification of kangaroo habitat distribution using three GIS models. *International Journal of Geographical Information Systems*, 10, 441–54.

Stohlgren, T.J., Jarnevich, C.S., Esaias, W.E., and Morisette, J.T. (2011). Bounding species distribution models. *Current Zoology*, 57, 642–47.

Strauss, R.E. (1979). Reliability estimates for Ivlev's electivity index, the forage ratio, and a proposed linear index of food selection. *Transactions of the American Fisheries Society*, 108, 344–52.

Thomas, D.L. and Taylor, E.J. (2006). Study designs and tests for comparing resource use and availability, part II. *Journal of Wildlife Management*, 70, 324–36.

Troy, J.R., Holmes, N.D., Veech, J.A., Raine, A.F., and Green, M.C. (2014). Habitat suitability modeling for the Newell's shearwater on Kauai. *Journal of Fish and Wildlife Management*, 5, 315–29.

Tsilimigras, M.C.B. and Fodor, A.A. (2016). Compositional data analysis of the microbiome: fundamentals, tools, and challenges. *Annals of Epidemiology*, 26, 330–35.

van Beest, F.M., Uzal, A., Vander Wal, E., et al. (2014). Increasing density leads to generalization in both coarse-grained habitat selection and fine-grained resource selection in a large mammal. *Journal of Animal Ecology*, 83, 147–56.

Verner, J., Morrison, M.L., and Ralph, C.J. (1986). *Wildlife 2000: Modeling Habitat Relationships of Terrestrial Vertebrates*. University of Wisconsin Press, Madison, Wis.

Wagner, W.E. (1998). Measuring female mating preferences. *Animal Behaviour*, 55, 1029–42.

Wecker, S.C. (1964). Habitat selection. *Scientific American*, 211, 109–17.

Willis, S.G., Foden, W., Baker, D.J., et al. (2015). Integrating climate change vulnerability assessments from species distribution models and trait-based approaches. *Biological Conservation*, 190, 167–78.

Yackulic, C.B., Chandler, R., Zipkin, E.F., et al. (2013). Presence-only modelling using Maxent: when can we trust the inferences? *Methods in Ecology and Evolution*, 4, 236–43.

CHAPTER 12

Conclusion

There is a long history of studying habitat in ecology and wildlife ecology and for good reason—habitat really matters to organisms. Even beyond the habitat associations of individual species, habitat and resource use drives everything in nature. Habitat as a particular kind of physical space and set of environmental conditions is crucial to survival and reproduction of the individual and persistence of the population. Habitat associations and more specifically habitat requirements of a species are the outcome of adaptation through natural selection. Given this overriding importance of habitat, we should expect that it might also strongly determine the spatial distribution of a species in nature. A given species will not be found everywhere there is habitat for it. Species do not typically saturate their habitat as there is almost always some amount of recruitment limitation. Nonetheless, a relatively simple probabilistic model of dispersal, settlement, and establishment (Chapter 3) might prove useful in understanding how an individual organism colonizes a new area with subsequent establishment of a sustainable population. Further, the dispersal–settlement–establishment (DSE) process might largely be governed by environmental cues that the dispersing individual responds to when selecting habitat. More specifically, the environmental cues may often be physical features or characteristics of the habitat. This line of thinking leads to the habitat-cue hypothesis of species distribution.

The habitat-cue hypothesis generates many predictions that can be tested (Chapter 3). My hope is that I have motivated some readers to further test the hypothesis and also think more broadly about the role of habitat in other ecological patterns and processes that we continue to study. To this end, an implicit goal of this book has been to more thoroughly formalize an area of study that could be referred to as habitat ecology. In Chapter 1, I defined it as the study of the habitat requirements of species and effects of habitat on individual survival, population persistence, and spatial distribution. To some extent, this is nothing new. In particular, wildlife ecologists have been studying species habitat requirements and characterizing habitat for over seven decades. Prior to wildlife ecology developing as an academic discipline and applied practice in the 1930s and 1940s, the study of habitat emerged slowly as the new field of ecology was born around 1900. Natural history writers of the late nineteenth century knew about habitat but typically used the term to designate a locality and perhaps describe the general environment where a species was found. Our modern concept of habitat developed as ecology itself was becoming established as a field of study between 1900 and the 1930s (Chapter 1). For various reasons, ecology and wildlife ecology as academic fields of study have followed separate historical trajectories despite a shared interest in habitat (Chapter 2). Given this historical legacy that still lingers today, one of my other goals in this book was to better connect these two disciplines. Someone that self-identifies as an ecologist and someone that self-identifies as a wildlife ecologist actually have a lot in common, particularly if they are interested in studying habitat and its importance to pattern and process in nature as well as the management and preservation of certain species and biodiversity in general.

But why study habitat? And why conduct a habitat analysis? There are many reasons, both pure and applied, for analytically examining the habitat associations of a species (Chapter 4). Briefly, we can

Habitat Ecology and Analysis. Joseph A. Veech, Oxford University Press (2021). © Joseph A. Veech. DOI: 10.1093/oso/9780198829287.003.0012

better understand a species' autecology as well as its influence in ecological communities and ecosystems by learning more about its habitat requirements and the way that it uses resources within the habitat. Further, such knowledge is also mandatory for proper management and preservation of the species and entire ecological communities. Conducting a habitat analysis (Chapters 5, 8, and 9) gives us the statistical confidence that the identified habitat associations truly do represent the habitat requirements of the species.

In Chapters 8 and 9, I was careful not to show too much favor toward any one of the six methods of habitat analysis that I discussed. For a given study, dataset, and research questions, the most appropriate method of analysis will depend on various factors related to characteristics of the species, study design and data collection protocol, and even peculiarities of the final dataset that may not have been initially anticipated prior to collecting the data (Chapters 6 and 7). None of the methods is "one size fits all." However, my own preference is multiple logistic regression. For a species response variable, it only requires presence–absence data or equivalent binary data, rather than abundance data. The latter can be more difficult to collect and presumably are more prone to error. Multiple logistic regression can handle many types of environmental predictor variables. The output or predicted y-values from a multiple logistic regression model is always constrained between 0 and 1. This is a beneficial property in that it sets bounds for the expected (predicted) effect of the environmental variables on the response variable. Moreover, the predicted y-values are often asymptotic at 0 and 1 as x approaches either its maximum or minimum observed value. This reflects ecological reality; likely no habitat characteristic has an *unlimited* effect on increasing either the abundance or "probability" of species presence at a location. One negative aspect of multiple logistic regression is that it ignores the information in abundance data, assuming the researcher has collected abundance data and converted it to presence–absence. Thus, logistic regression may not reveal a species association with a particular habitat variable(s) if that variable primarily influences abundance rather than the occurrence of the species. However, this is not overly problematic in that the researcher could also employ multiple linear regression to the non-zero abundance data to better reveal the effect of the habitat variable(s).

I also think it is often very worthwhile to compare multiple statistical models using either an Akaike information criterion (AIC)-based approach or by examining and comparing model fit in some other way (Chapter 10). Rather than being a set of environmental factors that *independently* influence a species' (or individual's) settlement, establishment, survival, and reproduction, the habitat of a species is best thought of as a multi-faceted whole composed of potentially many factors (characteristics) that are integrated with one another. Statistical analysis of habitat should reflect one's perspective on habitat. Thus, to characterize the habitat of a given species, I recommend including at least five potentially important environmental (predictor) variables and comparing logistic regression models composed of different subsets of variables. Among five predictor variables, there are 31 possible models. Ideally, to lessen the overall chance of Type I error (see below), the researcher will not conduct every possible model but rather use a specified strategy to decide which models to examine and compare (Chapter 10). As an upper limit, there is likely rapidly declining "return on investment" when including more than 30 predictor variables in a habitat analysis. Indeed, this many variables in any single regression model would require a large sample size or number of survey units; one rule of thumb is that sample size should be at least five times the number of predictor variables and some statisticians actually recommend eight times. There is also time, effort, and expense involved in collecting or compiling data on that many environmental variables. Note that there are over 1.07×10^9 unique model combinations when there are 30 predictor variables—a very carefully planned strategy would be needed to greatly limit the number of models conducted. Nonetheless, my main point is that predictor variables should be analyzed in combination with each other because they likely have synergistic or at least integrated effects on the species response variable. As habitat characteristics, the variables are holistic, not individualistic.

As mentioned in Chapter 10, AIC model comparison in the information-theoretic framework has often been promoted as an alternative to frequentist significance testing. That is, strict adherents to the information-theoretic approach might not like to see P-values incorporated into a model comparison table (or see them anywhere else in the Results section of a research paper). I do not follow this prohibition. It is perfectly appropriate to combine AIC model comparison with significance testing of the overall fit (R^2 value) of a multiple regression model and of the estimated values of the regression coefficients. Significance testing can complement or provide further support for the top models identified by the model comparison. In addition, significance testing allows the researcher to assess the possibility of study-wide Type I error. Briefly, a Type I error occurs when the null hypothesis is rejected because the P-value is less than the alpha level, even though the result (e.g., regression coefficient being significantly less than or greater than zero) is actually not truly significant. A Type I error is the finding of false significance or incorrect rejection of the null hypothesis. With regard to an entire study and multiple tests of significance, some of the tests with a significant result may actually represent Type I error at a rate that is equal to the alpha level. Thus, if 30 regression models are conducted and tested at $\alpha = 0.05$, then one or two are expected to be found significant due to Type I error ($30 \times 0.05 = 1.5$). It is impossible to know which models are Type I errors. However, there are methods for adjusting P-values to control for Type I error (consult any introductory statistics textbook for details).

Regardless of the statistical method employed to analyze habitat, it is important to recognize that the habitat must contain certain resources for the individual to survive and the population to persist. From the organism's perspective, habitat and resources are not separable. Further, a dispersing organism must first find the correct kind of habitat. As stated by Mannan and Steidl (2013, p. 231): "The ultimate challenge of characterizing habitat and understanding the process of habitat selection involves identifying the environmental features that trigger an individual's decision to settle in an area and the resources required for successful survival and reproduction." I'll end this chapter and the book with an example. Mountain plovers (*Charadrius montanus*) are a type of shorebird but during the summer they live and breed in grassland habitats far from water and from mountains. They are native to the vast grasslands of the western Great Plains in North America. Historically, they may have been closely associated with the very short grass and bare-ground habitat features created by grazing bison herds, prairie dog colonies, and fire (Goguen 2012). Prairie dog (*Cynomys ludovicianus*) colonies are still considered important breeding habitat for mountain plovers, particularly in northern areas where the grass away from colonies grows too thick and tall (Goguen 2012). Mountain plovers likely seek out and cue in on prairie dog colonies when migrating to their breeding grounds.

In a study of habitat at three spatial scales, colony, territory, and nest site, Goguen (2012) used multiple logistic regression to compare 14 prairie dog colonies having plovers to 30 without plovers. He found that plover presence was associated with shorter grass height, more bare ground, and less shrub cover at all three spatial scales. At the finest scale of nest site, plover nests tended to be situated on more steeply sloped ground and further from the entrance to a prairie dog burrow than were random points. Not surprisingly, some of the predictor variables, namely grass height, bare ground, and shrub cover were correlated with each other. Rather than being problematic for the statistical analysis, the multicollinearity among these variables represented their synergistic effect as habitat characteristics. The variables were correlated not because any of them have a strong cause and effect on the other two but instead because they are all the outcome of a common factor, prairie dogs, modifying the environment and creating the plover habitat. Goguen (2012) also explained how short grass and bare ground facilitate predator detection and foraging behavior of the plovers. Clearly, mountain plovers use prairie dog colonies as habitat given that such habitat promotes survival and reproduction, not to mention that mountain plovers evolved to use this unique type of habitat. Another informative aspect of this example is that many conservation organizations and government agencies consider mountain plovers to have threatened status due to habitat destruction. The species requires conservation action and perhaps active habitat management. There is a real need to characterize their habitat as thoroughly as possible and then put that knowledge toward preserving the species.

References

Goguen, C.B. (2012). Habitat use by mountain plovers in prairie dog colonies in northeastern New Mexico. *Journal of Field Ornithology*, 83, 154–65.

Mannan, R.W. and R.J. Steidl (2013). Habitat. Pages 229–45 in *Wildlife Management and Conservation: Contemporary Principles and Practices*. Krausman, P.R. and Cain, J.W. (editors). Johns Hopkins University Press, Baltimore, Md.

Index